# Nonisotopic DNA Probe Techniques

Edited by

**Larry J. Kricka**

Department of Pathology
and Laboratory Medicine
University of Pennsylvania
Philadelphia, Pennsylvania

**Academic Press, Inc.**
**Harcourt Brace Jovanovich, Publishers**
San Diego New York Boston London Sydney Tokyo Toronto

*Front cover photograph:* Color enchanced digitized image of a DNA sequence obtained using the chemiluminescent substrate CSPD to visualize bound alkaline phosphatase conjugate. This illustration was kindly provided by Irena Bronstein and Chris Martin of Tropix, Inc.

This book is printed on acid-free paper. ♾



Academic Press, Inc.
1250 Sixth Avenue, San Diego, California 92101-4311

*United Kingdom Edition published by*
Academic Press Limited
24–28 Oval Road, London NW1 7DX

Library of Congress Cataloging-in-Publication Data

Nonisotopic DNA probe techniques / edited by Larry J. Kricka.
p. cm.
Includes bibliographical references and index.
ISBN 0-12-426295-3 (hardcover). -- ISBN 0-12-426296-1 (pbk.)
1. DNA probes. I. Kricka, Larry J., date
[DNLM: 1. DNA Probes. 2. Nucleic Acid Hybridization. QU 58 N8124]
QP624.5D73N66 1992
574.87'3282--dc20
DNLM/DLC
for Library of Congress 91-41379
CIP

PRINTED IN THE UNITED STATES OF AMERICA
92 93 94 95 96 97 BC 9 8 7 6 5 4 3 2 1

# Nonisotopic DNA Probe Techniques

# CONTENTS

# PART TWO
# Detection Methods

## 3. Detection of Alkaline Phosphatase by Time-Resolved Fluorescence

*Eva Gudgin Templeton, Hector E. Wong, and Alfred Pollak*



## 4. Detection of Alkaline Phosphatase by Bioluminescence

*Reinhard Erich Geiger*



## 5. Detection of DNA on Membranes with Alkaline Phosphatase-Labeled Probes and Chemiluminescent AMPPD Substrate

*Annette Tumolo, Owen J. Murphy, Quan Nguyen, John C. Voyta, Frank Witney, and Irena Bronstein*



## 6. Detection of Alkaline Phosphatase by Colorimetry

*Ayoub Rashtchian*

# CONTRIBUTORS

*Numbers in parentheses indicate the pages on which the authors' contributions begin.*

**Lyle J. Arnold, Jr.** (275), Genta, Inc., San Diego, California 92121
**Patrick Balaguer** (203), INSERM, Unité 58, 34100 Montpellier, France
**Anne-Marie Boussioux** (203), INSERM, Unité 58, 34100 Montpellier, France
**Irena Bronstein** (127), Tropix, Inc., Bedford, Massachusetts 01730
**Theodore K. Christopoulos** (263), Department of Clinical Biochemistry, Toronto Western Hospital, Toronto, Ontario M5T 2S8, Canada
**Patrik Dahlén** (227), Wallac Biochemical Laboratory, SF-20101 Turku, Finland
**Eleftherios P. Diamandis** (263), Department of Clinical Biochemistry, Toronto Western Hospital, Toronto, Ontario M5T 2S8, Canada and Department of Clinical Biochemistry, University of Toronto, Toronto, Ontario M5G 1L5, Canada
**Ian Durrant** (167), Research and Development, Amersham International, Amersham, Buckinghamshire, HP7 9LL, England
**Reinhard Erich Geiger** (113), University of Munich, D-9000 Munich, Germany
**Pertti Hurskainen** (227), Wallac Biochemical Laboratory, SF-20101 Turku, Finland
**Christoph Kessler** (29), Boehringer Mannheim GmbH, Biochemical Research Center, Genetic Department, D-8122 Penzberg, Germany
**Larry J. Kricka** (3), Department of Pathology and Laboratory Medicine, University of Pennsylvania, Philadelphia, Pennsylvania 19104
**Timo Lövgren** (227), Department of Biochemistry, University of Turku, SF-20500 Turku, Finland
**Larry E. Morrison** (311), Amoco Technology Company, Naperville, Illinois 60566
**Owen J. Murphy** (127), Tropix, Inc., Bedford, Massachusetts 01730
**Norman C. Nelson** (275), Gen-Probe, Inc., San Diego, California 92121
**Quan Nguyen** (127), Genetic Systems Division, Bio-Rad Laboratories, Richmond, California 94806

**Jean-Claude Nicolas** (203), INSERM Unité 58, 34100 Montepellier, France
**Alfred Pollak** (95), Kronem Systems, Inc., Mississauga, Ontario L4V 1P1, Canada
**Ayoub Rashtchian** (147), Molecular Biology Research and Development, Life Technologies, Inc., Gaithersburg, Maryland 20898
**Mark A. Reynolds** (275), Genta, Inc., San Diego, California 92121
**Eva Gudgin Templeton** (95), Kronem Systems, Inc., Mississauga, Ontario L4V 1P1, Canada
**Béatrice Térouanne** (203), INSERM Unité 58, 34100 Montpellier, France
**Annette Tumolo** (127), Genetic Systems Division, Bio-Rad Laboratories, Richmond, California 94806
**Peter C. Verlander** (185), Department of Investigative Dematology, The Rockefeller University, 1230 York Avenue, New York, New York 10021
**Marie Agnès Villebrun** (203), INSERM Unité 58, 34100 Montpellier, France
**John C. Voyta** (127), TROPIX, Inc., Bedford, Massachusetts 01730
**Frank Witney** (127), Genetic Systems Division, Bio-Rad Laboratories, Richmond, California 94806
**Hector E. Wong** (95), Kronem Systems, Inc., Mississauga, Ontario L4V 1P1, Canada

# PREFACE

Numerous nonisotopic methods have now been developed as replacements for radioactive labels such as $^{32}$phosphorous and $^{125}$iodine in DNA probe hybridization assays. Most have been developed within the last five years; the range of nonisotopic methods is now so extensive that it is difficult to determine the relative merits and demerits for particular applications.

The objective of this book is to bring together descriptions of the principal nonisotopic methods for DNA hybridization assays, together with experimental details of the methods, including labeling and detection of the label. This book contains descriptions of bioluminescent, chemiluminescent, fluorescent, and time-resolved fluorescent detection methods. It covers the following combinations of label and detection reaction: acridinium esters/chemiluminescence; alkaline phosphatase/bioluminescence, colorimetry, chemiluminescence, time-resolved fluorescence; lanthanide chelates/time-resolved fluorescence; glucose 6-phosphate dehydrogenase/bioluminescence; fluorescence/fluorescence; and horseradish peroxidase/enhanced chemiluminescence/colorimetry. Non-separation DNA probe assay strategies based on selective hydrolysis of acridinium esters and energy transfer involving pairs of probes, one labeled with a chemiluminescent molecule and the other labeled with a fluorophore, are also presented.

Each chapter has been prepared by the inventor or developer of a particular nonisotopic method and thus provides an expert account of the method. Practical details for a range of applications are presented in step-by-step experimental procedures that provide a valuable source of authoritative information.

This book is intended to give research workers and assay developers a single source of information on nonisotopic procedures for DNA hybridization based assays.

Larry J. Kricka

# PART ONE
# Introduction

# Nucleic Acid Hybridization Test Formats: Strategies and Applications

Larry J. Kricka
Department of Pathology and Laboratory Medicine
University of Pennsylvania
Philadelphia, Pennsylvania



## I. INTRODUCTION

Nucleic acid hybridization tests for the detection of specific DNA and RNA sequences are now extensively used in research and routine laboratories (Diamandis, 1990; Leary and Ruth, 1989; Matthews and Kricka, 1988; Pollard-Knight, 1991). Hybridization assays have diverse applications in medicine and forensics, and some representative examples of these applications are listed in Table I. Labeled nucleic acid probes are utilized in a variety of assay formats including dot blots, Southern blots (DNA target), Northern blot (RNA target), *in situ* hybridization, plaque hybridization, and colony hybridization. An important aspect of nucleic hybridization assays is the choice of the substance used to label a nucleic acid probe and the label detection method. As yet there is no consensus on which substance is the ideal label for nucleic acid probes for use in the various assay formats. The first assays used a radioactive $^{32}$phosphorus label. However, this label has the major disadvantage of a relatively short

*Nonisotopic DNA Probe Techniques*

**Table I**
Applications of Nucleic Acid Hybridization Assays

| Application | Reference |
|---|---|
| Arteriosclerosis | Williams (1985) |
| Cell line authentication | Thacker *et al.* (1988) |
| Forensics | Budowle *et al.* (1990); Cawood (1989); Thornton (1989) |
| Blood stains | Gill *et al.* (1985) |
| Inherited disorders | Dawson (1990); Ropers (1987) |
| Cystic fibrosis | Kerem *et al.* (1989); Riordan *et al.* (1989) |
| Duchenne muscular dystrophy | Kunkel *et al.* (1989) |
| Phenylketonuria | DiLella *et al.* (1986); Woo *et al.* (1983) |
| Sickle cell anemia | Saiki *et al.* (1985) |
| Microbiology | Buck (1989); McGowan (1989); Wolfe (1988) |
| *E coli* | Miller *et al.* (1988) |
| *Neisseria gonorrhoeae* | Sanchez-Pescador *et al.* (1988) |
| *Legionella* | Wilkinson *et al.* (1986) |
| *Mycoplasma pneumoniae* | Dular *et al.* (1988) |
| Oncology | Knudson (1986) |
| Leukemia | Lovell (1989) |
| *Neu* oncogene | Slamon *et al.* (1987) |
| Paternity testing | Odelberg *et al.* (1988) |
| Virology | Landry (1990) |
| Cytomegalovirus | Spector and Spector (1985) |
| Hepatitis B | Kam *et al.* (1982) |
| Rotavirus | Flores *et al.* (1983) |

half-life (14.2 d) (cf. $^{125}$iodine used in immunoassay has a half-life of 60 d). Thus nucleic acid hybridization probes have a very short shelf-life. This has placed severe limitations on the routine use and commercialization of probe tests; hence, there are extensive efforts to develop and implement alternatives to the radioactive $^{32}$phosphorus label. Many different substances have been tested as nonisotopic replacements for $^{32}$phosphorus, and subsequent chapters of this book provide background and practical details of the application of various nonisotopic labels.

## II. NUCLEIC ACID LABELS

The majority of the substances used as labels for nucleic acid hybridization probes have been tested previously in immunoassay. Nonisotopic labels have been the focus of development because of the limitations of radioactive labels such as $^{32}$phosphorus (Kricka, 1985). These limitations are principally (1) a short half-life that restricts the shelf life of labeled probes and hence hybridization assay kits, (2) possible health hazards during preparation and use of the labeled nucleic acid, and (3) disposal of radioactive waste from the assay. The ideal label for a nucleic acid hybridization probe would have the following properties.

1. Easy to attach to a nucleic acid using a simple and reproducible labeling procedure;
2. Stable under nucleic acid hybridization conditions, typically temperatures up to 80° C, and exposure to solutions containing detergents and solvents such as formamide;
3. Detectable at very low concentrations using a simple analytical procedure and noncomplex instrumentation;
4. Nonobstructive on the nucleic acid hybridization reaction;
5. Applicable to solution or solid-phase hybridizations. In a solid-phase application, e.g., membrane-based assay, the label must produce a long-lived signal (e.g., enzyme label detected chemiluminescently or by time-resolved fluorescence);
6. Nondestructive. The label must be easy to remove for successive reprobing of membranes. Generally, reprobing is not problematic for $^{32}$phosphorus labels, but it is less straightforward for some nonisotopic labels (e.g., insoluble diformazan product of 5-bromo-4-chloro-3-indolyl-phosphate (BICP)-nitroblue tetrazolium (NBT)-alkaline phosphatase reaction has to be removed from a membrane with hot formamide);
7. Adaptable to nonseparation (homogenous) formats. Hybridization of labeled DNA probe to its complementary DNA sequence should modulate a property of the label so that it is detectable and distinguishable from unhybridized probe;
8. Stable during storage, providing longer shelf-life for commercial hybridization assay kits; and
9. Compatible with automated analysis. Widespread and large-scale applications of hybridization assays will lead to the need for automated analyzers. The label and the assay for the label must be compatible with a high throughput analyzer (rapid detection using the minimum number of reagents and analytical steps).

None of the labels listed in Table II fulfills all of these criteria and, just as in the case of immunoassays, there is still no agreement on the most appropriate nonisotopic label. Enzymes, such as horseradish peroxidase and alkaline phosphatase, have become particularly popular in recent years as a range of sensitive detection methods has evolved. Alkaline phosphatase, for example, can be detected using chemiluminescent, bioluminescent, and time-resolved fluorescent methods.

**Table II**
Direct Labels for Nucleic Acid Hybridization Assays

| |
|---|
| Chemiluminescent compounds |
| Acridinium ester |
| Isoluminol |
| Luminol |
| Enzymes |
| Alkaline phosphatase |
| Bacterial luciferase |
| Firefly luciferase |
| Glucose oxidase |
| Glucose 6-phosphate dehydrogenase |
| Hexokinase |
| Horseradish peroxidase |
| Microperoxidase |
| Papain |
| Fluorescent compounds |
| Fluorescein |
| Bimane |
| Ethidium |
| Methylcoumarin |
| Nitrobenzofuran |
| Pyrenebutyrate |
| Rhodamine |
| Terbium chelate |
| Tetramethylrhodamine |
| Texas Red |
| Miscellaneous |
| Latex particle |
| PolyAMP |
| Pyrene |
| Radioluminescent |
| $^{125}$Iodine |
| $^{32}$Phosphorus |
| $^{35}$Sulfur |
| Tritium |

# III. NUCLEIC ACID LABELING PROCEDURES

Detection of probe : nucleic acid target hybrids can be accomplished by direct or indirect labeling methods. In the former case, a label is attached directly to the nucleic acid by a covalent bond, or the label intercalates noncovalently between the double strand of the probe : nucleic acid target complex. The latter method, indirect labeling, employs a hapten (e.g., biotin) attached to the nucleic acid probe. The hapten is detected using a labeled specific binding protein (e.g., antibiotin, avidin, or streptavidin) (Table III). A slightly more complex format uses an intermediate binding protein to bridge between the hapten and the labeled binding protein (Table IV). Alternatively, a binding protein specific for double-stranded DNA can be used (e.g., monoclonal anti-dsDNA), and complexes are then detected using a labeled antispecies antibody (Mantero *et al.*, 1991). More complex indirect procedures have been developed to improve assay sensitivity (Wilchek and Bayer, 1990). In one design, a biotin-labeled probe is hybridized to the target DNA, followed by reaction of the biotinylated probe with streptavidin. The remaining binding sites on tetravalent streptavidin are then reacted with a biotinylated poly(alkaline phosphatase) to obtain a cluster of alkaline phosphatase labels around the bound biotinylated probe (Leary *et al.*, 1983).

Procedures for the direct labeling of a nucleic acid probe with a hapten or a direct label can be categorized into chemical, enzymatic, and synthetic procedures (Keller and Manak, 1989; Leary and Ruth, 1989; Matthews and Kricka, 1988). One of the goals of a labeling method is to

**Table III**
Indirect Labels for Nucleic Acid Hybridization Assays

| Hapten | Binding protein | Label |
|---|---|---|
| Biotin | Antibiotin | Gold colloid |
| | Avidin | Alkaline phosphatase |
| | | β-Galactosidase |
| | | Ferritin |
| | | Fluorescein |
| | | Horseradish peroxidase |
| | Streptavidin | β-Galactosidase |
| | | b-Phycoerythrin |
| Digoxigenin | Antidigoxigenin | Alkaline phosphatase |
| Ethidium | Antiethidium-DNA | β-Galactosidase |
| Glucosyl | Concanavilin A | Acid phosphatase |
| | | Glucose oxidase |
| IgG | Antispecies IgG | Horseradish peroxidase |
| IgG, Fab fragment | Antispecies IgG | Horseradish peroxidase |
| Lacoperon DNA | *Lac* repressor protein | Fluorescein |
| Poly(dA) | Poly(dT)-DNA | Horseradish peroxidase |
| Poly(dT) | Poly(dA)-DNA | Horseradish peroxidase |
| Protein A | IgG | Horseradish peroxidase |
| Protein G | IgG | Horseradish peroxidase |
| Sulfone | Antisulfone | Europium chelate |
| | Anti-RNA : DNA hybrid | Fluorescein |
| | Histone | $^{125}$Iodine |

**Table IV**
Indirect Labels for Nucleic Acid Hybridization Assays[a]

| Label | Binding protein | Binding protein-label |
|---|---|---|
| Biotin | Antibiotin | Antispecies IgG-horseradish peroxidase |
| | | Protein A-colloidal gold |
| 5-Bromodeoxyuridine | Anti-5-bromodeoxyuridine | Antispecies IgG-alkaline phosphatase |
| DNP | Anti-DNP | Antispecies IgG-pyruvate kinase |
| *N*-2-Acetylaminofluorene | | Anti-(*N*-2-guanosinyl) acetylaminofluorene : antispecies IgG-alkaline phosphatase |
| Streptavidin | Biotin-IgG | Streptavidin-alkaline phosphatase |
| Sulfone | Antisulfone | Antispecies IgG-horseradish peroxidase |
| | | Antispecies IgG-glucose 6-phosphate dehydrogenase |
| | Antidouble-stranded DNA | Antispecies IgG-alkaline phosphatase |

[a] That use an intermediate binding protein.

confine the label to sites on the nucleic acid that are not involved in the hydrogen bonding necessary for hybrid formation. However, in practice, hydrogen bonding amino groups are used as labeling sites. This is effective only because the labeling density necessary to provide a detectable probe is low (10–30 modified bases/1000 bases); hence, it does not adversely influenced hybrid stability.

A nucleic acid probe has a limited range of functional groups suitable for attachment of a label; these are listed in Table V. Various enzymatic labeling procedures have been developed using deoxyribonuclease-DNA polymerase I (Rigby *et al.*, 1977), terminal transferase (Riley *et al.*, 1986), $T_4$ polynucleotide kinase (Maxam and Gilbert, 1977), and the Klenow fragment of *Escherichia coli* DNA (random hexanucleotide priming reaction) (Feinberg and Vogelstein, 1983). Labeled probes can also be made by cloning the probe sequence into an M13 bacteriophage vector (Hu and Messing, 1982). Polymerase chain reaction (PCR) has also been adapted for labeling, thus allowing simultaneous amplification and labeling (Lion and Haas, 1990; Schowalter and Sommer, 1989). An alternative approach to labeling is to introduce labels or reactive groups for the subsequent attachment of labels during oligomer synthesis (Ruth, 1984; Ruth and Bryan, 1984). Appropriately activated nucleotides can be incorporated either internally (Haralambidis *et al.*, 1987; Ruth *et al.*, 1985) or at the 5′-end (Smith *et al.*, 1985; Sproat *et al.*, 1987). The various types of labeling methods are discussed in detail in Chapter 2.

**Table V**
Chemical Labeling of Nucleic Acids

| Site | Reaction | Reference |
|---|---|---|
| Adenine | | |
| 6-Amino | Bifunctional coupling reagents (e.g., glutaraldehyde) | Matthews and Kricka (1988) |
| Cytosine | | |
| 4-Amino | Bisulfite-catalyzed transamination | Draper and Gold (1980); Viscidi *et al.* (1986) |
| | Thiolation | Malcom and Nicolas (1984) |
| 5-Carbon | Mercuration | Hopman *et al.* (1986) |
| 6-Carbon | Sulfonylation | Lebacq *et al.* (1988) |
| Guanine | | |
| 8-Carbon | *N*-acetoxy-*N*-2-acetyl-aminofluorene | Tchen *et al.* (1984) |
| | Bromination (*N*-bromosuccinimide) | Keller *et al.* (1988) |
| Thymine | | |
| 5,6-Bond | Cycloaddition | Cimino *et al.* (1985) |
| Uracil | | |
| 5-Carbon | Mercuration | Hopman *et al.* (1986) |
| 5,6-Bond | Cycloaddition | Cimino *et al.* (1985) |
| 3′-End hydroxyl | Oxidation | Bauman *et al.* (1981); Broker *et al.* (1978) |
| 5′-End phosphate | Aminoalkylation | Chu *et al.* (1983); Chu and Orgel (1985) |
| Nonspecific | Nitrene C—C, C—H bond insertion | Forster *et al.* (1985) |
| | Cross-linking (glutaraldehyde) | Al-Hakim and Hull (1986); Renz (1983); Renz and Kurz (1984); Sodja and Davidson (1978); Syvanen *et al.* (1985) |
| | (UV irradiation) | Czichos *et al.* (1989); Oser *et al.* (1990) |
| Noncovalent | Intercalation | Al-Hakeem and Sommer (1987) |

# IV. DETECTION OF LABELS AND NUCLEIC ACID HYBRIDIZATION ASSAY SENSITIVITY

The sensitivity of a nucleic acid hybridization assay is determined primarily by the detection limit of the label. This is owing to the sandwich assay design of a hybridization assay and the high binding constant for complementary nucleic acid interaction; thus, low losses due to hybrid dissociation will be encountered. The amount of a specific target DNA sequence in a cell is very small. For example, a mammalian cell contains approximately 6 pg of DNA, comprising 3 billion base pairs or 3 million kB of DNA sequence (a bacterial cell only contains 0.1 pg of DNA). An assay for a typical 1-kB gene using a 1-kB probe requires a detection limit of 1 attogram of nucleic acid. The extreme sensitivity required for the detection of single copy genes has generated several different analytical strategies.

1. Amplifying labels that initiate a cascade reaction, and multiple labeling strategies. Those based on the biotin : avidin system are especially effective at increasing detection sensitivity. However, sensitivity can be degraded by background from assay reagents, nonspecific binding of labeled probe, and contaminants.

2. Background rejection via the use of time-resolved fluorescent labels (terbium and europium chelates) (Syvanen *et al.*, 1986) and substrates (e.g., salicyl phosphate–europium chelate) (Evangelista *et al.*, 1991). These end-points utilize long-lived fluorescent chelates activated by a pulse of excitation light. The fluorescent signal is measured after the short-lived background fluorescence has decayed. Interference can also be minimized by appropriate design of the hybridization assay. Thompson *et al.* (1989) developed a reversible target capture assay, which uses a 3′-poly(dA)-tailed capture probe, a label probe, and an oligo(dT)-labeled paramagnetic particle. The target nucleic acid hybridizes to the probes, and the hybrids are captured via interaction of the dA tail on the capture probe and the dT tail on the paramagnetic particle. The particles are washed, and then the relatively weak bond between the dA and dT tails is disrupted using guanidine thiocyanate. The particles are replaced with fresh particles, and the cycle of capture, washing, and release is repeated. In this way any interferents in solution or bound to the particles are removed before the measurement step.

3. Selective amplification of DNA or RNA target. This strategy is becoming increasingly popular. A DNA or RNA target can be selectively amplified, thus reducing the requirement for ultrasensitive detection techniques. Several different methods have been developed. The PCR amplifies a DNA target sequence 10 million-fold after 30 cycles (Mullis, 1987; Mullis and Faloona, 1987), and a transcription-based amplification system will produce a 2–5 million-fold amplification of RNA target after 4 cycles (Lizardi *et al.*, 1988). Increasing the amount of analyte reduces the requirement for an ultrasensitive detection technique, but it does add extra steps to the overall analytical procedure.

4. Probe amplification using the Q-beta replicase system. This is an exponential amplification method for RNA probes based on Q-beta RNA polymerase (Kwoh *et al.*, 1989). In this method a RNA probe is hybridized to its target, isolated as the hybrid, and then denatured. The released RNA

probe is amplified a billion-fold in a single step, and then the vast number of probe copies is quantitated in a secondary hybridization step.

5. Probe networks provide a further method of introducing an amplification factor into a hybridization assay. This assay utilizes three types of probes (Urdea *et al.*, 1987). A set of short primary probes binds to the target, and these are subsequently reacted with a polymerized second probe, which in turn hybridizes with a multiple enzyme-labeled probe. Amplification of 100-fold can be achieved compared to the signal from a single oligomer probe.

Table VI compares the detection limits for various direct and indirect labels. Some of the most sensitive detection schemes involve enzyme labels in combination with either a chemiluminescent or bioluminescent reaction. For example, the detection limit for an alkaline phosphatase label using the adamantyl dioxetane phosphate (AMPPD) is 0.001 attomoles (Bronstein *et al.*, 1990). Results of comparative studies to determine the most effective label-detection reaction combination for a hybridization assay can vary considerably. Urdea *et al.* (1988) compared the detection limits of seven different labels in a sandwich hybridization assay. The lowest detection limits were obtained using $^{32}$phosphorus and horseradish peroxidase (enhanced chemiluminescent end-point). Another study (Balaguer *et al.*, 1991) compared luminescent (chemiluminescent or bioluminescent) detection of four different streptavidin–enzyme conjugates. The following signal : blank ratios were obtained after a 5-min incuba-

**Table VI**
Detection Limits for Nucleic Acid Labels[a]

| Detection method | CL | BL | FL | COL | RL | EL |
|---|---|---|---|---|---|---|
| Acridinium ester | 0.5 | — | — | — | — | — |
| Alkaline phosphatase | 0.001 | 0.01 | 0.1 | 50 | — | — |
| β-D-galactosidase | — | 5 | — | — | — | — |
| Europium chelate | — | — | 10 | — | — | — |
| Glucose 6-phosphate dehydrogenase | — | 0.1 | — | — | — | — |
| Horseradish peroxidase | 25 | 0.4 | 10 | 100 | — | — |
| Isoluminol | 1000 | — | — | — | — | — |
| $^{32}$Phosphorus | — | — | — | — | 50 | — |
| Rhodamine | 5000 | — | — | — | — | — |
| Ruthenium *tris* (bipyridyl) | | | | | | 20 |
| Texas Red | 20,000 | — | — | — | — | — |
| Xanthine oxidase | 3 | — | — | — | — | — |

[a] Detection limits in attomoles. BL, bioluminescence; CL, chemiluminescence; COL, colorimetry; EL, electrochemiluminescence; FL, fluorimetry; RL, radioluminescence. From Arnold *et al.* (1989); Blackburn *et al.* (1991); Bronstein and Kricka (1989); Diamandis (1990); Urdea *et al.* (1988).

tion in an assay to detect 1 pg of biotinylated pBR 322 spotted on a membrane: glucose 6-phosphate dehydrogenase, 7.0; alkaline phosphatase, 4.5; xanthine oxidase, 2.8; and peroxidase, 1.8. Other investigators have shown different relative detection limits, and more definitive studies are still needed to clarify which is the most effective label and detection reaction for a particular application (e.g., dot blot, *in situ* hybridization).

## A. Detection of Nonisotopic Labels

The principal types of detection methods for nonisotopic labels are bioluminescence, chemiluminescence, colorimetry, electrochemiluminescence, fluorescence, and time-resolved fluorescence. For some labels several detection methods are possible; these are summarized in Table VII.

### 1. Chemiluminescence and Bioluminescence

Chemiluminescence is the emission of light that occurs in certain chemical reactions because of decay of chemiexcited molecules to the electronic ground state. Bioluminescence is the chemiluminescence of nature (e.g., the firefly, *Photinus pyralis*), which involves luciferin substrates and luciferase enzymes, or photoproteins. Both methods are very sensitive (zeptomole amounts, $10^{-21}$ moles), rapid, and versatile [adaptable for hybridizations in solution and on membranes and monitoring with charge coupled device (CCD) cameras] (Hooper and Ansorge, 1990; Wick, 1989).

### 2. Colorimetry

Colorimetric assays produce soluble colored products and are relatively insensitive compared to luminescent assays (chemiluminescence, fluorescence, etc). Most attention has focused on reactions to produce insoluble colored products for locating hybrids on solid phases. These have the advantage of a simple visual read-out and a permanent record, especially for membrane-based assays.

### 3. Electrochemiluminescence

In this process, an electrochemical reaction produces excited-state species that decay to produce a ground state product and light. A disadvantage of this detection reaction is the need for specialized equipment that combines electrochemical-generation and light-detection capabilities.

**Table VII**
Assays for Selected Nucleic Acid Probe Direct and Indirect Labels

**Alkaline phosphatase (AP)**

Bioluminescent assay

$$\text{firefly luciferin-}O\text{-phosphate} \xrightarrow{\text{AP}} \text{firefly luciferin}$$

$$\text{firefly luciferin} \xrightarrow{\text{firefly luciferase + ATP + } Mg^{2+}} \text{light}$$

Chemiluminescent assay

$$\text{AMPPD} \xrightarrow{\text{AP}} \text{light}$$

Time-resolved fluorescent assay

$$\text{salicyl phosphate} \xrightarrow{\text{AP}} \text{salicylic acid} \xrightarrow{Eu^{3+}} Eu^{3+} \text{ chelate}$$

**Glucose 6-phosphate dehydrogenase (G6PDH)**

Bioluminescent assay

$$\text{glucose 6-phosphate + } NADP^{+} \xrightarrow{\text{G6PDH}} \text{glucose + NADPH}$$

$$\text{NADPH + FMN} \xrightarrow{\text{NAD(P)H:FMN oxidoreductase}} \text{NADP + flavin mononucleotide}H_2$$

$$FMNH_2 + \text{decanal} + O_2 \xrightarrow{\text{marine bacterial luciferase}} \text{light}$$

**Horseradish peroxidase (HRP)**

Chemiluminescent assay

$$\text{luminol} + H_2O_2 + \textit{para}\text{-iodophenol} \xrightarrow{\text{HRP}} \text{light}$$

**Xanthine oxidase**

Chemiluminescent assay

$$\text{xanthine} + O_2 \xrightarrow{\text{xanthine oxidase}} \text{hypoxanthine} + H_2O_2$$

$$H_2O_2 + \text{luminol + Fe-EDTA} \longrightarrow \text{light}$$

## 4. Fluorescence and Time-Resolved Fluorescence

Fluorescence measurements are capable of detecting single molecules of fluorescein (Mathies and Stryer, 1986). In practice, however, fluorescence is plagued by background signal due to nonspecific fluorescence present in biological samples. Since the background fluoresence tends to be short-lived, it can be avoided by using a long-lived fluorophore that is excited by a rapid pulse of excitation light. The fluorescence emission is then measured after the short-lived background has decayed, thus eliminating interference.

Detection methods for the most popular labels are summarized in the following sections, and further information, together with experimental protocols, is provided in other chapters of this book.

- *Acridinium esters*

  Oxidation of an acridinium ester by a mixture of sodium hydroxide and hydrogen peroxide produces a rapid flash of light. The detection limit for an acridinium ester-labeled probe is 0.5 amol (Arnold *et al.*, 1989; Weeks *et al.*, 1983). This flash reaction kinetics makes application of acridinium ester labels difficult for membrane-based assays.
- *Aequorin*

  Apoaequorin can now be produced by recombinant DNA techniques (Prasher *et al.*, 1985; Stults *et al.*, 1991). It is converted to aequorin (the photoprotein present in the hydrozoan jellyfish *Aequorea*) by reaction with coelentrazine, and light emission from aequorin is triggered by calcium ions. Light is emitted as a flash (469 nm); hence, this type of label is difficult to configure in membrane-based applications.
- *Alkaline phosphatase*

  There is now a series of highly sensitive detection methods for alkaline phosphatase labels. The bioluminescent method uses firefly D-luciferin-*O*-phosphate as a substrate. An alkaline phosphatase label dephosphorylates the substrate to liberate D-luciferin, which, unlike the phosphate, is a substrate for the bioluminescent firefly luciferase reaction (Hauber *et al.*, 1989; Miska and Geiger, 1987, 1990), see Chapter 4 .

Several chemiluminescent assays have also been developed for this enzyme. The most sensitive and widely investigated is based on an adamantyl 1,2-dioxetane aryl phosphate (AMPPD; see Chapter 5). This substance is dephosphorylated to a phenoxide intermediate, which decomposes to form adamantanone and an excited-state aryl ester, which emits light at 477 nm as a protracted glow (>1 hr) (Bronstein *et al.*, 1990, 1991). Light emission can be enhanced using a variety of compounds including water-soluble polymers {e.g., poly[vinylbenzyl(benzyldimethly ammonium chloride)]} (Bronstein, 1990), and fluoresceinated detergents (e.g., cetyltrimethylammonium bromide) (Schaap *et al.*, 1989). A second generation of dioxetane substrates is now available in which a 5-substituent on the adamantyl ring (e.g., 5-chloro, 5-bromo) prevents aggregation of these molecules (Bronstein *et al.*, 1991), thus reducing reagent background due to thermal degradation.

One of the most popular colorimetric methods for alkaline phosphatase is based on the formation of an insoluble purple dye by sequential dephosphorylation and reduction of BCIP (McGadey, 1970). An alternative substrate is naphthol AS-TR phosphate, which in combination with Fast Red RC, reacts with alkaline phosphatase to produce a red precipitate (Chapter 6).

A salicyl phosphate substrate has been used in a time-resolved assay for alkaline phosphatase. Enzymatic cleavage of the phosphate group produces salicylic acid, which forms a long-lived fluorescent chelate with a europium or terbium ion (Evangelista *et al.*, 1991). Soluble or insoluble chelates can be produced depending on the substituent on the salicylic acid—a fluoro substituent produces a soluble chelate; a branched-chain alkyl group produces an insoluble chelate (Chapter 3).

- *Europium chelates*
  Lanthanides, e.g., europium$^{3+}$ and terbium$^{3+}$, form highly fluorescent chelates with naphthoyltrifluoroacetone or 4,7-*bis*(chlorosulfophenyl)-1, 10-phenanthroline-2,9-dicarboxylic acid (BCPDA) (Diamandis,1988; Soini and Lovgren, 1987) (see Chapters 9 and 10). These lanthanide chelates have a long-lived fluorescence (100–1000 us) and are useful as time-resolved fluorescent labels. Sensitivity enhancement via multiple labeling can be achieved by coupling the BCPDA to streptavidin and using the multiple-labeled streptavidin in conjunction with biotinylated probes (Dahlen, 1987; Oser *et al.*, 1990; Syvanen *et al.*, 1986).
- *Fluorescein*
  This fluorophore (fluorescence quantum yield, >0.85; excitation 492 nm; emission 520 nm) has been used effectively in nonseparation energy transfer probe assays with other fluorophore labels and with chemiluminescent labels (isoluminol) (Morrison *et al.*, 1989) (see Chapter 12).
- *Glucose 6-phosphate dehydrogenase*
  A glucose 6-phosphate dehydrogenase label can be measured via the coupled marine bacterial luciferase NAD(P)H : FMN oxidoreductase reaction (Balaguer *et al.*, 1989) (see Chapter 8) (Table VII). The signal, emitted as a glow, is well suited to membrane-based applications.
- *Horseradish peroxidase*
  The enzyme horseradish peroxidase (HRP) catalyzes the chemiluminescent oxidation of luminol and, in the presence of small amounts of certain phenols (*para*-iodophenol), naphthols (1-bromo-2-naphthol) and amines (*para*-anisidine) (*enhancers*), the analytical features of this reaction are significantly improved (Kricka *et al.*, 1988a; Thorpe and Kricka, 1986). The intensity of the light emission is increased by several orders of magnitude, and background light emission from the luminol-peroxide assay reagent is greatly reduced, which leads to a dramatic increase in the signal : background ratio. The light emission from this reaction is a long-lived glow (>30 min), suitable for membrane-based assays (Durrant, 1990; Durrant *et al.*, 1990; Matthews *et al.*, 1985; Schneppenheim and Rautenberg, 1987) (see Chapter 7). A dioxetane substrate for HRP has been also described (Urdea and Warner, 1990). HRP cleaves a 2-methylnaphthoxy substituent from the dioxetane to produce AMPPD, which is detected via its alkaline phosphatase-catalyzed chemiluminescent decomposition (*vide supra*).

  Several substrates that produce insoluble colored products are in use, including 3-amino-9-ethylcarbazole (red product), 4-chloro-1-naphthol (blue product), and 3,3′-diaminobenzidine (brown product). 3,3′,5,5′-Tetramethybenzidine (TMB) is oxidized by HRP to a *semisoluble* blue cationic dye, which can be trapped by treating the membrane with dextran sulfate (Sheldon *et al.*, 1987).

- *Luminol and isoluminol*

Metal ions and peroxidases (e.g., cobalt, microperoxidase) catalyze the chemiluminescent oxidation of luminol and isoluminol. Labeling is most convenient via the aryl amino group, but the chemiluminescence from the luminol label is reduced 10-fold when this amino group is substituted (Schroeder *et al.*, 1978). In contrast, substitution of the aryl amino group of the less efficient isoluminol (chemiluminescence quantum yield: luminol, 0.01, isoluminol, 0.001) causes a 10-fold increase in the quantum yield. Hence isoluminol and its derivatives have been preferred as labels and applied mainly in conjunction with fluorophores in energy-transfer assays (see Chapter 12).

- *Renilla luciferase*

Recombinant *Renilla* (sea pansy) luciferase catalyzes the bioluminescent oxidation of coelenterazine. Light is emitted as a glow (at 480 nm) and, in the presence of green fluorescent protein, energy transfer occurs with a shift in light emission to 508 nm (Ward and Cormier, 1978). This enzyme has been used mainly as a secondary label (biotin conjugate) (Stults *et al.*, 1991).

- *Ruthenium bipyridyl complexes*

Ruthenium and osmium complexes, e.g., ruthenium(II) *tris*(bipyridyl) [$(Ru(bpy)_3{}^{2+})$], can be used as electrochemiluminescent labels and detected by reaction with tripropylamine radicals (Blackburn *et al.*, 1991; Massey *et al.*, 1987). $Ru(bpy)_3^{2+}$ is oxidized at the electrode surface to form $Ru(bpy)^{3+}$. This reacts with tripropylamine cation radicals, to produce excited state $Ru(bpy)_3^{2+}$. This species decays to its ground state, and light is emitted at 620 nm.

- *Xanthine oxidase*

Xanthine oxidase catalyzes the luminescent oxidation of luminol. The long-lived light emission (lasting for several days) is enhanced by an iron–ethylenediaminetetraacetic acid (EDTA) complex via hydroxyl radical production (Balaguer *et al.*, 1991; Baret *et al.*, 1990).

## V. PATENTS

Many aspects of nucleic acid technology (labels, detection of labels, assay formats, amplification, sample treatment) have been patented. Currently, several hundred patents delineate the bounds of proprietary nucleic acid technology. Table VIII provides a brief survey of the scope of some recent patents and patent applications. The intellectual property for which patent protection either has been obtained, or is currently being sought, is divided into eight categories representing different aspects of nucleic acid hybridization assays. Some of the earlier patents directed principally to nonisotopic immunoassays have importance for nucleic acid hybridization because broad claims were allowed to ligands, binding partners, and binding assays (Buckler and Schroeder, 1980), thus anticipating the application of nonisotopic labels in nucleic acid hybridization assays. In the past, the patent literature has represented a very specialized area of interest. However,

**Table VIII**
Selected Patents and Patent Applications Relating to Nucleic Acid Technology

| Technology | Patent | Reference |
|---|---|---|
| Amplification | | |
| Target | WO 8910979 | Adler *et al.* (1989) |
| | WO 9001068 | Berninger (1990) |
| | WO 8901050 | Burg *et al.* (1989) |
| | US 4683202 | Mullis (1987) |
| | WO 8909835 | Orgel (1989) |
| | EP 373960 | Gingeras *et al.* (1990) |
| Probe | WO 9002820 | Axelrod *et al.* (1990) |
| | WO 9002819 | Chu *et al.* (1990) |
| Assay format | | |
| Amplifier probes | WO 9013667 | Urdea *et al.* (1990a) |
| Complementary probes | WO 8801302 | Gingeras *et al.* (1988) |
| Immobilized probe | EP 163220 | Carrico (1985) |
| Immobilized probe—antihybrid | EP 339686 | Carrico (1989) |
| Interacting labeled probes | WO 8603227 | Taub (1986) |
| Multiple probes | EP 320308 | Backman and Wang (1989) |
| | EP 318245 | Hogan and Milliman (1989) |
| | WO 8903891 | Urdea *et al.* (1990b) |
| Devices | | |
| Liquid-impervious receptacle | EP 295069 | Matkovitch (1988) |
| Porous track | EP 387696 | McMahon and Gordon (1990) |
| Reagent-delivery system | EP 299359 | Korom *et al.* (1989) |
| Test sheets | US 4613566 | Potter (1986) |
| Labeling | | |
| Cationic detergent | US 4873187 | Taub (1990) |
| Functionalized polynucleotide | WO 8607361 | Stabinsky (1986) |
| Label and detection system | | |
| Acridinium ester | EP 309230 | Arnold and Nelson (1987) |
| Agglutination | WO 8705334 | Gefter and Holland (1987) |
| Alkaline phosphatase—bacterial bioluminescence | EP 386691 | Watanabe and Shibata (1990) |
| Antibody to intercalation complex | US 4563417 | Albarella & Anderson (1986) |

*(continues)*

**Table VIII** (*continued*)

| Technology | Patent | Reference |
|---|---|---|
| Antihybrid antibody | EP 209702 | Carrico (1987) |
| | EP 144913 | Albarella *et al.* (1985) |
| ATP-encapsulated liposome—bioluminescence | US 4704355 | Bernstein (1988) |
| Arabinose | EP 227459 | McCormick (1987) |
| Avidin-biotin | EP 184701 | Wong (1986) |
| Benzidine meriquinone | EP 219286 | Bloch *et al.* (1987) |
| Carbonic anhydrase inhibitor | EP 210021 | Musso *et al.* (1987a) |
| Catalase | JP 61009300 | Kasahara and Ashihara (1986) |
| Chelator | WO 8702708 | Musso *et al.* (1987b) |
| Dioxetanes | US 4978614 | Bronstein, 1990 |
| | WO 8906226 | Edwards and Bronstein (1989) |
| | CA 2006222 | Akhaven-Tafti and Schaap (1990) |
| | EP 401001 | Urdea and Warner (1990) |
| Electroluminescence | WO 8706706 | Massey *et al.* (1987) |
| Electron transfer (tetra-thiafulvalene mediator) | EP 234938 | Turner *et al.* (1987) |
| Energy transfer | EP 229943 | Heller and Jablonski (1987) |
| | US 4822733 | Morrison (1989) |
| | EP 232967 | Morrison (1987) |
| Enzyme—bacterial bioluminescence | EP 362042 | Nicolas *et al.* (1990) |
| Fluorescence polarization | EP 382433 | Garman and Moore (1990) |
| Horseradish peroxidase—chemiluminescence | US 4598044 | Kricka *et al.* (1986) |
| | US 4729950 | Kricka *et al.* (1988b) |
| HPLC separation | JP 1046650 | Toyo Soda (1989) |
| Labeled cross-linkers | WO 8502628 | Yabucaki *et al.* (1985) |
| Lanthanides | WO 8904375 | Musso *et al.* (1989) |
| Oxidases—chemiluminescence | EP 256932 | Baret (1988) |
| Ruthenium and rhodium chelates | WO 9005732 | Barton (1990) |
| Sample preparation | | |
| Chromatography | WO 8808036 | Tedesco and Petersen (1988) |
| Polycationic support | EP 281390 | Arnold *et al.* (1988) |
| Sonication | EP 337690 | Li *et al.* (1989) |
| Solid supports and separation procedures | | |
| Electrophoresis | EP 244207 | Murao and Hosaka (1987) |
| Gel exclusion | EP 278220 | Kuhns (1988) |
| Magnetic particles | WO 9006042 | Hornes and Korsnes (1990b) |
| | WO 8605815 | Hill *et al.* (1986) |
| Microbeads | EP 304845 | Wang (1989) |
| Nylon (amidinated) | EP 221308 | Carrico and Patterson (1987) |
| Photochemically reactive solid support | EP 130523 | Dattagupta and Crothers (1985) |
| Solid carrier—microfiltration | EP 200113 | Jolley and Ekenberg (1986) |
| Superparamagnetic particles | WO 9006045 | Hornes and Korsnes (1990a) |
| Transparent controlled-pore glass particles | US 4780423 | Bluestein *et al.* (1988) |
| Waveguide | EP 245206 | Sutherland *et al.* (1987) |

*(continued)*

**Table VIII** (*continued*)

| Technology | Patent | Reference |
|---|---|---|
| Specific probes | | |
| *Entamoeba histolytica* | GB 2223307 | Huber *et al.* (1990) |
| *Listeria monocytogenes* | US 7411965 | Datta (1990) |
| *Mycobacteria* | EP 398677 | Wirth and Patel (1990) |
| *Neisseria gonorrhoeae* | EP 353985 | Woods *et al.* (1990) |
| *Neisseria* strains | EP 337896 | Rossau and Vanheuvers (1989) |
| Non-A, non-B hepatitis | EP 377303 | Mishiro and Nakamura (1990) |
| Papillomavirus | US 4908306 | Lorincz (1990) |
| Type I diabetes mellitus | US 4965189 | Owerbach (1990) |
| *Yersinia enterocolitica* | EP 339783 | Shah *et al.* (1989) |

recent developments indicate that a general appreciation of patents and their scope will become increasingly important in view of the trend to the broad enforcement of issued patents (e.g., PCR).

## VI. CONCLUSIONS

Nucleic acid hybridization assays are still under active development; this is most evident in the diversity of nonisotopic labels and end-points. No single label or labeling, and label detection method has gained supremacy. The following chapters describe aspects of current nonisotopic replacements for the $^{32}$phosphorus label in nucleic acid hybridization assays.

## REFERENCES

Adler, K. E., Greene, R. A., amd Kasila, P. A. (1989). Enzymatic amplification and detection of nucleic acids. *WO Patent* 8910979.

Akhavan-Tafti, H., and Schaap, A. P. (1990). New chemiluminescent 1,2-dioxetane compounds. *Can. Patent* 2006222.

Albarella, J. P., and Anderson, L. H. D. (1986). Nucleic acid hybridization assay using antibodies to intercalating complexes with double-stranded nucleic acid. *U.S. Patent* 4563417.

Albarella, J. P., Anderson, L. H. D., and Carrico, R. J. (1985). Determination of polynucleotide sequence in sample by using labeled probe and antihybrid. *Eur. Patent* 144913.

Al-Hakeem, M., and Sommer, S. S. (1987). Terbium identifies double-stranded RNA on gels by quenching the fluorescence of intercalated ethidium bromide. *Anal. Biochem.* **163,** 433–439.

Al-Hakim, A. H., and Hull, R. (1986). Studies towards the development of chemically synthesized nonradioactive biotinylated hybridization probes. *Nucleic Acids Res.* **14,** 9965–9976.

Arnold, L. J., Jr., and Nelson, N. C. (1987). Homogeneous protection assay. *Eur. Patent Appl.* 309230.

Arnold, L. J., Jr., Nelson, N. C., Reynolds, M. A., and Waldrop, A. A. (1988). Separating polynucleotide from solution by binding to polycationic support. *Eur. Patent* 281390.

Arnold, L. J., Jr., Hammond, P. W., Wiese, W. A., and Nelson, N. C. (1989). Assay formats involving acridinium-ester-labeled DNA probes. *Clin. Chem.* **35,** 1588–1594.

Axelrod, V. D., Kramer, F. R., Zardi, P. M., and Mills, D. R. (1990). Oligonucleotide-promoter DNA molecular adduct. *WO Patent* 9002820.

Backman, K. C., and Wang, C. N. J. (1989). Detecting target nucleic acid in a sample. *Eur. Patent* 320308.

Balaguer, P., Terouanne, B., Boussioux, A. M., and Nicolas, J. C. (1989). Use of bioluminescence in nucleic acid hybridization reactions. *J. Biolumin. Chemilumin.* **4,** 302–309.

Balaguer, P., Bonafe, N., Terouanne, B., Boussioux, A. M., Baret, A., and Nicolas, J. C. (1991). Luminescent detection of DNA probes: A comparative study of four luminescent detection systems. *In* "Bioluminescence and Chemiluminescence: Current Status" (P. E. Stanley and L. J. Kricka, eds.), pp. 187–190. Wiley, Chichester, England.

Baret, A. (1988). Signal reagent for chemiluminescent assay. *Eur. Patent* 256932.

Baret, A., Fert, V., and Aumaille, J. (1990). Application of a long-term enhanced xanthine oxidase-induced luminescence in solid-phase immunoassays. *Anal. Biochem.* **187,** 20–26.

Barton, J. K. (1990). New transition metal complexes. *WO Patent* 9005732.

Bauman, J. G. J., Wiegent, J., and Van Duijn, P. (1981). Cytochemical hybridization with fluorochrome-labeled RNA. *J. Histochem. Cytochem.* **29,** 227–237.

Berninger, M. (1990). Sequence-specific assay for detection of nucleic acids. *WO Patent* 9001068.

Bernstein, D. (1988). Assay for specific binding pair analyte. *U.S. Patent* 4704355.

Blackburn, G. F., Shah, H. S., Kenten, J. H., Leland, J., Kamin, R. A., Link, J., Peterman, J., Shah, A., Talley, D. B., Tyagi, S. K., Wilkins, E., Wu, T.-G., and Massey R. J. (1991). Electrochemiluminescence detection for development of immunoassays and DNA-probe assays for clinical diagnostics. *Clin. Chem.,* **37,** 1534–1539.

Bloch, W., Sheridan, P. J., and Goodson, R. J. (1987). Composition for visualization of biological materials. *Eur. Patent* 219286.

Bluestein, B. I., Famulare, A. J., and Worthy, T. E. (1989). Fluorescent assay for ligand-ligand-binding partner complex. *U.S. Patent* 4780423.

Broker, T. R., Angerer, L. M., Yen, P. H., Hershey, N. D., and Davidson, N. (1978). Electron microscopic visualization of tRNA genes with ferritin-avidin : biotin labels. *Nucleic Acids Res.* **5,** 363–384.

Bronstein, I. (1990). Method of detecting a substance using enzymatically induced decomposition of dioxetanes. *U.S. Patent* 4978614.

Bronstein, I., and Kricka, L. J. (1989). Clinical applications of luminescent assays for enzymes and enzyme labels. *J. Clin. Lab. Anal.* **3,** 316–322.

Bronstein, I., Edwards, B., and Voyta, J. C. (1990). 1,2-Dioxetanes: Novel chemiluminescent enzyme substrates. *J. Biolumin. Chemilumin.* **4,** 99–111.

Bronstein, I., Juo, R. R., Voyta, J. C., and Edwards, B. (1991). Novel chemiluminescent adamantyl 1,2-dioxetane enzyme substrates. *In* "Bioluminescence and Chemiluminescence: Current Status" (P. E. Stanley and L. J. Kricka, eds.), pp. 73–82. Wiley, Chichester, England.

Buck, G. E. (1989). Nonculture methods for detection and identification of microorganisms in clinical specimens. *Selected Top. Ped. Pathol.* **36,** 95–112.

Buckler, R. T., and Schroeder, H. R. (1980). Chemiluminescent naphthalene-1,2-dicarboxylic acid hydrazide-labeled polypeptides and proteins. *U.S. Patent* 4225485.

Budowle, B., Baechtel, F. S., Guisti, A. M., and Monson, K. L. (1990). Applying highly polymorphic variable number of tandem repeats loci genetic markers to identity testing. *Clin. Biochem.* **23,** 287–293.

Burg, L. J., Pouletty, P., and Broothroyd, J. C. (1989). Amplification of target polynucleotide sequences. *WO Patent* 8901050.

Carrico, R. J. (1985). Determination of specific polynucleotide sequence in test medium by adding immobilized probe containing complementary sequence and detecting duplex formation. *Eur. Patent* 163220.
Carrico, R. J. (1987). Solid-phase hybridization assay using polynucleotide probe and detecting hybridized probe using antihybrid antibodies. *Eur. Patent* 209702.
Carrico, R. J. (1989). Nucleic acid hybridization assay. *Eur. Patent* 339686.
Carrico, R. J., and Patterson, W. L. (1987). Immobilized nucleic acid on nylon support carrying amidine residues. *Eur. Patent* 221308.
Cawood, A. H. (1989). DNA fingerprinting. *Clin. Chem.* **35,** 1832–1837.
Chu, B. C. F., Wahl, G. M., and Orgel, L. E. (1983). Derivatization of unprotected oligonucleotides. *Nucleic Acids Res.* **11,** 6513–6528.
Chu, B. C. F., and Orgel, L. E. (1985). Detection of specific DNA sequences with short biotin-labeled probes. *DNA* **4,** 327–331
Chu, B. C., Yoyce, G. F., and Orgel, L. E. (1990). Adduct of linked moieties of oligonucleotide sequence. *WO Patent* 9002819.
Cimino, G. D., Gamper, H. D., Isaacs, S. T., and Hearst, J. E. (1985). Psoralens as photoactive probes of nucleic acid structure and function: Organic chemistry, photochemistry, and biochemistry. *Annu. Rev. Biochem.* **54,** 1151–1193.
Czichos, J., Kohler, M., Reckmann, B., and Renz, M. (1989). Protein–DNA conjugates produced by UV irradiation and their use as probes for hybridization. *Nucleic Acids Res.* **17,** 1563–1572.
Dahlen, P. (1987). Detection of biotinylated DNA probes by using $Eu^{3+}$-labeled streptavidin and time-resolved fluorimetry. *Anal. Biochem.* **164,** 78–83.
Datta, A. (1990). Synthetic *Listeria monocytogenes* oligonucleotide probes. *U.S. Patent* 7411965.
Dattagupta, N., and Crothers, D. (1985). Solid support with photochemically linked nucleic acid. *Eur. Patent* 130523.
Dawson, D. B. (1990). Use of nucleic acid probes in genetic tests. *Clin. Biochem.* **23,** 279–285.
Diamandis, E. P. (1988). Immunoassays with time-resolved fluorescence spectroscopy; principles and applications. *Clin. Biochem.* **21,** 139–150.
Diamandis, E. P. (1990). Analytical methodology for immunoassays and DNA hybridization assays—current status and selected systems. *Clin. Chim. Acta* **194,** 19–50.
DiLella, A. G., Marvit, J., Lidsky, A. S., Guettler, F., Woo, S. L. C. (1986). Tight linkage between a splicing mutation and a specific DNA haplotype in phenylketonuria. *Nature* **322,** 799–803.
Draper, D. E., and Gold, L. M. (1980). A method for linking fluorescent labels to polynucleotides: Application to studies of ribosome–ribonucleic acid interactions. *Biochemistry* **19,** 1774–1781.
Dular, R., Kajioka, R., and Kasatiya, S. (1988). Comparison of Gen-Probe commercial kit and culture technique for the diagnosis of *Mycoplasma pneumoniae* infection. *J. Clin. Microbiol.* **26,** 2240–2245.
Durrant, I. (1990). Light-based detection of biomolecules. *Nature* **346,** 297–298.
Durrant, I., Benge, L. C. A., Sturrock, C., Devenish, A. T., Howe, R., Roe, S., Moore, M., Scozzafava, G., Proudfoot, L. M. F., Richardson, T. C., McFarthing, K. G. (1990). The application of enhanced chemiluminescence to membrane-based nucleic acid detection. *BioTechniques* **8,** 564–570.
Edwards, B., and Bronstein, I. Y. (1989). New cycloalkyl substituted fluorescent chromophore compounds. *WO Patent* 8906226.
Evangelista, R. A., Pollack, A., and Gudgin Templeton, E. F. (1991). Enzyme-amplified lanthanide luminescence for enzyme detection in bioanalytical assays. *Anl. Biochem.* **197,** 213–224.

Feinberg, A. P., and Vogelstein, B. (1983). A technique for radiolabeling DNA restriction endonuclease fragments to high specific activity. *Anal. Biochem.* **132,** 6–13.

Flores, J., Purcell, R. H., Perez, T., Wyatt, R. G., Boeggeman, E., Sereno, M., White, L., Chanock, R. M., and Kapikian, A. Z. (1983). A dot hybridization assay for detection of rotavirus. *Lancet* **i,** 555–559.

Forster, A. C., McInnes, J. L., Skingle, D. C., and Symons, R. H. (1985). Nonradioactive hybridization probes prepared by the chemical labeling of DNA and RNA with a novel reagent, photobiotin. *Nucleic Acids Res.* **13,** 745–762.

Garman, A. J., and Moore, R. S. (1990). Detecting nucleic acid sequence using fluorescence polarization. *Eur. Patent* 382433.

Gefter, M. L., and Holland, C. A. (1987). Detection of nucleic acid sequences. *WO Patent* 8705334.

Gill, P., Jeffreys, A. J., and Werrett, D. J. (1985). Forensic applications of DNA "fingerprints." *Nature* **318,** 577–579.

Gingeras, T. R., Ghosh, S. S., Davis, G. R., Kwoh, D. Y., and Musso, G. F. (1988). Nucleic acid probe assay. *WO Patent* 8801302.

Gingeras, T. R., Guatelli, J. C., and Whitfield, K. M. (1990). Amplification of RNA sequences. *Eur. Patent* 373960.

Haralambidis, J., Chai, M., and Tregear, G. (1987). Preparation of base-modified nucleosides suitable for nonradioactive label attachment and their incorporation into synthetic oligodeoxyribonucleotides. *Nucleic Acids Res.* **15,** 4857–4876.

Hauber R., Miska W., Schleinkofer L., and Geiger R. (1989). New, sensitive, radioactive-free bioluminescence-enhanced detection system in protein blotting and nucleic acid hybridization. *J. Biolumin. Chemilumin.* **4,** 367–372.

Heller, M. J., and Jablonski, E. J. (1987). Single-strand polynucleotide sequences fluorescent hybridization assay. *Eur. Patent* 229943.

Hill, H. A. O., Gear, M. J., Williams, S. C., and Green, M. J. (1986). Magnetized nucleic acid sequence comprising single- or double-stranded nucleic acid linked to magnetic or magnetizable substance. *WO Patent* 8605815.

Hogan, J. J., and Milliman, C. L. (1989). Enhancing nucleic acid hybridization. *Eur. Patent* 318245.

Hopman, A. H. N., Wiegant, J., Tesser, G. I., and Van Duijn, P. (1986). A nonradioactive *in situ* hybridization method based on mercurated nucleic acid probes and sulfhydryl-hapten ligands. *Nucleic Acids Res.* **14,** 6471–6488.

Hooper, C. E., and Ansorge, R. E. (1990). Quantitative luminescence imaging in the biosciences using the CCD camera: Analysis of macro and microsamples. *Trends Anal. Chem.* **9,** 269–277.

Hornes, E., and Korsnes, L. (1990a). Novel nucleic acid probes. *WO Patent* 9006045.

Hornes, E., and Korsnes, L. (1990b). Detection and quantitative determination of target RNA or DNA. *WO Patent* 9006042.

Hu, N., and Messing, J. (1982). The making of strand-specific M13 probes. *Gene* **17,** 271–277.

Huber, M., Gitler, C., Revel, M., Mirelman, D., Garfinkel, L. I., and Giladi, M. (1990). DNA probes for *Entamoeba histolytica* detection and differentiation. *G.B. Patent* 2223307.

Jolley, M. E., and Ekenberg, S. J. (1986). Solid phase nucleic acid hybridization assay involves hybridization whilst solid carrier is suspended for use in DNA or RNA assay. *Eur. Patent* 200113.

Kam, W., Rall, L. B., Smuckler, E. A., Schmid, R., and Rutter, W. J. (1982). Hepatitis B viral DNA in liver and serum of asymptomatic carriers. *Proc. Natl. Acad. Sci. U.S.A.* **79,** 7522–7526.

Kasahara, Y., and Ashihara, Y. (1986). Heat-resistant enzyme-labeled polynucleotide for polynucleotide determination. *Jap. Patent* 61009300.

Keller, G. H., and Manak, M. M. (1989). "DNA Probes." Stockton Press, New York.
Keller, G. H., Cumming, C. U., Huang, D. P., Manak, M. M., and Ting, R. (1988). A chemical method for introducing haptens onto DNA probes. *Anal. Biochem.* **170,** 441–450.
Kerem, B.-S., Rommens, J. M., Buchanan, J. A., Markiewicz, D., Cox, T. K., Chakravarti, A., Buchwald, M., Tsui, L.-C. (1989). Identification of the cystic fibrosis gene: Genetic analysis. *Science* **245,** 1073–1080.
Knudson, A. G. (1986). Genetics of human cancer. *Annu. Rev. Genet.* **20,** 231–251.
Korom, G. K., Caplan, D. E., Eng, K. K. M., and Hansa, J. G. (1989). Reagent delivery system for solid-phase assay device. *Eur. Patent* 299359.
Kricka, L. J. (1985). "Ligand-Binder Assays." Dekker, New York.
Kricka, L. J., Stott, R. A. W., and Thorpe, G. H. G. (1988a). Enhanced chemiluminescence enzyme immunoassays. *In* "Complementary Immunoassays" (W. P. Collins, ed.), pp. 169–179. Wiley, Chichester, England.
Kricka, L. J., O'Toole, M., Thorpe, G. H. G. H., and Whitehead, T. P. (1988b). Enhanced chemiluminescent or luminometric assay. *U.S. Patent* 4729950.
Kricka, L. J., Thorpe, G. H. G. H., and Whitehead, T. P. (1986). Enhanced chemiluminescent or luminometric assay. *U.S. Patent* 4598044.
Kunkel, L. M., Beggs, A. H., and Hoffman, E. P. (1989). Molecular genetics of Duchenne and Becker muscular dystrophy; emphasis on improved diagnosis. *Clin. Chem.* **35,** B21–B24.
Kuhns, M. C. (1988). Target nucleic acid sequences detection. *Eur. Patent* 278220.
Kwoh, D. Y., Davis, G. R., Whitfield, K. M., Chappelle, H. L., DiMichelle, L. J., and Gingeras, T. S. (1989). Transcription-based amplification system detection of amplified human deficiency virus type I with a bead-based sandwich hybridization format. *Proc. Natl. Acad. Sci. U.S.A.* **86,** 1173–1177.
Landry, M. L. (1990). Nucleic acid hybridization in viral diagnosis. *Clin. Biochem.* **23,** 267–77.
Leary, J. J., Brigati, D. J., and Ward, D. C. (1983). Rapid and sensitive colorimetric method for visualizing biotin-labeled DNA probes hybridized to DNA or RNA on nitrocellulose: Bio-blots. *Proc. Natl. Acad. Sci. U.S.A.* **80,** 4045–4049.
Leary, J. J., and Ruth, J. L. (1989). Nonradioactive labeling of nucleic acid probes. *In* "Nucleic Acid and Monoclonal Antibody Probes" (B. Swaminathan and G. Prakash, eds.), pp. 33–57. Dekker, New York.
Lebacq, P., Squalli, D., Duchenne, M., Pouletty, P., and Joannes, M. (1988). A new sensitive nonisotopic method using sulfonated probes to detect picogram quantities of specific DNA sequences on dot blot hybridizations. *J. Biochem. Biophys. Methods* **15,** 255–266.
Li, M. K., McLaugglin, D., Palome, E., and Kessler, J. (1989). Release of nucleic acid from sample. *Eur. Patent* 337690.
Lion, T., and Haas, O. A. (1990). Nonradioactive labeling of probe with digoxigenin by polymerase chain reaction. *Anal. Biochem.* **188,** 335–337.
Lizardi, P. M., Guerra, C. E., Lomeli, H., Tussie-Luna, I., and Kramer, F. R. (1988). Exponential amplification of recombinant RNA hybridization probes. *Biotechnology* **6,** 1197–1202.
Lorincz, A. T. (1990). Nucleic acid hybridization probes. *U.S. Patent* 4908306.
Lovell, M. A. (1989). Molecular genetics of leukemia and lymphoma. *Clin. Chem.* **35,** B43–B47
Malcolm, A. D. B., and Nicolas, J. L. (1984). Method of detecting a polynucleotide sequence and labeled polynucleotides useful therein. *WO Patent* 8403520.
Mantero, G., Zonaro, A., Albertini, A., Bertolo, P., and Primi, D. (1991). DNA enzyme immunoassay: General method for detecting products of polymerase chain reaction. *Clin. Chem.* **37,** 422–429.

Massey, R. J., Powell, M. J., Mied, P. A., Feng, P., Della, C. L., Dressick, W. J., and Poonian, M. S. (1987). Electrochemiluminescent assays and kits using ruthenium and osmium bipyridyl complexes as labels. *WO Patent* 8706706.

Mathies, R. A., and Stryer, L. (1986). Single molecule fluorescence detection: A feasibility study using phycoerythrin. *In* "Applications of Fluorescence in the Biomedical Sciences" (D. Lansing Taylor, A. S. Waggoner, R. F. Murphy, F. Lanni, and R. R. Birge, eds.) pp. 129–140. Liss, New York.

Matkovich, V. I. (1989). Detection of components in liquid sample. *Eur. Patent* 295069.

Matthews, J. A., and Kricka, L. J. (1988). Analytical strategies for the use of DNA probes. *Anal. Biochem.* **169,** 1–25.

Matthews, J. A., Batki, A., Hynds, C., and Kricka, L. J. (1985). Enhanced chemiluminescent method for the detection of DNA dot-hybridization assays. *Anal. Biochem.* **151,** 205–209.

Maxam, A., and Gilbert, W. (1977). A new method for sequencing DNA. *Proc. Natl. Acad. Sci. U.S.A.* **74,** 560–564.

McCormick, R. M. (1987). Arabinonucleic acid probes. *Eur. Patent* 227459.

McGadey, J. (1970). A tetrazolium method for nonspecific alkaline phosphatase. *Histochemie* **23,** 180–184.

McGowan, K. L. (1989). Infectious diseases: Diagnosis utilizing DNA probes. *Clin. Pediatr.* **28,** 157–162.

McMahon, M. E., and Gordon, J. (1990). Improved reaction kinetics in nucleic acid hybridization using porous chromatographic material as solid support. *Eur. Patent* 387696.

Miller, C. A., Patterson, W. L., Johnson, P. K., Swartzell, C. T., Wogoman, F., Albarella, J. P., and Carrico, R. J. (1988). Detection of bacteria by hybridization of rRNA with DNA-latex and immunodetection of hybrids. *J. Clin. Microbiol.* **26,** 1271–1276.

Mishiro, S., and Nakamura, T. (1990). Non-A, non-B hepatitis virus genome RNA. *Eur. Patent* 377303.

Miska, W., and Geiger, R. (1987). Synthesis and characterization of luciferin derivatives for use in bioluminescence-enhanced enzyme immunoassays. New ultrasensitive detection systems for enzyme immunoassays I. *J. Clin. Chem. Clin. Biochem.* **25,** 23–30.

Miska, W., and Geiger, R. (1990). Luciferin derivatives in bioluminescence-enhanced enzyme immunoassays. *J. Biolumin. Chemilumin.* **4,** 119–128.

Morrison, L. E. (1987). Competitive homogeneous assay. *Eur. Patent* 232967.

Morrison, L. E. (1989). Improved luminescent homogeneous assays. *U.S. Patent* 4822733.

Morrison, L. E., Halder, T. C., and Stols, L. M. (1989). Solution-phase detection of polynucleotides using interacting fluorescent labels and competitive hybridization. *Anal. Biochem.* **183,** 231–244.

Mullis, K. B. (1987). Process for amplifying nucleic acid sequences. *U.S. Patent* 4683202.

Mullis, K. B., and Faloona, F. A. (1987). Specific synthesis of DNA *in vitro* via a polymerase-catalyzed chain reaction. *Methods Enzymol.* **155,** 335–350.

Murao, Y., and Hosaka, S. (1987). Specific binding assay. *Eur. Patent* 244207.

Musso, G. F., Ghosh, G. F., Orgel, L. E., and Wahl, G. M. (1987a). Carbonic anhydrase inhibitor-tagged nucleic acid probes. *Eur. Patent* 210021.

Musso, G. F., Ghosh, G. F., and Gingeras, T. R. (1987b). New nucleic acid probes for hybridization assays. *WO Patent* 8702708.

Musso, G. F., Ghosh, G. F., and Gingeras, T. R. (1989). Lanthanide chelated tagged nucleic acid probes. *WO Patent* 8904375.

Nicolas, J. C., Balaguer, P., Terouanne, B., and Boussioux, A. M. (1990). Specific DNA or RNA sequence assay by bioluminescent reaction. *Eur. Patent* 362042.

Odelberg, S. J., Demers, D. B., Westin, E. H., and Hossaini, A. A. (1988). Establishing paternity using minisatellite DNA probes when the putative father is unavailable for testing. *J. Forens. Sci.* **33,** 921–928.

Orgel, L. E. (1989). Amplification of a target DNA segment of known sequence. *WO Patent* 8909835.

Oser, A., Collasius, M., and Valet, G. (1990). Multiple end labeling of oligonucleotides with terbium chelate-substituted psoralen for time-resolved fluorescence detection. *Anal. Biochem.* **191,** 295–301.

Owerbach, D. (1990). DQ beta gene oligonucleotide(s) for detection of proclivity in humans for development of type I diabetes mellitus. *U.S. Patent* 4965189.

Pollard-Knight, D. V. (1991) Rapid and sensitive luminescent detection methods for nucleic acid detection. *In* "Bioluminescence and Chemiluminescence: Current Status" (P. E. Stanley and L. J. Kricka, eds.), pp. 83–90. Wiley, Chichester, England.

Potter, H. (1986). Simultaneous hybridization assay of multiple RNA or DNA fragments by stacking several test sheets with single probe sheet containing bands of labeled material. *U.S. Patent* 4613566.

Prasher, D., McCann, R. O., and Cormier, M. J. (1985). Cloning and expression of the cDNA for Aequorin, a bioluminescent calcium-binding protein. *Biochem. Biophys. Res. Commun.* **126,** 1259–1268.

Renz, M. (1983). Polynucleotide-histone H-1 complexes as probes for blot hybridizations. *Eur. J. Mol. Biol.* **2,** 817–822.

Renz, M., and Kurz, C. (1984). A colorimetric method for DNA hybridization. *Nucleic Acids Res.* **12,** 3435–3444.

Richardson, R. W., and Gumport, R. I. (1983). Biotin and fluorescent labeling of RNA using $T_4$ RNA ligase. *Nucleic Acids Res.* **11,** 6167–6184.

Rigby, P. W. J., Dieckmann, M., Rhodes, C., and Berg, P. (1977). Labeling deoxyribonucleic acid to high specificity by nick translation with DNA polymerase I. *J. Mol. Biol.* **113,** 237–257.

Riley, L. K., Marshall, M. E., and Coleman, M. S. (1986). A method for biotinylating oligonucleotide probes for use in molecular hybridizations. *DNA* **5,** 333–337.

Riordan, J. R., Rommens, J. M., Kerem, B.-S., Alon, N., Rozmahel, R., Grzelczak, Z., Zielenski, J., Lok, S., Plavsic, N., Chou, J.-L., Drumm, M. L., Iannuzzi, M. C., Collins, F. C., Tsui, L.-C. (1989). Identification of the cystic fibrosis gene; cloning and characterization of complementary DNA. *Science* **245,** 1066–1073.

Ropers, H. H. (1987). Use of DNA probes for diagnosis and prevention of inherited disorders. *Eur. J. Clin. Invest.* **17,** 475–487.

Rossau, R., and Vanheuvers, H. (1989). Hybridization probes for Neisseria strains. *Eur. Patent* 337896.

Ruth, J. L. (1984). Chemical synthesis of nonradioactively labeled DNA hybridization probes. *DNA* **3,** 123.

Ruth, J. L., and Bryan, R. N. (1984). Chemical synthesis of modified oligonucleotides and their utility as nonradioactive hybridization probes. *Fed. Proc.* **43,** 2048.

Ruth, J. L., Morgan, C. A., and Pasko, A. (1985). Linker arm nucleotide analogs useful in oligonucleotide synthesis. *DNA* **4,** 93.

Saiki, R. K., Scharf, S., Faloona, F., Mullis, K. B., Horn, G. T., Erlich, H. A., and Arnhein, N. (1985). Enzymatic amplification of beta-globin genomic sequences and restriction site analysis for diagnosis of sickle-cell anemia. *Science* **230,** 1350–1354.

Sanchez-Pescador, R., Stempien, M. S., and Urdea, M. S. (1988). Rapid chemiluminescent nucleic acid assays for the detection of TEm-1 beta-lactamase-mediated penicillin resistance in *Neisseria gonorrhoeae*. *J. Clin. Microbiol.* **26,** 1934–1938.

Schaap, A. P., and Akhaven-Tafti, H. (1990). New chemiluminescent 1,2-dioxetane derivatives. *WO Patent* 9007511.

Schaap, A. P., Akhaven, H., and Romano, L. J. (1989). Chemiluminescent substrates for alkaline phosphatase: Applications to ultrasensitive enzyme-linked immunoassays and DNA probes. *Clin. Chem.* **35,** 1863–1864.

Schneppenheim, R., and Rautenberg, P. (1987). A luminescence Western blot with enhanced sensitivity for antibodies to human immunodeficiency virus. *Eur. J. Microbiol.* **6,** 49–51.

Schowalter, D. B., and Sommer, S. S. (1989). The generation of radiolabeled DNA and RNA probes with polymerase chain reaction. *Anal. Biochem.* **177,** 90–94.

Schroeder, H. R., Boguslaski, R. C., Carrico, R. J., and Buckler, R. T. (1978). Monitoring specific binding reactions with chemiluminescence. *Methods Enzymol.* **57,** 424–445.

Shah, J. S., Chan, S. W., Pitman, T. B., and Lane, D. J. (1989). Detection of *Yersinia enterolitica*. *Eur. Patent* 339783.

Sheldon, E., Kellogg, D. E., Levenson, C., Bloch, W., Aldwin, L., Birch, D., Goodson, R., Sheridan, P., Horn, G., Watson, R., and Erlich, H. A. (1987). Nonisotopic M13 probes for detecting the beta-globin gene: Application to diagnosis of sickle-cell anaemia. *Clin. Chem.* **33,** 1368–1371.

Slamon, D. J., Clark, G. M., Wong, S. G., Levin, W. J., Ullrich, A., and McGuire, W. (1987). Human breast cancer: Correlation of relapse and survival with amplification of the HER-2/*neu* oncogene. *Science* **235,** 177–182.

Smith, L. M., Fung, S., Hunkapiller, M. W., Hunkapiller, T. J., and Hood, L. (1985). The synthesis of oligonucleotides containing an aliphatic amino group at the 5′-terminus: Synthesis of fluorescent DNA primers for use in DNA sequence analysis. *Nucleic Acids Res.* **13,** 2399–2412.

Sodja, A., and Davidson, N. (1978). Gene mapping and gene enrichment by the avidin–biotin interaction: Use of cytochrome c as a polyamine bridge. *Nucleic Acids Res.* **5,** 385–401.

Soini, E., and Lovgren, T. (1987). Time-resolved fluorescence of lanthanide probes and applications in biotechnology. *CRC Crit. Rev. Anal. Chem.* **18,** 105–154.

Spector, S. A., and Spector, D. H. (1985). The use of DNA probes in studies of human cytomegalovirus. *Clin. Chem.* **31,** 1514–1520.

Sproat, B. A., Beijer, B., and Rider, P. (1987). The synthesis of protected 5′-amino-2′,5′-dideoxyribonucleoside 3′-*O*-phosphoramidites; applications of 5′-amino-oligoribonucleotides. *Nucleic Acids Res.* **15,** 6181–6196.

Stabinsky, Y. (1986). Preparation of functionalised polynucleotide using synthesis of polynucleotide on a hydroxylamine-linked support-bound carboxyl. *WO Patent* 8607361.

Stanley, C. J., and Johannsson, A. (1988). Device for biochemical assay. *WO Patent* 8804428.

Stults, N. L., Stocks, N. A., Cummings, R. D., Cormier, M. J., and Smith, D. F. (1991). Applications of recombinant bioluminescent proteins as probes for proteins and nucleic acids. *In* "Bioluminescence and Chemiluminescence: Current Status" (P. E. Stanley and L. J. Kricka, eds.), pp. 533–536. Wiley, Chichester, England.

Sutherland, R. M., Bromley, P., and Gentile, B. (1987). Determining specifically sequenced nucleic acid. *Eur. Patent* 245206.

Syvanen, A.-C., Alanen, M., and Soderlund, H. A. (1985). Complex of single-strand binding protein and M13 DNA as hybridization probe. *Nucleic Acids Res.* **13,** 2789–2802.

Syvanen, A.-C., Tchen, P., Ranki, M., and Soderlund, H. (1986). Time-resolved fluorimetry: A sensitive method to quantify DNA-hybrids. *Nucleic Acids Res.* **14,** 1017–1028.

Taub, F. (1986). Assay for nucleic acid sequences especially genetic lesions. *WO Patent* 8603227.

Taub, F. (1990). Conjugation of enzyme and nucleic acid molecule. *U.S. Patent* 4873187.

Tchen, P., Fuchs, R. P. P., Sage, E., and Leng, M. (1984). Chemically modified nucleic acids as immunodetectable probes in hybridization experiments. *Proc. Natl. Acad. Sci. U.S.A.* **81,** 3466–3470.

Tedesco, J. L., and Peterson, J. (1988). Isolating DNA and/or RNA from samples containing protein. *WO Patent* 8808036.

Thacker, J., Webb, M. B. T., and Debenham, P. G. (1988). Fingerprinting cell lines; use of human hypervariable DNA probes to characterize mammalian cell cultures. *Somatic Cell Mol. Genet.* **14,** 519–525.

Thompson, J. D., Decker, S., Haines, D., Collins, R. S., Feild, M., and Gillespie, D. (1989). Enzymatic amplification of RNA purified from crude cell lysate by reversible target capture. *Clin. Chem.* **35,** 1878–1881.

Thornton, J. I. (1989). DNA profiling. *Chem. Eng. News* **Nov 20,** 18–30.

Thorpe, G. H. G., and Kricka, L. J. (1986). Enhanced chemiluminescent reactions catalyzed by horseradish peroxidase. *Methods Enzymol.* **133,** 331–354.

Toyo Soda Manufacturing Company (1989). Nucleic acid hybridization assay in biological samples. *Jap. Patent* 1046650

Turner, A. P. F., Hendry, S. P., and Cardosi, M. F. (1987). Bioelectrochemical electron transfer process. *Eur. Patent* 234938.

Urdea, M. S., and Warner, B. D. (1990). Chemiluminescent doubletriggered 1,2-dioxetanes. *Eur. Patent Appl.* 401001

Urdea, M. S., Running, J. A., Horn, T., Clyne, J., Ku, L., and Warner, B. D. (1987). A novel method for the rapid detection of specific nucleotide sequences in crude biologic samples without blotting or radioactivity; application to the analysis of hepatitis B virus in human serum. *Gene* **61,** 253–264.

Urdea, M. S., Warner, B. D., Running, J. A., Stempien, M., Clyne, J. and Horn, T. (1988). A comparison of nonradioisotopic hybridization assay methods using fluorescent, chemiluminescent and enzyme-labeled synthetic oligodeoxyribonucleotide probes. *Nucleic Acids Res.* **16,** 4937–4956.

Urdea, M. S., Warner, B., Running, J. A., Kolberg, J. A., Clyne, J. M., Sanchez-Pescador, R., and Horn, T. (1990a). Amplified nucleic acid sandwich hybridization assay for hepatitis B virus using amplifier and capture probes. *WO Patent* 9013667

Urdea, M. S., Warner, B., Running, J. A., Kolberg, J. A., Clyne, J. M., Sanchez-Pescador, R., and Horn, T. (1990b). Nucleic acid multimer for hybridization assays. *WO Patent* 8903891

Viscidi, R. P., Connelly, C. J., and Yolken, R. H. (1986). Novel chemical method for the preparation of nucleic acids for nonradioactive hybridization. *J. Clin. Microbiol.* **23,** 311–317.

Wang, C. G. (1989). Assaying gene expressions. *Eur. Patent* 304845.

Ward, W. W., and Cormier, M. J. (1978). Energy transfer via protein–protein interaction in *Renilla* bioluminescence. *Photochem. Photobiol.* **27,** 389–396.

Watanabe, H., and Shibata, S. (1990). DNA probe assay with bacterial luciferase system. *Eur. Patent* 386691.

Weeks, I., Beheshti, I., McCapra, F., Campbell, A. K., and Woodhead, J. S. (1983). Acridinium esters as high-specific activity labels in immunoassay. *Clin. Chem.* **29,** 1474–1479.

Wick, R. A. (1989). Photon counting imaging: Applications in biomedical research. *BioTechniques* **7,** 262–268.

Wilchek, M., and Bayer, E. A. (eds.) (1990). Avidin-biotin technology. Methods in Enzymology, **184,** Academic Press, San Diego.

Wilkinson, H. W., Sampson, J. S., and Plikaytis, B. B. (1986). Evaluation of a commercial gene probe for identification of *Legionella* cultures. *J. Clin. Microbiol.* **23,** 217–220.

Williams, D. L. (1985). Molecular biology in arteriosclerosis research. *Arteriosclerosis* **5,** 213–227.

Wirth, D. F., and Patel, R. J. (1990). Identification of *mycobacteria* by hybridization using specific nucleic acid sequence. *Eur. Patent* 398677.

Wolfe, H. J. (1988). DNA probes in diagnostic pathology. *Am. J. Clin. Pathol.* **90,** 340–344.

Wong, G. (1986). Determination of a ligand using specific binding substance for ligand labeled with avidin and biotin-labeled enzyme. *Eur. Patent* 184701.

Woo, S. L. C., Lidsky, A. S., Guttler, F., Chandra, T., and Robson, K. (1983). Cloned human phenylalanine hydroxylase gene allows prenatal diagnosis and carrier detection of classical phenylketonuria. *Nature* **306,** 151–155.

Woods, D., Madonna, J. M., and Mulcahy, L. S. (1990). Nucleic acid probe. *Eur. Patent* 353985.

Yabucaki, K. K., Isaacs, S. T., and Gamper, H. B. (1985). Probe for specific nucleic acid base sequences in hybridization assay. *WO Patent* 8502628.

# 2 Nonradioactive Labeling Methods for Nucleic Acids

Christoph Kessler
Biochemical Research Center
Genetic Department
Germany

# I. OVERVIEW

Increased attempts have been made in the last decade to substitute radioisotopes like [$^{3}H$], [$^{14}C$], [$^{32}P$], [$^{35}S$], or [$^{125}J$] (Maitland *et al.*, 1987) by nonradioactive reporter molecules for the detection of nucleic acids like DNA, RNA, or oligonucleotides (Matthews and Kricka, 1988; Keller and Manak, 1989). The most frequently applied indicator systems are based on hybridization of the nucleic acid target molecules (analytes) with nonradioactively labeled complementary nucleic acids (probes). These probes are coupled with a variety of detector systems either directly by covalent binding or indirectly by additional specific high-affinity interaction. Table I lists currently described nonradioactive nucleic acid labeling and detection systems.

Most of the recently developed systems use enzymatic, photochemical, or chemical incorporation of the reporter group detected by optical, luminescence, fluorescence, metal-precipitating or electrochemical indicator systems (Meinkoth and Wahl, 1984; Pereira, 1986; Urdea *et al.*, 1987; Donovan *et al.*, 1987; Viscidi and Yolken, 1987; Höfler, 1987; Moench, 1987; Yolken, 1988; Kessler, 1991).

In the direct systems, the analyte-specific probes are directly and covalently linked with the signal-generating reporter group; thus, the detection of nucleic acids in direct systems consists of hybrid formation between analyte and labeled probe; and signal generation via the covalently coupled reporter group.

Frequently used direct reporter groups are fluorescent dyes like fluorescein and rhodamine (Lichter *et al.*, 1990) as well as marker enzymes like alkaline phosphatase (AP) (Jablonski *et al.*, 1986; Jablonski and Ruth, 1986) or horseradish peroxidase (HRP) (Pollard-Knight *et al.*, 1990).

In contrast, the indirect systems first require the modification of the analyte-specific probe by introduction of a particular modification group. This modification group binds through an additional, noncovalent interaction to a universal reporter group; thus, the signal-generating reporter group is linked indirectly to the hybrid by an additional interaction be-

**Table I**
Nonradioactive Nucleic Acid Detection Systems

| Mode of labeling | Mode of detection | Sensitivity [pg]/ copy number[a] | Development/availability | Reference(s) |
|---|---|---|---|---|
| *Systems on polynucleotide basis* | | | | |
| Enzymatic modification | | | | |
| Biotin-dUTP/Nick translation | Streptavidin-AP | 10/1 × $10^6$ | Enzo | Langer *et al.* (1981); Leary *et al.* (1983) |
| Biotin-dUTP/Tailing | Streptavidin-AP | 10/1 × $10^6$ | Enzo | Brakel and Engelhardt (1985) |
| Biotin-dATP/Nick translation | Streptavidin-AP | 0.5/5 × $10^4$ | LTI | Gebeyehu *et al.* (1987) |
| Digoxigenin-dUTP/Random priming | Anti-Digoxigenin-AP | 0.1/1 × $10^4$ | Boehringer Mannheim | Kessler *et al.* (1990); Höltke *et al.* (1990); Seibl *et al.* (1990); Mühlegger *et al.* (1990) |
| Digoxigenin-rUTP/Transcription | Anti-Digoxigenin-AP | 0.1/1 × $10^4$ | Boehringer Mannheim | Höltke and Kessler (1990) |
| Chemical modification | | | | |
| AAIF | Secondary AP-ab | 40/4 × $10^6$ | INSERM | Tchen *et al.* (1984) |
| | Secondary $Eu^{2+}$-ab | 2/2 × $10^5$ | Orion | Syvänen *et al.* (1986b) |
| Sulfone | Secondary ab | 2/2 × $10^5$ | Orgenics | Proverenny *et al.* (1979) |
| POD | Direct | 20/2 × $10^6$ | EMBO | Renz and Kurz (1984) |
| | Direct/Luminol | 5/5 × $10^5$ | Amersham | Renz and Kurz (1984); Pollard-Knight *et al.* (1990) |
| | Direct | Nd[b] | Digene | Taub (1986) |
| Biotin-Angelicin | Secondary ab | Nd | Miles | Albarella *et al.* (1989) |
| Biotin-psoralen | Streptavidin-POD TRF | 0.5/5 × $10^4$ | Cetus | Sheldon *et al.* (1986) |
| | | 5/5 × $10^5$ | MPI | Oser *et al.* (1988) |

*(continues)*

**Table I** (*continued*)

| Mode of labeling | Mode of detection | Sensitivity [pg]/ copy number[a] | Development/availability | Reference(s) |
|---|---|---|---|---|
| Photobiotin | Streptavidin-AP | $40/4 \times 10^6$ | BRESA/LTI | Forster *et al.* (1985) |
| | Streptavidin-$Eu^{2+}$ | $10/1 \times 10^6$ | Orion | Dahlen *et al.* (1987) |
| | Gold-ab | $6/6 \times 10^5$ | Miles | Tomlinson *et al.* (1988) |
| Photo-DNP | Secondary ab | $1/1 \times 10^5$ | Biotec Research | Keller *et al.* (1989) |
| Photodigoxigenin | Anti-digoxigenin-AP | $0.5/5 \times 10^4$ | Boehringer Mannheim | Mühlegger *et al.* (1990) |
| Biotin transamination | Streptavidin-AP | $0.5/5 \times 10^4$ | Johns Hopkins University | Viscidi *et al.* (1986) |
| Digoxigenin transamination | Anti-digoxigenin-$\beta$-Gal | $5/5 \times 10^5$ | Boehringer Mannheim | Graf and Lenz (1984) |
| $Eu^{2+}$ transamination | TRF | $30/3 \times 10^6$ | Wallac | Dahlen *et al.* (1988) |
| $Hg^{2+}$ derivatization | HS hapten/Secondary ab | $70/7 \times 10^6$ | Univ. Leiden | Hopman *et al.* (1986a; 1986b) |
| Bromo derivatization | Secondary ab | $8/8 \times 10^5$ | Biotech Res. | Keller *et al.* (1988) |
| Biotin hydrazide | Streptavidin-AP | $2/2 \times 10^5$ | Weizmann Inst. | Reisfeld *et al.* (1987) |
| | | $10/1 \times 10^6$ | Showa Univ. | Takahashi *et al.* (1989) |
| Diazobiotin | Streptavidin-AP | Nd | Weizmann Inst. | Rothenberg and Wilchek (1988); Nur *et al.* (1989) |
| Biotin-DNA-binding protein | Streptavidin acid phosphatase | $10/1 \times 10^6$ | Orion | Syvänen *et al.* (1985) |
| *Systems on oligonucleotide basis* | | | | |
| Enzymatic modification | | | | |
| Biotin-dUTP/tailing | Streptavidin-AP | $100/1 \times 10^7$ | INSERM/LTI | Kumar *et al.* (1988) |
| Digoxigenin-dUTP/tailing | Antidigoxigenin-AP | $0.5/5 \times 10^4$ | Boehringer Mannheim | Schmitz *et al.* (1991) |
| Chemical modification | | | | |
| Aminocytosine-oligo-marker enzyme | Direct | $30/3 \times 10^6$ | Molecular Biosystems | Jablonski *et al.* (1986); Yamakawa *et al.* (1989) |
| | | $100/1 \times 10^7$ | Chiron | Urdea *et al.* (1988) |
| 5′-Amino-oligo-marker enzyme | Direct | $30/3 \times 10^6$ | Univ. Adelaide | Li *et al.* (1987) |
| | | Nd | EMBO | Sproat *et al.* (1987) |
| HS-Oligo-marker enzyme | | Nd | Salk Inst. | Chu and Orgel (1988) |

[a] 1000 bp fragment

[b] Nd, not determined

tween the modification group and a high-affinity binding partner coupled with the reporter group. The detection of nucleic acids in indirect systems is therefore divided into three reaction steps: hybrid formation between analyte and modified probe; specific noncovalent interaction between the modified probe and the binding partner coupled with the reporter group; and signal generation via the reporter group indirectly bound to the analyte.

A variety of interaction pairs between modification group and binding partner have already been realized (Table II). Aside from the well-known systems using a vitamin [e.g., biotin (Langer *et al.*, 1981; Forster *et al.*, 1985; Chu and Orgel, 1985a; Reisfeld *et al.*, 1987)] or a hapten [e.g., digoxigenin (Kessler, 1990, 1991; Kessler *et al.*, 1990; Höltke *et al.*, 1990, 1991; Seible *et al.*, 1990; Mühlegger *et al.*, 1990; Höltke and Kessler, 1990; Schmitz *et al.*, 1991;], alternative kinds of interaction with the respective modification groups have been established by either the selective binding of heavy metal ions [e.g., mercury (Bauman *et al.*, 1983; Hopman *et al.*, 1986a, 1986b)], specific recognition of particular nucleic acid sequences [e.g., promoter or repressor sequences (Paau *et al.*, 1983; Dattagupta *et al.*, 1988)], or nucleic acid conformations [e.g., DNA/RNA hybrids (Van Prooijen-Knegt *et al.*, 1982; Stollar and Rashtchian, 1987; Rashtchian *et al.*, 1987; Coutlee *et al.*, 1989a; 1989b; 1989c)].

Among the indirect systems, only the biotin and digoxigenin systems are characterized by subpicogram sensitivities; all other systems are of less sensitivity. Even though the biotin system is characterized by high sensitivity and a wide range of applications, the disadvantage of this system is that an endogenous ubiquitous vitamin, vitamin H, is used as modification group (Lardy and Peanasky, 1953; Chaiet and Wolf, 1964; Greene, 1975). This results in higher background reactions especially with biological samples. The digoxigenin system, characterized by analogous sensitivity, circumvents these background side-reactions by using the cardenolide digoxigenin occurring only in *Digitalis* plants (Pataki *et al.*, 1953; Reichstein and Weiss, 1962; Hegnauer, 1971).

Besides various methods for linking fluorescent dyes of marker enzymes directly to nucleic acid probes, a broad repertoire of enzymatic, photochemical, and chemical labeling methods can be applied for introduction of modification groups into nucleic acid probes (Table III). The enzymatic reactions are catalyzed by a number of DNA-dependent DNA (DNAP) or RNA (RNAP) polymerases, RNA-dependent DNA polymerases (RT) or terminal transferases (TdT) (Langer *et al.*, 1981; Ausubel *et al.*, 1987; Kessler *et al.*, 1990; Höltke and Kessler, 1990; Schmitz *et al.*, 1991). The photochemical derivatization of nucleic acids can be accomplished using azide-containing compounds by the photochemical dissociation of nitrogen and subsequent reaction of the intermediate resulting nitrene (Forster *et al.*, 1985; Mühlegger *et al.*, 1990). Chemical modification can be per-

**Table II**
Binding Partners of Modified Nucleic Acids

| Modification of analyte : probe | Example(s) | Reference(s) |
|---|---|---|
| Vitamins | | |
| Vitamin : Binding protein | Biotin: Avidin/Streptavidin | Bayer and Wilchek (1980); Langer *et al.* (1981); Wilchek and Bayer (1988; 1990) |
| Haptens | | |
| Hapten : Hapten-specific antibody | Digoxigenin (DIG): >DIG | Kessler *et al.* (1990); Höltke *et al.* (1990); Seibl *et al.* (1990); Mühlegger *et al.* (1990); Höltke and Kessler (1990); Schmitz *et al.* (1991) |
| | Dinitrophenyl (DNP): >DNP | Keller *et al.* (1988; 1989) |
| | Fluorescein isothiocyanate (FITC): >FITC | Serke and Pachmann (1988); Parsons (1988) |
| | Biotin (Bio): >Bio | Langer-Safer *et al.* (1982); Agrawal *et al.* (1986); Binder (1987) |
| | *N*-2-Acetylaminofluoren (AAF): >AAF | Tchen *et al.* (1984); Landegent *et al.* (1984, 1985); Cremers *et al.* (1987) |
| | *N*-2-Acetylamino-7-iodofluoren (AAIF): >AAIF | Tchen *et al.* (1984); Syvänen *et al.* (1986b) |
| | 5C-Bromo-desoxyuridine (Br-dU): >Br-dU | Traincard *et al.* (1983); Porstman *et al.* (1985); Sakamoto *et al.* (1987) |
| | 5C-Sulfite-desoxycytidine ($SO_3$-dC): >$SO_3$-dC | Herzberg (1984); Pezzella *et al.* (1987); Hyman *et al.* (1987) |
| | Ethidium (Et): >Et | Albarella and Anderson (1985a; 1985b); Dattagupta *et al.* (1985) |

| | | |
|---|---|---|
| Conformation of nucleic acid | | |
| Nucleic acid hybrid : Conformation-specific antibody | RNA/DNA: >RNA/DNA | Van Prooijen-Knegt *et al.* (1982); Stollar and Rashtchian (1987); Rashtchian *et al.* (1987); Coutlee *et al.* (1989a, 1989b, 1989c) |
| | RNA/DNA: >RNA/DNA | Coutlee *et al.* (1989b) |
| | DNA: >DNA | McKnabb *et al.* (1989) |
| Sequence of nucleic acid | | |
| Nucleic acid sequence : Binding protein | ssDNA: *E. coli* ssb protein | Syvänen *et al.* (1985); McKnabb *et al.* (1989) |
| | dsDNA: histone | Renz (1983); Bulow and Link (1986) |
| | T7 promoters A1/A2/A3: *E. coli* RNA polymerase | Paau *et al.* (1983) |
| | 5-Aza-dC: DNA methyltransferase | Reckmann and Rieke (1987) |
| | *lac* operon : *lac* repressor | Dattagupta *et al.* (1988) |
| | Protein A-NS : IgG-Fc | Dattagupta *et al.* (1984); Czichos *et al.* (1989) |
| | S peptide-NS : S protein | Rabin *et al.* (1985) |
| Modification of heavy metal ions | | |
| Heavy metal : mercaptane | $Hg^{2+}$ : Glutathione-TNB: >TNB | Baumann *et al.* (1983); Hopman *et al.* (1986a; 1986b) |
| Polyadenylation | | |
| Polyadenylation : polythymidine ends | $(dA)_x$ : $(dT)_x$ | Woodhead and Malcolm (1984); Kumar *et al.* (1988); Parsons (1988) |
| Polyadenylation : polynucleotide phosphorylase | $(dA)_x$ : PNP pyruvate kinase/ATP-coupled luciferase reaction | Vary *et al.* (1986); Gillam (1987) |

**Table III**
Possible Modes of Modification of Nucleic Acid Probes

| Reactive groups of binding partner: Modifying agent | Reference(s) |
|---|---|
| Enzymatic modification | |
| dsDNA : M-dNTP/*E. coli* DNA polymerase I | Rigby *et al.* (1977); Höltke *et al.* (1990) |
| ssDNA : M-dNTP/primer/Klenow | Langer *et al.* (1981); Gregersen *et al.* (1987); Höltke *et al.* (1990) |
| Polymerase | |
| 3′-OH-ssDNA/RNA : M-dNTP, M-ddNTP/ terminal transferase | Riley *et al.* (1986); Pitcher *et al.* (1987); Schmitz *et al.* (1991) |
| ssRNA : M-dNTP/primer/reverse | Vary *et al.* (1986) |
| Transcriptase | |
| dsDNA transcription unit : M-NTP/ SP6,T7,T3 RNA polymerase | McCracken (1989); Theissen *et al.* (1989); Höltke and Kessler (1990) |
| Photochemical modification | |
| ss,dsDNA/ss,dsRNA : M-Azidobenzoyl/h | Ben-Hur and Song (1984); Forster *et al.* (1985); Cimino (1985); Mühlegger *et al.* (1990) |
| dsDNS/dsRNA : M-NS-Intercalator[a]/h | Brown *et al.* (1982); Dattagupta and Crothers (1984); Sheldon *et al.* (1985, 1987); Dattagupta *et al.* (1989) |
| Chemical modification | |
| DNA·CHO : M-hydrazide | Reisfeld *et al.* (1987) |
| DNA·$Hg^{2+}$ : M-mercaptane | Dale *et al.* (1975); Bergstrom and Ruth (1977); Langer *et al.* (1981); Ward *et al.* (1982); Baumann *et al.* (1983); Hopman *et al.* (1986a; 1986b) |
| DNA·SH : M-amine | Renz and Kurz (1984); Landes 1985; Al-Hakim and Hull (1986) |
| DNA·$NH_2$ : M-amine/$SO_3^{2-}$ | Draper and Gold (1980); Viscidi *et al.* (1986); Gillam and Tener (1986) |
| Allylamine-oligonucleotide : M-*N*-Hydroxys-uccinimide ester | Langer *et al.* (1981); Cook *et al.* (1988); Urdea *et al.* (1988) |
| 5′-*P*-Oligonucleotide : M-*N*-Hydroxysuccinimide ester/diamino-ethyl, -hexyl | Agrawal *et al.* (1986) |
| 3′-MF-CPG-Oligonucleotide : M-*N*-Hydroxy-succinimide ester | Kempe *et al.* (1985); Nelson *et al.* (1989a; 1989b) |

[a] Psoralen, angelicin, acridin, Ethidium.

formed by a number of alternative reactions: e.g., sulphite-catalyzed transamination, mercury and bromine derivatization, substitution of amino groups with bifunctional reagents, as well as mercaptane derivatization of amine residues of nucleotide bases (for review see Matthews and Kricka, 1988).

Enzymatic labeling reactions are especially useful because they result in highly labeled nucleic acid probes that can be applied for high-sensitivity nucleic acid detection.

The most frequently used sites of base modification are the C-5 position of uracil and cytosine, the C-6 of thymine, and the C-8 of guanine and adenine; these positions are not involved in hydrogen bonding. In addition, the $N^4$-position of cytosine and the $N^6$-position of adenine have been used for base modification. However, these sites are involved in hydrogen bonding. Oligonucleotides may also be modified at either the 5′- or 3′-terminus; internal labeling at the 2-position of the deoxyribose as well as at the phosphate diester bridge is also possible.

## II. METHODS FOR ENZYMATIC LABELING

Enzymatic labeling procedures result in probes mediating highly sensitive nucleic acid detection. The various labeling methods comprise labeling of single- or double-stranded DNA, single-stranded RNA, and oligodeoxynucleotides as target molecules (see also Chan and McGee, 1990; Pollard-Knight, 1990).

In these enzymatic labeling procedures, nucleotide analogs modified with particular haptens like biotin, digoxigenin, or fluorescein are used instead of or in combination with their nonmodified counterparts. In case of DNA fragments and oligodeoxynucleotides, hapten-dUTP, hapten-dCTP or hapten-dATP are applied as modified substrates (Langer *et al.*, 1981; Gebeyhu *et al.*, 1987); for labeling RNA, hapten-UTP is used as nucleotide analog (Hóltke and Kessler, 1990).

Most of the labeling reactions are based on standard reactions (Ausubel *et al.*, 1987; Sambrook *et al.*, 1989a) adapted to modified nucleotides.

In the case of DNA, homogeneous labeling can be obtained by random priming with Klenow polymerase, nick translation with *Escherichia coli* DNA polymerase I or by the polymerase chain reaction with *Taq* DNA polymerase. In the latter reaction, vector-free double- or single-stranded probes may be synthesized. DNA end-labeling can be achieved by the tailing reaction with terminal transferase preferentially at 3′-protruding ends, by the fill-in reaction with Klenow polymerase at 5′-protruding ends, or a T4 DNA polymerase replacement reaction by sequential action of 3′ → 5′ exonuclease and 5′ → 3′ polymerase at 3′-protruding fragment termini guided by the absence or presence of the respective nucleotides.

Starting from RNA as template, labeled DNA probes can be synthesized with viral reverse transcriptase (AMV RT; Mo-MLV RT) by oligodeoxynucleotide-primed RNA-dependent DNA synthesis (reverse transcription) using hapten-modified deoxynucleotides (e.g., bio-dUTP or DIG-dUTP) in addition to nonmodified deoxynucleotides as substrates.

RNA can be homogeneously labeled by synthesizing "run-off" transcripts catalyzed by the phage-coded DNA-dependent SP6, T7, or T3

RNA polymerases. The promoter-directed RNA probe synthesis results in vector-free hybridization probes (cf. PCR-generated probes).

Oligodeoxynucleotides can be enzymatically 3′-end-labeled by the tailing reaction with terminal transferase; internal enzymatic labeling is obtained with Klenow polymerase starting from short primers that bind to complementary sequences within the 3′-region of the template oligodeoxynucleotide strand.

## A. DNA Labeling Techniques

In this chapter the three main labeling reactions will be discussed: random-primed labeling, nick translation, and synthesis of double- or single-stranded probes via PCR. Whereas nick translation is characterized by the exchange of unlabeled nucleotides by labeled nucleotides (replacement synthesis), both random-primed and PCR-mediated probe synthesis result in the new DNA sequences (net synthesis). With PCR, the novel sequences are synthesized in large excess over template DNA by the repeated amplification cycles.

### 1. Random-Primed Labeling

Besides labeling hybridization probes with radioisotopes, the random-primed probe synthesis was adapted to a variety of nonradioactive modification groups; e.g., biotin or digoxigenin (Kessler *et al.*, 1990; Seibl *et al.*, 1990). The reaction scheme is outlined in Fig. 1.

Random-primed labeling was developed by Feinberg and Vogelstein (1983; 1984); labeling is catalyzed by Klenow polymerase (PolIk), the large fragment of *E. coli* DNA polymerase I (PolI) obtained by proteolytic subtilisin cleavage of PolI. The Klenow polymerase lacks the 5′ → 3′ exonuclease activity of the holoenzyme but still contains the 5′ → 3′ polymerase as well as the 3′ → 5′ exonuclease proofreading activity.

For random priming, linear double-stranded DNA is first denatured (e.g., by heat treatment) and then chilled on ice to stabilize the single-stranded DNA chains. Supercoiled DNA is labeled to a lesser extent because of faster reannealing rates observed with this configuration; thus, circular DNA molecules have to be linearized before the labeling reaction to obtain high labeling densities.

A random mixture of hexa-deoxynucleotides are annealed to the dena-

**Fig. 1** Random-primed labeling. (a) reaction scheme; (b) structure of hapten-modified deoxynucleoside triphosphate (DIG-[11]-dUTP).

a

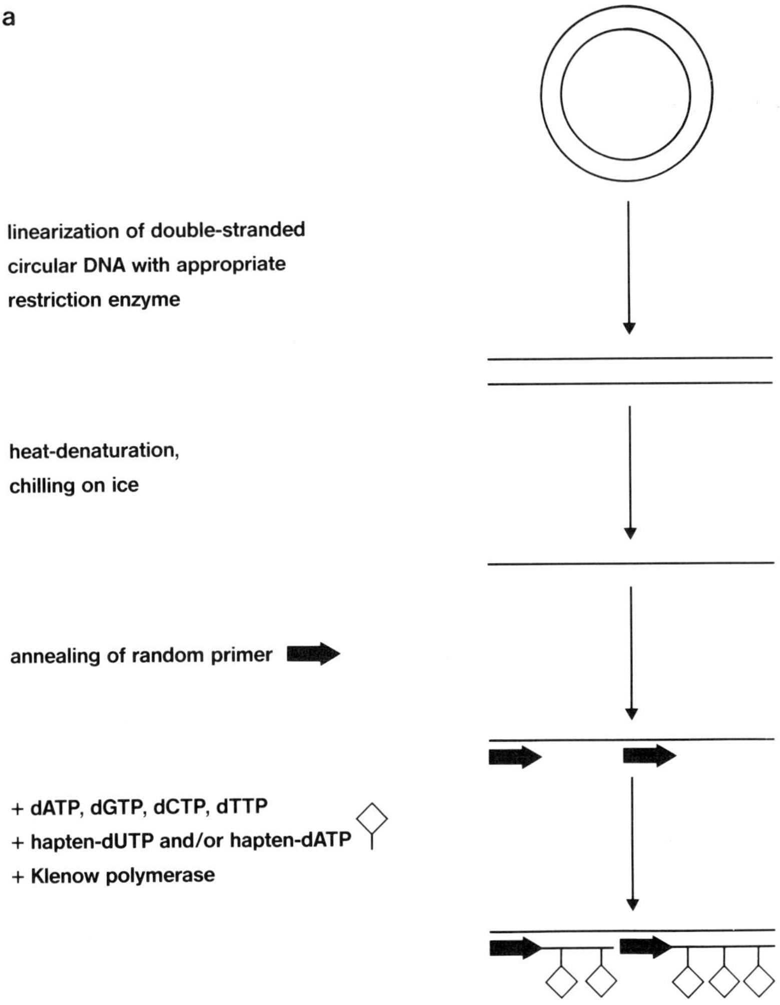

linearization of double-stranded circular DNA with appropriate restriction enzyme

heat-denaturation, chilling on ice

annealing of random primer

+ dATP, dGTP, dCTP, dTTP
+ hapten-dUTP and/or hapten-dATP
+ Klenow polymerase

b

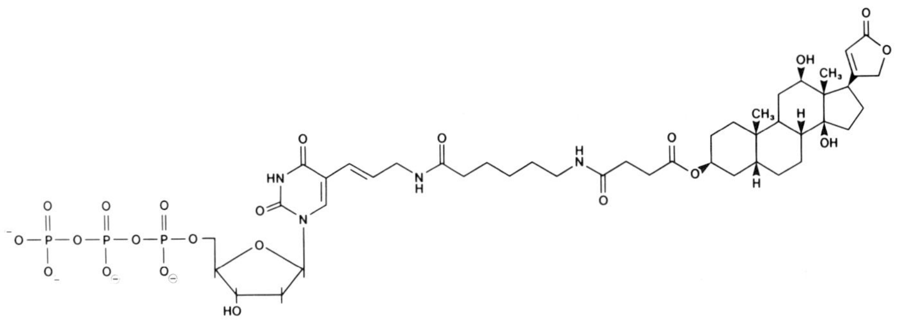

tured single-stranded templates in a statistical manner; these annealed oligonucleotides serves as 3′-OH primers in the subsequent polymerase reaction. The probe synthesis is started by adding Klenow polymerase and all four deoxynucleotides, of which at least one (mostly dUTP) is modified with a hapten. Since the primers have random specificity (random primers) they are able to bind to all possible target sequences; this means that the primers bind to the target molecule in a statistical manner. During the polymerization reaction, Klenow polymerase incorporates not only the nonmodified deoxynucleotides but also the hapten-modified substrates (e.g., bio-dUTP or DIG-dUTP); this results in highly labeled hybridization probes. If vector-free probes are to be synthesized, the vector sequences have to be removed before labeling by restriction enzyme treatment and subsequent fragment separation.

During random-primed synthesis, both complementary target strands may act as templates after strand denaturation. Reannealing of both template strands is prevented by fixing the single-stranded configuration by low temperatures and by high primer concentrations. The ratio between primer and template DNA also inversely controls the length of the newly synthesized labeled probe; the higher the primer concentration, the shorter the synthesized probes.

The resulting labeled probe is therefore a statistical mixture of individual probe chains complementary to different sequences of the template; in addition, the labeled DNA chains are markedly shorter than those of the input template DNA. Because both DNA strands fo the original DNA act as templates, the sequences of the labeled probe chains are at least partially complementary, so that denaturation before hybridization is necessary.

Labeling density depends on the ratio between hapten-modified and non-modified dNTP; best results in terms of sensitivity of detection of labeled hybrids are obtained if mixtures of unmodified and hapten-modified deoxynucleotides are applied. In the case of digoxigenin incorporation, a mixture of 35% DIG-dUTP and 65% dTTP gives the best results (Höltke *et al.*, 1990). It should be noted that the optimal ratio is different using different haptens; thus, this ratio has to be determined for every labeling system.

Random-primed labeling is very useful for labeling DNA fragments isolated from gels (ng-amounts); however, larger amounts up to several $\mu$g can be labeled with this method. The incorporation reaches a plateau after 30 to 60 min; during prolonged incubation, the incorporated label remains stable. The sensitivities obtained with probes modified with haptens such as biotin or digoxigenin via random-priming are in the range of 100 fg on blots (Kessler *et al.*, 1990; Höltke *et al.*, 1990).

## 2. Nick Translation

Nick translation was first developed for labeling hybridization probes with radioisotopes (Rigby *et al.*, 1977). Subsequently the incorporation of modification groups into DNA by nick translation was reported (e.g., biotin: Langer *et al.*, 1981; digoxigenin: Höltke *et al.*, 1990). The reaction scheme of nick translation is shown in Fig. 2.

Nick translation involves the simultaneous action of pancreatic deoxyribonuclease (DNaseI) and *E. coli* DNA polymerase I holoenzyme (Kornberg polymerase: PolI) on double-stranded DNA. DNaseI is a double-strand specific endonuclease hydrolyzing phosphodiester bonds in DNA chains, resulting in short oligonucleotides characterized by 5′-phosphate groups. In the presence of $Mg^{2+}$ ions, the action of DNaseI is restricted to the generation of single-stranded nicks because of the independent action of the enzyme on each DNA strand. The rate of introducing nicks into the single-stranded chains depends on the concentration of DNaseI in the incubation mixture. Nick translation requires low concentrations of DNaseI; using these conditions, only a few nicks are introduced in each DNA strand. The generated free phosphorylated 5′ ends within the nicks trigger the action of the PolI holoenzyme. PolI is characterized by three independent catalytic activities: 5′ → 3′ polymerase; 5′ → 3′ exonuclease; and 3′ → 5′ exonuclease (proofreading activity). The 5′ → 3′ exonuclease activity of PolI progressively removes deoxynucleotides from the phosphorylated 5′ end; the 5′ → 3′ polymerase activity of PolI synchronously adds new deoxynucleoside triphosphates to the opposite free 3′-hydroxyl ends of the nick. If nonmodified deoxynucleoside triphosphates and hapten-modified deoxynucleoside triphosphates are in the reaction mixture, the hapten is incorporated during this polymerization reaction. Because PolI acts in a strict template-dependent manner, the nucleotide sequence remains unchanged during nick translation.

As a result of the syntergistic action of the two PolI activities, the nick moves along the DNA fragment in 5′ → 3′ direction, terminating at the 3′ fragment end (*nick translation*). The net effect is the replacement of unlabeled nucleotides by labeled nucleotides in the double-stranded molecule. Because DNaseI acts in a statistical manner, nicks are introduced into both DNA strands; thus, both strands are labeled during nick translation, resulting in nearly fully labeled double-stranded molecules. Therefore, like probes prepared by random-priming, nick-translated probes have to be denatured before hybridization.

A critical parameter of nick translation is the ratio between DNaseI and PolI (Sambrook *et al.*, 1989b). The ratio is not a fixed value but dependent on the amount, length, and configuration of the DNA to be labeled. There-

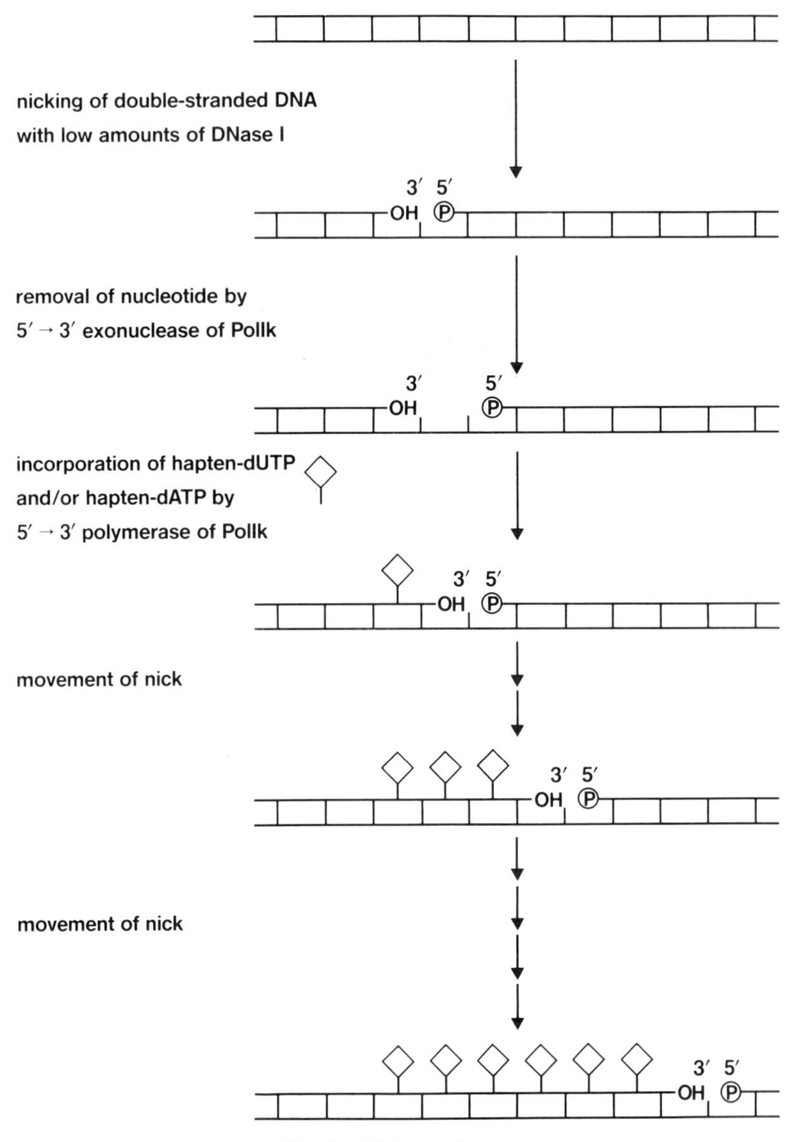


**Fig. 2** Nick translation.

fore the actual amounts of both enzymes needed for optimal nick translation should be determined when different kinds of DNA are used as targets for the labeling reaction. In order to determine the optimal enzyme concentrations, different amounts of both enzymes have to be tested with respect to nick-translation efficiency. The combination of both enzyme concentrations that gives best results is then used for subsequent preparative labeling reactions. Because hapten-modified probes are stable at −20° C for months or even years, large quantities of DNA probes can be synthesized, for comparable and reproducible results.

As with random priming, the best labeling density depends on the optimal ratio between hapten-modified and unmodified deoxynucleotides, and this must be determined for every system. Because labeling conditions have to be adapted to the nature of the DNA to be labeled, the nick-translation technique is suitable for preparative labeling of DNA probes. The incorporation is optimal after 1 to 2 hr; longer incubation periods result in a slight decrease of the labeling product, caused by the input DNaseI and the exonuclease activities of PolI. The observed sensitivities using nick-translated biotin or digoxigenin probes are in the range of 1001 to 500 fg on blots (Leary *et al.*, 1983; Chan *et al.*, 1985; Gebeyehu *et al.*, 1987; Höltke *et al.*, 1990).

### 3. Polymerase Chain Reaction

A powerful technique for labeling vector-free DNA probes is the polymerase chain reaction (PCR). PCR was developed by Saiki *et al.* (1985b) and adapted to a variety of applications including amplification, sequencing, cloning, and probe synthesis (for review see Erlich, 1989; Erlich *et al.*, 1989; Innis *et al.*, 1990). Probes can be labeled by modification of either the two primers or the deoxynucleoside triphosphates used for polymerization. The reaction scheme of PCR is shown in Fig. 3. The PCR is catalyzed by the thermostable enzyme *Taq* DNA polymerase isolated from the thermophilic eubacterium *Thermus aquaticus*. The whole amplification reaction consists of up to 60 sequential temperature cycles and is performed as follows:

1. Denaturation of the target DNA into single strands by heat (90–95° C);
2. Hybridization of two flanking antiparallel primers to the DNA single strands (40–60° C); and
3. Copying the two single strands by elongation of the primers using the heat-stable *Taq* DNA polymerase (70–75° C).

Since both single strands are copied simultaneously, the result is an exponential amplification of the target sequence. If RNA is the starting material, a reverse transcription step has to precede the amplification reaction.

In the first labeling approach, nonradioactively labeled PCR probes are generated by the use of 5′-terminal labeled primers. The most commonly employed 5′-terminal labels are biotin and a variety of fluorescent dyes (Chollet and Kawashima, 1985; Smith *et al.*, 1985, 1986, 1987; Adarichev *et al.*, 1987; Brosalina and Grachev, 1986; Ansorge *et al.*, 1986, 1987). It is

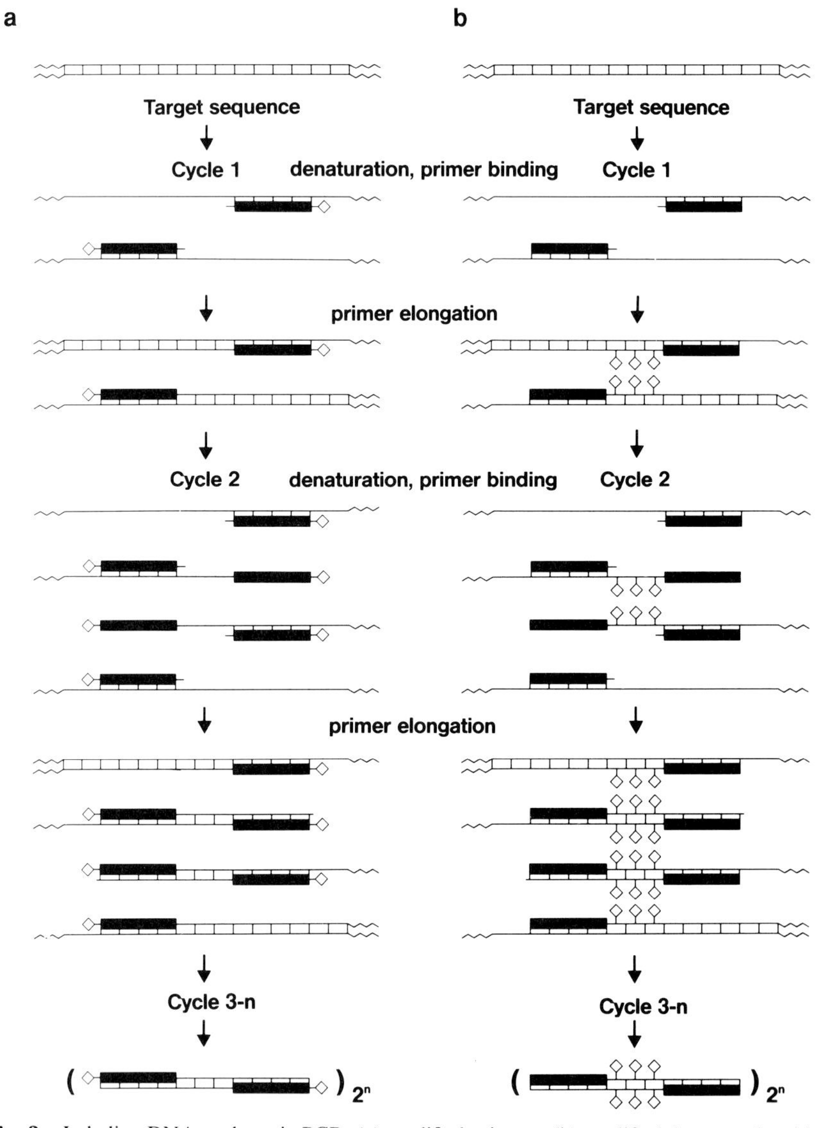


**Fig. 3** Labeling DNA probes via PCR. (a) modified primers; (b) modified deoxynucleoside triphosphates.

also possible to label PCR primers directly with marker enzymes, e.g. horseradish peroxidase or alkaline phosphatase (Levenson and Chang, 1990). Using HRP, 1 : 1 adducts of enzyme to nucleic acid are formed.

In the alternative labeling approach, the modification group is incorporated into the PCR product by modified deoxynucleoside triphosphates, resulting in homogeneously labeled vector-free hybridization probes. This approach has been described for biotin (Lo *et al.*, 1988) as well as for digoxigenin (Seibl *et al.*, 1990). With this alternative approach $\mu$g amounts of vector-free probes can be synthesized in a few hours. In the case of biotin, the optimal ratio between dTTP and bio-dUTP is 3 : 1, the respective ratio between dTTP and DIG-dUTP is 2 : 1.

Both labeling procedures can be used to generate probes of cloned inserts in which the sequence of the insert is still unknown. In this case primers directly flanking the cloning site have to be used. Applying biotin- or digoxigenin-labeled PCR probes, sub-pg- amounts of target sequences can be detected on blots (Lo *et al.*, 1988; Seibl *et al.*, 1990).

This method results in probes containing both complementary strands in a labeled form; thus, before labeling, a denaturation step has to be included. In an alternative protocol, defined single-stranded DNA hybridization probes containing up to 5000 nucleotides are synthesized by a *run off* synthesis (Stürzl and Roth, 1990). This method is based on asymmetric PCR cycles using only one PCR primer and cutting the amplified sequence with an appropriate restriction enzyme at a site located at a defined distance from the primer binding site. The resulting hybridization probes are as sensitive as probes obtained by nick translation, but they hybridize in a strand-specific manner. The single-stranded DNA probes can be handled under the same conditions as nick-translated probes and offer the benefits of single-stranded RNA run-off probes and the stability of DNA probes.

If the sequence of interest is cloned into a multiple cloning site flanked by two different, antiparallel primer sites [e.g., SP6 and T7 primers in pSPT18/19 (Pfeiffer and Gilbert, 1988)], single-stranded sense and antisense DNA probes can be synthesized using either the SP6 or T7 primer after cutting the insert for asymmetric PCR.

## B. RNA Labeling Techniques

RNA probes are commonly synthesized with the RNA polymerases from bacteriophages SP6, T7, or T3 by *in vitro* transcription of DNA, which is cloned into appropriate transcription vectors like the pSPT-vectors containing highly specific SP6-, T7-, or T3-specific promoters in front of a multiple cloning site (Melton *et al.*, 1984; Morris *et al.*, 1986; Krieg and Melton, 1987). Cloning particular DNA fragments in one of these restriction sites results in complete transcription units acting as templates for DNA-dependent RNA synthesis. The reaction scheme of *in vitro* RNA synthesis is shown in Fig. 4.

The RNA polymerase-catalyzed reaction commonly starts by initiation of the transcription reaction with a purine ribonucleoside triphosphate at a fixed position downstream of the promoter sequence (Butler and Cham-

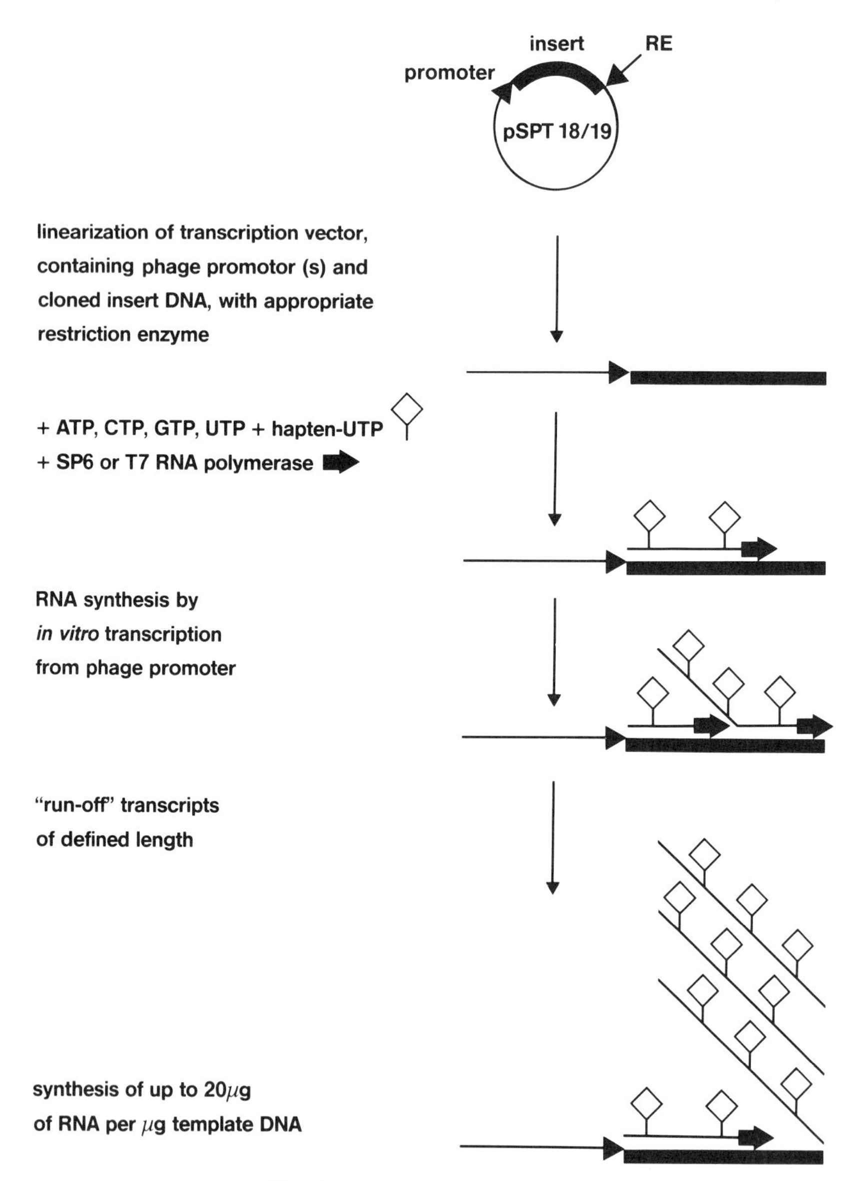


**Fig. 4** "Run-off" RNA synthesis.

berlain, 1982). If a modified ribonucleotide is added to the nonmodified nucleotide mixture, the growing RNA chain is labeled by integration of the modified ribonucleotide. Mostly hapten-labeled UTP (e.g., bio-UTP; DIG-UTP) is used as labeling reagent (Theissen *et al.*, 1989; Höltke and Kessler, 1990). The transcription reaction is terminated at the end of the linearized transcription unit; thus hapten-modified probe molecules of defined length and sequence are synthesized. The completion of the transcription reaction requires 1–2 hr depending on the desired labeling density; e.g., with digoxigenin, labeling with 35% digoxigenin-modified UTP and 65% UTP reaches an optimum after 2 hr with T7 RNA polymerase. Using 1 μg template DNA and 1 m*M* ATP, GTP and CTP each, 0.65 m*M* UTP and 0.35 m*M* DIG-UTP in 20 μl transcription buffer including 20 units RNase inhibitor, up to 20 μg labeled transcripts may be obtained. Thus, *in vitro* labeling with RNA polymerases may be used for preparative RNA probe synthesis.

After the transcription reaction, the template DNA can be removed by digestion with RNase-free DNase. The resulting labeled RNA probe contains no vector sequences, so that cross-hybridization caused by unspecific interaction between vector and target sequences is absent during hybrid formation.

Although the core region of phage-specific promoters is only 17 nucleotides in length and differs in only a few nucleotides between SP6, T7, and T3 promoters (PFeiffer and Gilbert, 1988), the promoters are highly specific for the respective RNA polymerase (Butler and Chamberlain, 1982; Kassavetis *et al.*, 1982). After linearization of the recombinant clone directly downstream of the transcription unit, run-off transcripts of defined length are synthesized. By positioning two promoters of different specificity and polarity at both sides of the transcription unit, transcription of both sense (coding) and antisense (noncoding) RNA can be initiated.

The advantages of RNA probes (e.g., single-strandedness, non-self-complementarity, defined length, and higher stability of the RNA/DNA hybrids have been widely exploited in different hybridization applications. In most cases radioactively labeled probes have been used and these have all the pitfalls of instability, low resolution, and all the disadvantages of handling isotopes. Applying nonradioactively labeled RNA probes in blot experiments as little as 0.1 pg homologous DNA or RNA can be detected in dot, Southern or Northern blots. This sensitivity is comparable with that obtained with radioactively labeled RNA probes.

## C. Labeling of Oligodeoxynucleotides

The method of choice for enzymatic labeling of oligodeoxynucleotides is the tailing reaction catalyzed by terminal transferase. This enzymatic approach complements the chemical labeling methods described in the next section (see also Kempe *et al.*, 1985).

Synthetic oligonucleotides are currently applied to almost every field of molecular biology as well as in diagnostic and forensic applications. These include hybridization of oligodeoxynucleotides to cloned DNA and genomic digests, screening of gene libraries, restriction-site mapping of cloned DNA, gel-shift assays, and a variety of *in situ* hybridization approaches. The recently developed *South-Western* hybridization technique for the isolation of DNA-binding proteins and characterization of the DNA-binding sequences also makes use of synthetic oligodeoxynucleotides.

The terminal transferase-catalyzed tailing reaction is based on the property of this enzyme to add deoxy- or dideoxyribonucleotides to 3′ ends of DNA in a template-independent manner (Roychoudhury *et al.*, 1979; Tu and Cohen, 1980). Repeated enzymatic addition of a particular nucleotide results in the generation of homopolymeric tails that do not hybridize with the target DNA sequence. The nature of the incorporated nucleotides is dependent only on the nature of the deoxyribonucleoside triphosphates employed. However, the reaction conditions differ for the incorporation of the various nucleotides. In the tailing reaction, the oligodeoxynucleotide is labeled only within the tail sequence; labeling of the probe sequence within the hybridizing region does not occur. The outline of the tailing reaction is shown in Fig. 5.

If biotin-labeled deoxynucleotides are used, the best sensitivity is obtained with a 1 : 2 mixture of bio-UTP/UTP (Riley *et al.*, 1986; Keller and Manak, 1989); if digoxigenin is used as modification, group tailing of oligodeoxynucleotides with a 1 : 10 mixture of DIG-dUTP/dATP results in labeling at every 10–12 nucleotides within a total tail length of 40 to 50 nucleotides. Unspecific cross-hybridization by these DIG-dUMP-labeled oligo(dA) tails is not observed. The addition of fluorescein-labeled ddCMP residues to DNA has also been reported (Trainor and Jensen, 1988).

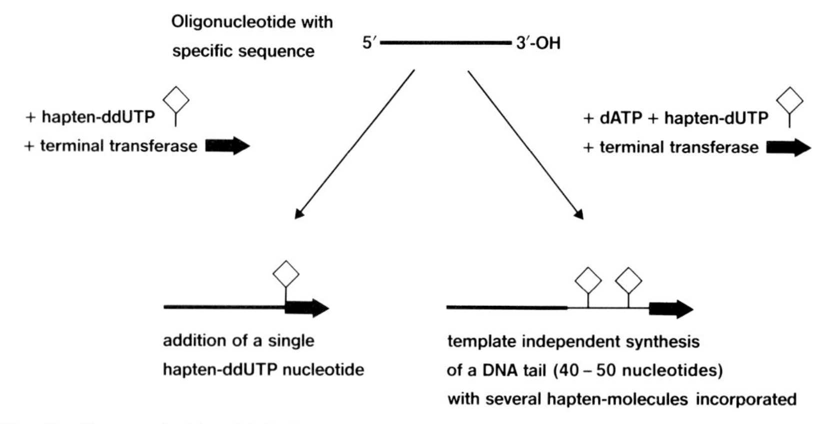


**Fig. 5** Enzymatic 3′-end-labeling of oligodeoxynucleotides via the tailing reaction with modified dNTPs or ddNTPs.

With DIG-labeled 30mers, as low as 50 pg DNA can be detected on blots. Compared with labeled DNA or RNA probes, the reduced sensitivity reflects the reduced length of hybridization region. Labeling of the 3′-end of the same 30mer oligodeoxynucleotide with DIG-2′,3′-ddUTP results in the incorporation of a single modified nucleotide because of the chain termination reaction. The application of these monolabeled oligodeoxynucleotide probes leads to a sensitivity of 500 pg DNA on blots (Schmitz *et al.*, 1991).

Because there is no shift in $T_m$ as compared with the nonmodified oligodeoxynucleotide, the tailing reaction is suitable for labeling not only oligodeoxynucleotides but also short DNA fragments (100–200 bp in length) for which $T_m$ is a critical factor.

# III. METHODS FOR CHEMICAL LABELING

A variety of chemical procedures for modification of nucleic acids have been described. These chemical labeling procedures can be categorized into light-induced labeling reactions, direct derivatization of nucleic acids with detectable tags, as well as modification of nucleic acids by activated linker molecules. In oligonucleotides these compounds can be labeled either after oligonucleotide synthesis at both 5′ or 3′ ends, or internally during oligonucleotide synthesis by integration of allylamine-substituted nucleotide compounds and further reaction with *N*-hydroxysuccinimide esters carrying the modification group.

## A. Light-Activated Labeling Techniques

Labeling with azido compounds and photoactive furocoumarins lead to covalently linked modification groups. The difference between the types of modification is that with the azide compounds, a random binding with the aromatic ring systems has been observed, whereas with the furocoumarins (e.g., psoralen), distinct cycloaddition reactions with thymine or uridine occur after intercalation.

### 1. Labeling with Azide Compounds

The most commonly used photolabeling reagents are aryl azide compounds. Aryl azides are stable in the dark and can be photoactivated *in situ* by illumination with UV light ($\lambda = 320$ nm) to generate highly reactive aryl nitrenes. These bind to the aromatic bases of ribosomal RNA by cross-linking experiments. 3-Nitrophenyl azide compounds substituted at C-4 with a linker molecule carrying the nonradioactive modification group have the advantage that they can be activated with long-wave UV light, avoiding damage of the nucleic acid probe.

The resulting modified probes are full-length molecules; thus they can be used both as nonradioactively labeled molecular weight markers and as homogeneously labeled hybridization probes (Forster *et al.*, 1985; Dahlen *et al.*, 1987; Tomlinson *et al.*, 1988; Keller *et al.*, 1989; Mühlegger *et al.*, 1990). The resulting linkages are stable at alkaline pH, elevated temperature, or irradiation with UV light; therefore, no loss of label is observed under Southern blot conditions.

Both single- and double-stranded DNA and RNA molecules can be labeled in small or large amounts; the labeling density is in the range of one modification group per 200 to 400 nucleotides. The reaction time for labeling is typically 2 to 3 min.

A variety of haptens have been coupled to 3-nitro-aryl azides (Fig. 6). These haptens include biotin (Forster *et al.*, 1985), digoxigenin (Mühlegger *et al.*, 1990), dinitrophenyl (DNP) (Keller *et al.*, 1989), and ethidium (Albarella and Anderson, 1985; Albarella *et al.*, 1985); most often the photoreactive aryl azide group is coupled with the modification group via a linear spacer 9 to 11 atoms in length.

In an alternative approach, azido groups are first coupled to DNA by

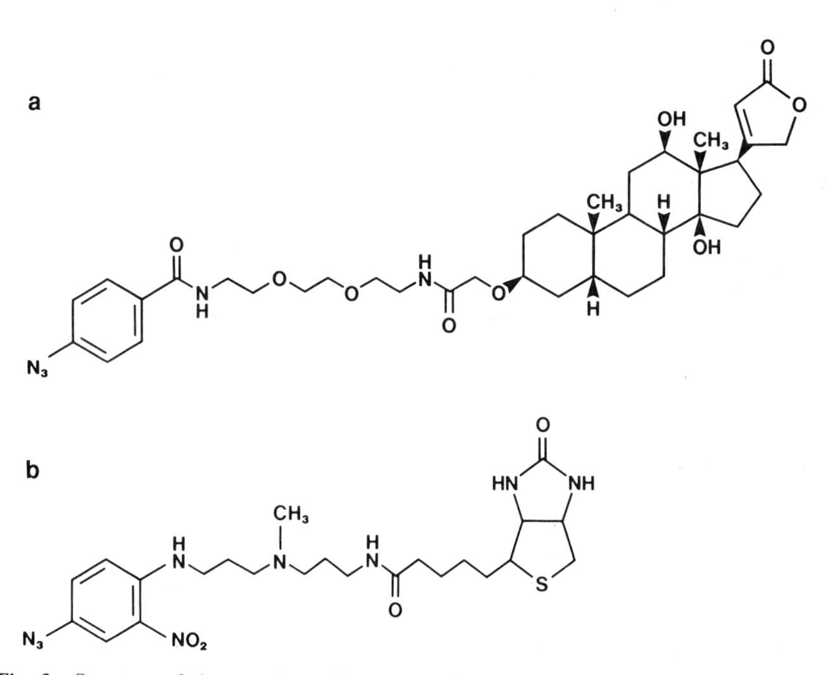


**Fig. 6** Structure of photoractive azide compounds. (a) photodigoxigenin; (b) photobiotin.

reacting with *p*-azidophenyl glyoxal to modify terminal guanosine residues (Heller and Morrison, 1985). These DNA-bound azido groups are used to react with modification groups after UV illumination via nitrene intermediates.

## 2. Labeling with Furocoumarin Compounds

A number of photoreactive substances will intercalate into nucleic acids, and then in a second step, covalently link with flanking bases through a photoreaction. These substances include furocoumarin compounds (psoralen, angelicin), acridine dyes (acridine orange), phenanthrolines (ethidium bromide), phenazines, phenothiazines, and quinones (Ben-Hur and Song, 1984; Dattagupta and Crothers, 1984; Sheldon *et al.*, 1985).

The most important intercalating compound is the bifunctional photoreagent psoralen (Cimino *et al.*, 1985), forming either mono- or diadducts with pyrimidine bases of nucleic acids after intercalation and photoactivation (Fig. 7). Structurally, psoralens are tricyclic compounds formed by the linear fusion of a furan ring with a coumarin. The formation of the photoadducts is mediated by either the furan and/or the coumarin heterocycle. The number of adducts is limited and dependent on the geometry of the intercalation complex.

Photoaddition of psoralen to nucleic acids occurs with incident light in the range of $\lambda = 320 - 400$ nm. Binding of both ends of psoralen (diadduct formation) to opposite strands of a nucleic acid helix results in the formation of a covalent interstrand cross-link. Psoralen reacts primarily with thymine (DNA) and uridine (RNA), whereas cytidine residues undergo only minor reactions. Mono- as well as diadducts are chemically stable;

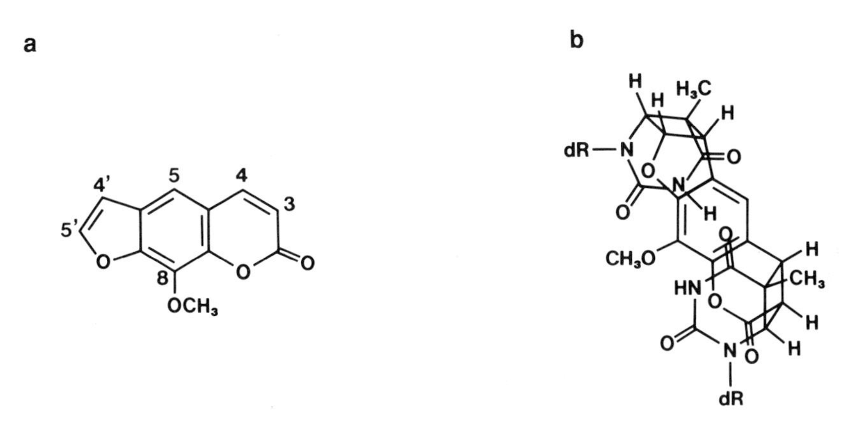


**Fig. 7** (a) Structure of 8-methoxypsoralen; (b) structure of the dT-8-MOP-dT-diadduct.

therefore, these adducts can be used under a variety of reaction conditions (elevated temperatures and ionic strengths, absence of presence of divalent cations or organic solvents).

Photochemical covalent cycloaddition of psoralen or angelicin has been used to modify nucleic acids with biotin or fluorescein by coupling these nonradioactive haptens with psoralen via a linear linker, similar to those applied with the azide compounds or the modified nucleotides used for enzymatic labeling (Sheldon *et al.*, 1986; Oser *et al.*, 1988; Albarella *et al.*, 1989). Linkage of the haptens is obtained by reaction of 4,5′,8-trimethylpsoralen at the C-4 position or by reaction of the activated linker derivatized with the nonradioactive hapten with 4′-aminomethyl-4,5′,8-trimethylpsoralen.

## B. Direct Derivatization with Chemical Tags

Labeling nucleic acids with directly detectable tags results either in noncovalent binding of the tag (e.g., ethidium bormide reacting with double-stranded RNA) or in covalent bond formation [e.g., sulfonation of C-4 positions of pyrimidines or derivatization of C-8 positions of purines with fluoren derivatives like *N*-acetoxy-*N*-2-acetylaminofluorene (AAF) or its 7-iodo derivative (AAIF)].

### 1. Noncovalent Labeling with Ethidium Bromide

This labeling system is based on the observation that the lanthanide cation terbium quenches the fluorescence of ethidium bromide by 40-fold when bound to double-stranded RNA (Al-Hakeem and Sommer, 1987). In contrast, quenching of double- or single-stranded DNA is only 2.5-fold, and quenching of single-stranded RNA is less than 5-fold. This selective quenching permits specific detection of double-stranded RNA in yeast or some viruses.

Terbium ions exhibit a low level of intrinsic fluorescence in aqueous solution. This is greatly enhanced when it is chelated to organic ligands, such as the aromatic ring systems of the nucleic acid bases, to form charge transfer complexes (Gross and Simpkins, 1981). Ethidium bromide exhibits marked enhancement of fluorescence on intercalation into double-stranded nucleic acids but only weak fluorescence with single strands. Terbium ions selectively quench the fluorescence of the ethidium intercalation complex, particularly with double-stranded RNA. This allows the detection of this particular nucleic acid in agarose gels by selective quenching the orange fluorescence of ethidium bromide. An alternative detection system is based on the recognition of intercalating ethidium bromide by ethidium bromide-specific antibodies (Albarella and Anderson, 1985; Dattagupta *et al.*, 1985).

## 2. Derivatization with Sulphone Groups

With this labeling approach, a sulphone group is introduced as hapten into the nucleic acid by sulphonation (Verdlov *et al.*, 1974). Sulphonation is obtained with a high concentration of sodium bisulphite at position C-6 of cytidine residues. The resulting sulphone derivative is relatively unstable, but can be stabilized by the substitution of the amino group at C-4 of the cytosine base with the nucleophilic reagent methylhydroxylamine. Cytosines are transformed by this reaction into $N^4$-methoxy-5.6-dihydrocytosine-6-sulphonate derivatives. These cytosine derivatives of nucleic acid probes can directly detected by antisulphone specific antibodies coupled with optical reporter groups (Herzberg, 1984; Nur *et al.*, 1989).

## 3. Derivatization with Fluorene Derivatives

This method is based on the chemical modification of guanine residues in DNA and RNA at position C-8 using *N*-acetoxy-*N*-2-acetylaminofluorene (AAF) or its 7-iodo derivative (AAIF) (Tchen *et al.*, 1984). After heat denaturation, a three-fold excess of AAF or AAIF is added to the probe and the reaction proceeds at 37° C for 2 hr. Cross-linking is avoided by treatment with alkali; unreacted fluorene derivatives are removed by repeated extraction with ethanol. Detection of fluorene-derivatized nucleic acids can be obtained with an enzyme-linked immunosorbent assay (ELISA)-type detection reaction using antifluorene antibodies (Tchen *et* al., 1984; Landgenet *et al.*, 1984, 1985; Syvänen *et al.*, 1986b; Cremers *et al.*, 1987).

## 4. Derivatization with Mercury

Cytosine and uracil nucleotides are readily mercurated by heating at 37 to 50° C with mercuric acetate in buffered aqueous solutions (pH 5.0–8.0) (Dale *et al.*, 1975). Polynucleotides can be mercurated under similar conditions. Cytosine and uracil bases are modified in RNA, whereas only cytosine residues in DNA are substituted. There is little, if any, reaction with adenine, thymine, or guanine bases. The rate of nucleic acid mercuration is influenced by ionic strength; the lower the ionic strength, the faster the reaction. The mild reaction conditions give minimal strand-breakage and do not produce pyrimidine hydrates.

The position of mercuration of cytosine and uracil is C-5 (Bergstrom and

Ruth, 1977); in uracil, this modification mimics the methyl group of thymine. Although sufficiently stable to permit biochemical studies, the mercury–carbon bond is extremely sensitive to cleavage by electrophiles and reducing agents. Treatment of mercurated polynucleutides with $I_2$, *N*-bromo-succinimide has been found to generate the corresponding iodinated and brominated nucleic acids rapidly. Therefore, the mercurated, iodinated, or brominated nucleic acids can be either detected directly or further substituted with activated linker arms carrying a detectable hapten such as biotin (Langer *et al.*, 1981; Hopman *et al.*, 1986a, 1986b; Keller *et al.*, 1988).

## C. Derivatization via Activated Linker Arms Carrying Detectable Haptens

Besides direct modification of nucleic acids with chemical tags and direct detection of these tags, alternative methods are available in which the detectable haptens are introduced into the nucleic acids via an activated linker arm. Primary modification of the nucleic acids is obtained either by transamination, mercuration, bromination, thiolation, or amine substitution with bifunctional reagents, followed by condensation with activated linker arms carrying a detectable hapten.

### 1. Transamination

Bisulphite catalyzes transamination between cytidine derivatives and primary amines. Aqueous bisulfite causes deamination of cytidine derivatives; however, in the presence of primary amines, it can catalyze transamination to give $N^4$-substituted cytidines with deamination only as a side reaction (Viscidi *et al.*, 1986). If the reaction is performed at neutral pH (pH 7.3), deamination is minimized, although the rate of transamination is reduced in comparison with lower pH.

If bifunctional diamines (e.g., ethylenediamine or 1,6-diaminohexane) are used as primary amines, then the amino group not attached to the cytosine residue may act as a recipient for an activated linker carrying a detectable hapten such as biotin or digoxigenin (Graf and Lenz, 1984; Viscidi *et al.*, 1986; Gillam and Tener, 1986). The linkers can be activated using *N*-hydroxysuccinimide ester. The overall reaction comprises two parts (Fig. 8): bisulfite-catalyzed transamination with one amino group of diamines; and Reaction of the second free amino group with activated linkers.

A one-step modified protocol uses biotin hydrazides for transamination (Reisfeld *et al.*, 1987). In this reaction, the biotin hapten is directly coupled

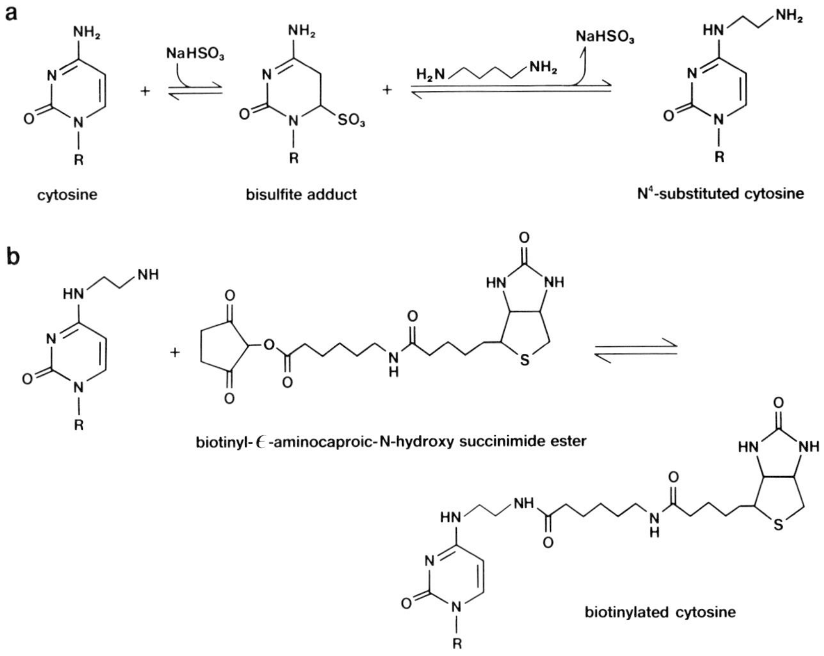

**Fig. 8** Chemical DNA labeling by transamination. (a) transamination reation; (b) reaction with activated hapten (biotin).

to the cytidine nucleotide; however, the linker region between the nucleotide and biotin is relatively short, leading to reduced sensitivities.

Critical in the transamination reaction is that a balance exits between transamination and deamination; the ratio between both reactions is dependent on the pH value of the labeling solution and the pK of the applied diamine. Low pK values like that of ethylenediamine (pK 7.6) favor transamination using a pH value of 7.3 for the transamination reaction. Furthermore, high transamination rates are obtained only after prolonged incubation up to several days at elevated temperatures under a nitrogen atmosphere.

In addition, the reactive $N$-$^4$-position of cytidine is involved in hydrogen bonding, resulting in disturbed hybrid formation and thus low sensitivities.

## 2. Allylamine Derivatization

Derivatization of nucleic acids with mercuric acetate results in mercurated nucleic acids (Dale *et al.*, 1975); predominantly the C-5 positions of uridine and cytidine bases are modified (see also Section III,B). These mercurated derivatives can be used to react with allylamine in the presence of palladium catalysts. This reaction results in C-5-allylamin-substituted bases and can be performed either with nucleotides or with polynucleotides.

The C-5-allylamine-substituted compounds can react with *n*-hydroxysuccinimide ester carrying a detectable hapten like biotin or di-goxigenin. This condensation reaction results in biotin or digoxigenin haptens coupled to the C-5 position of the cytosine or uridine base via a linear linker (Langer *et al.*, 1981; Hopman *et al.*, 1986a; Mühlegger *et al.*, 1990). The C-5 position of the pyrimidine bases is not involved in hydrogen bonding; thus the formation of double-stranded hybrids is not disturbed, allowing high sensitivity detection. The length of the linker can be enlarged by introduction of γ-aminobutyryl and/or ε-aminocaproic acid residues as additional linker components, which reduces unspecific interactions between probe and hapten and suppresses steric hindrance effects during hapten detection (Mühlegger *et al.*, 1990).

### 3. Bromo Derivatization

The bases of both DNA and RNA are readily brominated under mild conditions using dilute aqueous bromine or *N*-bromosuccinimide. Bromination occurs at the C-8 position of purines as well as C-5 of cytosine and (possibly) C-6 of thymine. The brominated intermediates can be detected directly (Br-dU) (Traincard *et al.*, 1983; Porstman *et al.*, 1985; Sakamoto *et al.*, 1987) or by reaction with amines such as 1,6-diaminohexan pre-bound with detectable haptens like DNP via one of the two amino functions (Keller *et al.*, 1988) (Fig. 9).

In contrast to the transamination reaction the modified sites are not involved in hydrogen bonding; thus, base pairing is not disturbed. Cross-linking is avoided without using a vast excess of linker arm because only one end of the arm contains a free amino group; the other is pre-attached to the hapten. Furthermore, it is also of advanatage that the bromine-mediated labeling reaction is rapid; it requires only 2 hr, although coupling is at elevated temperatures (50° C).

### 4. Thiolation

Amino groups of pyrimidine and purine nucleotides can be thiolated by reaction of nucleic acids with thiolating reagents such as *N* succimidyl-3-(2-pyridyldithio)propionate (Malcolm and Nicolas, 1984) or *N*-acetyl-*N*-(p-glyoxylbenzoyl)cysteamine (Rabin *et al.*, 1985). The thiol derivatives can react with mercaptanes to form disulfide bridges linking detectable haptens via the disulfide bond to the nucleotide.

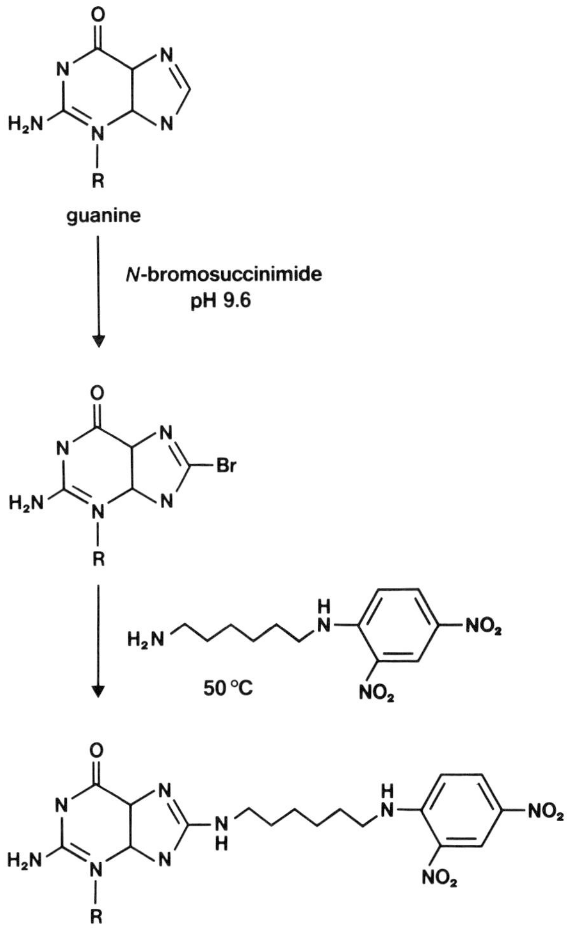

guanine

N-bromosuccinimide
pH 9.6

50 °C

**Fig. 9** Chemical DNA labeling by bromination.

## 5. Substitution of Amino Groups with Bifunctional Reagents

Another approach for chemically labeling nucleic acids is the substitution of amino groups of pyrimidines or purines with bifunctional reagents (Landes, 1985). The first reactive group reacts with the amino functions of the nucleotides, and the second reactive group couples basic macromolecules carrying a detectable hapten. A variety of bifunctional cross-linking reagents have been described: glutaraldehyde, 1,2,7,8-diepoxyoctane, *bis*(succinimidyl)suberate or *bis* (sulfonosuccinimidyl)suberate. Suitable basic macromolecules include polyethylenimine (Al-Hakim and Hull, 1986), cytochrome C (Sodja and Davidson, 1978), or histon H1 (Renz, 1983), and these are prelabeled with detectable haptens such as biotin via biotin bridges (Al-Hakim and Hull, 1986). In these more complex derivatization reactions, the complete side chain comprises the bivalent cross-linking bridge, the basic macromolecule, the biotin bridge, and biotin as detectable hapten; the derivative is obtained by cross-linking the activated nucleotides with the prelabeled macromolecules (Fig. 10).

The long side chain favors high sensitivities, whereas the high molecular mass of the basic macromolecule decreases sensitivity. Polyethylenimine (60 kDa) is superior to histone H1 (23 kDa) or cytochrome C (12.3 kDa). The contrary effects on sensitivity are explained by steric hindrance: the larger the cross-linking reagents, the more basic macromolecules can be linked per unit length of nucleic acid, but the more bulky the basic macromolecules, the fewer molecules can be linked to the DNA.

## D. Chemical Labeling of Oligonucleotides

Chemical labeling of oligonucleotides can be performed internally or at both termini during chemical oligonucleotide synthesis by the incorporation of allylamine-modified protected synthesis components (Haralambidis *et al.*, 1987; Cook *et al.*, 1988). These allylamine-derivatized oligonucleotides react with *N*-hydroxysuccinimide esters coupled with the haptens of interest. End-labeling can be obtained at the 5′ end by reaction with activated ethyl- or hexyl-derivatives (Agrawal *et al.*, 1986), by direct labeling 5′-phosphorylated oligonucleotides with functionalized hapten derivatives (Kempe *et al.*, 1985), or by end-labeling at the 3′-end by use of multifunctional CPG carriers (MF-CPG: multifunctional controlled pore glass) (Nelson *et al.*, 1989a; 1989b). 5′-End-labeling via $NH_2$- or SH-groups has also been reported (Chollet and Kawashima, 1985; Chu and Orgel, 1985a; Smith *et al.*, 1985; Conolly and Rider, 1985; Ansorge *et al.*, 1986; Wachter *et al.*, 1986; Coull *et al.*, 1986; Connolly, 1987; Tous *et al.*, 1988). Finally, the introduction of suitable marker groups into oligonucleotides via modified phosphodiester linkages is also possible (Asseline *et al.*, 1984; Chu and Orgel, 1985b; Thuong *et al.*, 1987; Jäger *et al.*, 1988; Froehler *et al.*, 1988).

The three most important methods for internal or 5′- and 3′-end-labeling are described in more detail in the next sections.

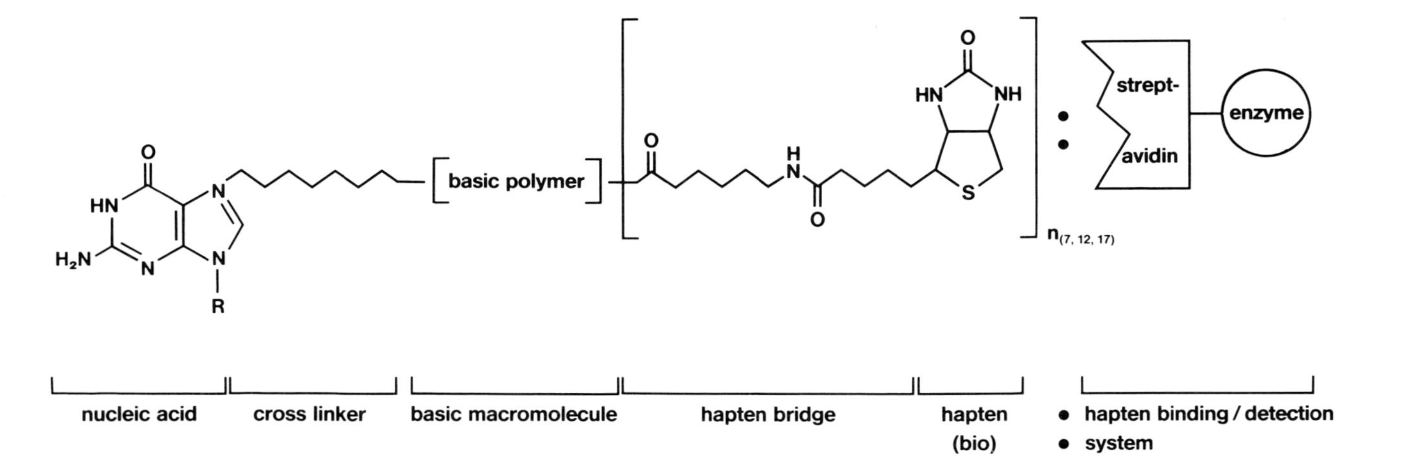


**Fig. 10** Hapten-coupling via basic macromolecules.

### 1. Chemical Labeling during Oligonucleotide Synthesis with Allylamine-Modified Protected Synthesis Components

This labeling reaction uses allylamine-modified protected synthesis components during oligonucleotide synthesis (5′-dimthoxytrityl-5(3-trifluoracetylaminopropenyl)-2′-deoxyuridine-3′-N,N-diisopropylamino-cyanoethoxyphosphoamidite) (Haralambidis *et al.*, 1987; Cook *et al.*, 1988) (Fig. 11).

The required derivatives are available for all nucleotides; pyrimidine bases are modified at the C-5 position, and purine bases are derivatized at C-8. These derivatizations do not interfere with hydrogen bonding; therefore, hybrid formation with the modified probes is not hindered. After incorporation of the allylamine derivatives and removal of the protecting groups, the terminal amino functions can be reacted with activated linker moieties carrying a hapten; e.g., biotin- or digoxigenin-*N*-hydroxysuccinimide ester. Because integration of allylamine-substituted monomers into the growing oligonucleotide can be obtained at every position, this labeling approach is very flexible. Another advantage is that labeling of oligonucleotides can be performed during routine automated oligonucleotide synthesis.

### 2. Chemical 5′-End-Labeling with Activated Ethyl- and Hexyl-Derivatives (Aminolink I/II)

As an alternative to the incorporation of allylamine compounds during chemical oligonucleotide synthesis, 5′-terminal phosphate residues synthetic oligonucleotides can be labeled with activated ethyl- or hexyl-derivatives (Agrawal *et al.*, 1986) (Fig. 12):

The oligonucleotide reacts with aminolink I (ethyl-) or aminolink II (hexyl-) derivatives after removal of the protection groups, but is still bound to the solid support. Reaction takes place by a oxidative condensation of the 5′-terminal phosphate and the phosphate moiety of the aminolink compound. Liberation from the solid support and removal of the trifluoracetyl protection group results in a 5′-terminal amino function, and this amino group can react with *N*-hydroxysuccinimide ester-carrying haptens (e.g., biotin, digoxigenin). These 5′-end-labeled oligonucleotides can hybridize without any interaction with the hybridizing sequence, resulting in good sensitivities.

### 3. Chemical 3′-End-Labeling by Synthesis with Multifunctional Controlled Pore Glass Carriers (MF-CPG)

With this method the functional amino group needed for coupling activated esters is already present in the linker arm introduced into the grow-

allylamine-modified nucleotide

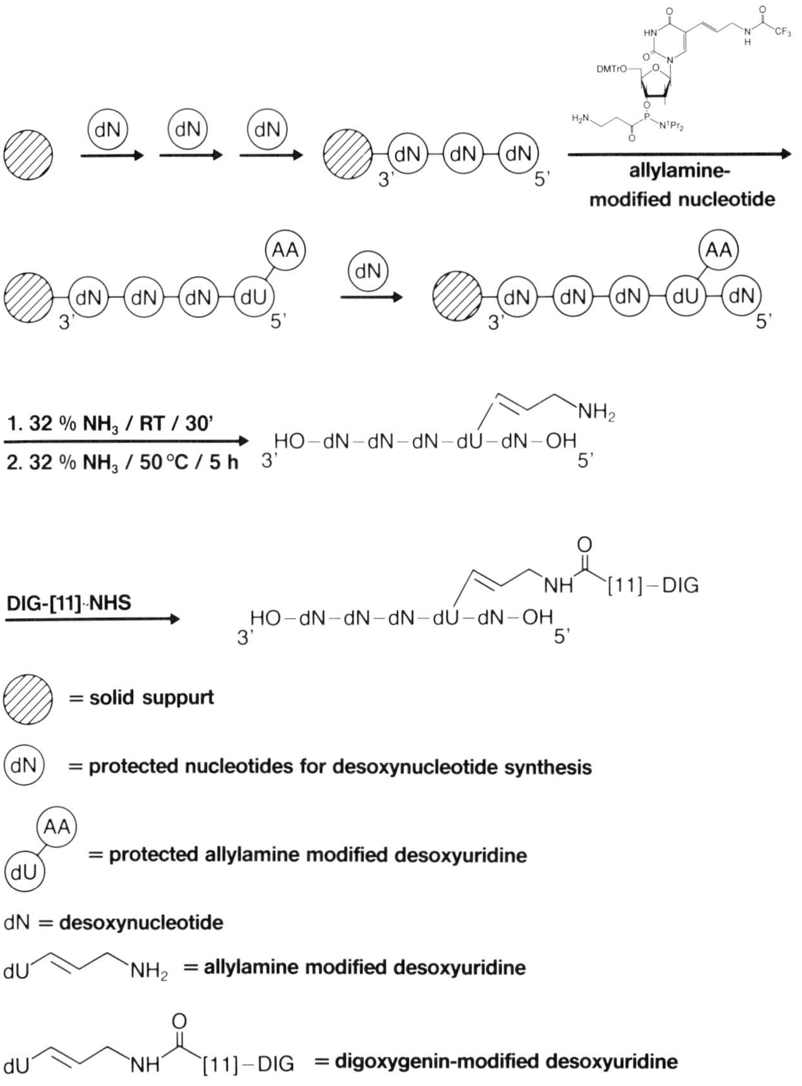

1. 32 % $NH_3$ / RT / 30'

2. 32 % $NH_3$ / 50 °C / 5 h

3' HO–dN–dN–dN–dU–dN–OH 5'

DIG-[11]-NHS

3' HO–dN–dN–dN–dU–dN–OH 5'

= solid suppurt

(dN) = protected nucleotides for desoxynucleotide synthesis

(dU)(AA) = protected allylamine modified desoxyuridine

dN = desoxynucleotide

dU–CH=CH–CH2–$NH_2$ = allylamine modified desoxyuridine

dU–CH=CH–CH2–NH–C(=O)–[11]–DIG = digoxygenin-modified desoxyuridine

**Fig. 11** Chemical internal labeling of oligodeoxynucleotides.

ing oligonucleotide during oligonucleotide synthesis on the solid support. After normal oligonucleotide synthesis, the oligonucleotide is solubilized from the solid support by hydrolysis of a 3′-terminal carboxyl moiety (Kempe *et al.*, 1985; Nelson *et al.*, 1989a, 1989b) (Fig. 13).

The resulting 3′-terminal coupling group contains an amino function, and as with the 5′-end-labeling reaction, condensation of this amino func-

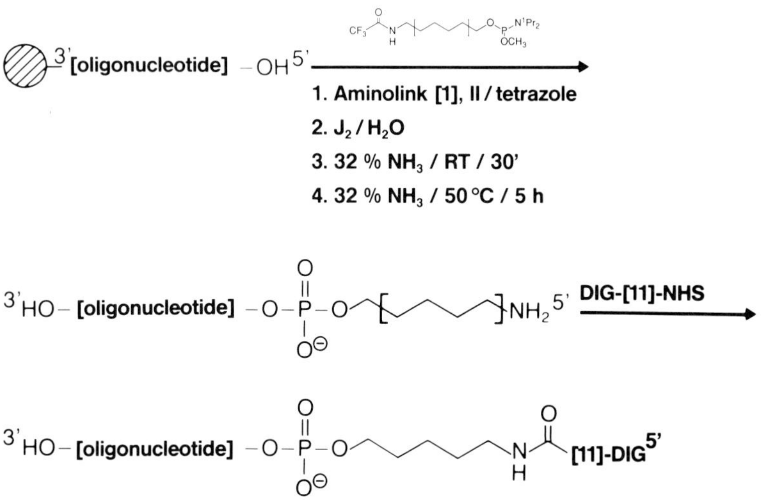

3' [oligonucleotide] –OH 5'

1. Aminolink [1], II / tetrazole
2. $J_2 / H_2O$
3. 32 % $NH_3$ / RT / 30'
4. 32 % $NH_3$ / 50 °C / 5 h

3' HO– [oligonucleotide] –O–P–O–[...]–$NH_2$ 5'  DIG-[11]-NHS

3' HO– [oligonucleotide] –O–P–O–...–NH–C(O)–[11]-DIG 5'

**Fig. 12** Chemical 5′-end-labeling of oligodeoxynucleotides.

tion with biotin- or digoxigenin-*O*-hydroxysuccinyl esters may occur. Analogous to the 5′-end-labeled oligonucleotides, the 3′-end-labeled oligonucleotides hybridize without interaction with the hybridizing region, again resulting in good sensitivities.

# IV. METHODS FOR CHEMICAL LABELING OF DNA, RNA, AND OLIGODEOXYNUCLEOTIDES WITH MARKER ENZYMES

In addition to the various methods for labeling of nucleic acids with chemical tags or haptens, a variety of procedures are described for direct covalent chemical coupling of marker enzymes to oligodeoxynucleotide probes.

These oligodeoxynucleotide : enzyme conjugates can be directly applied for generation of labeled hybridization complexes without additional binding steps between a modification group and a respective binding component; therefore, these direct detection systems are simpler and more convenient. However, because flexibility of application of directly enzyme-labeled probes is not so high as with indirectly labeled hybridization probes, directly labeled oligonucleotide probes are often used in standardized hybridization approaches; e.g., standardized sequences of midivariant regions applied as fingerprinting probes, labeled universal

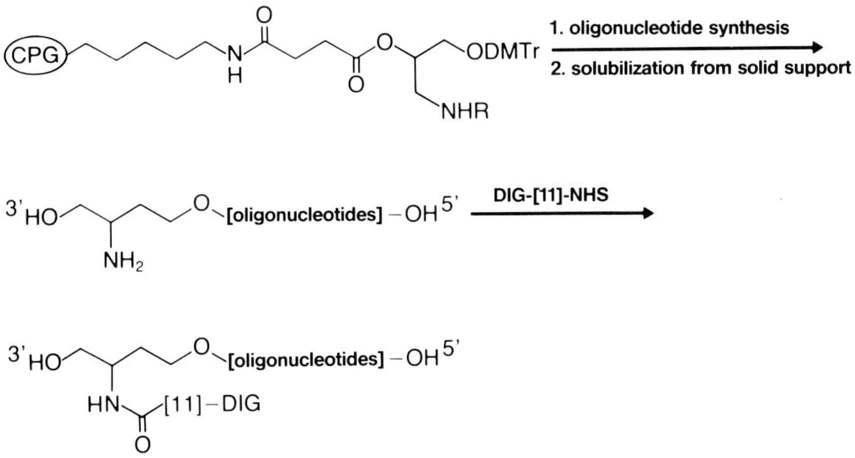

CPG — N(H) — O — ODMTr, NHR
1. oligonucleotide synthesis
2. solubilization from solid support

3' HO — O — [oligonucleotides] – OH 5', $NH_2$
DIG-[11]-NHS

3' HO — O — [oligonucleotides] – OH 5', HN — [11] – DIG

**Fig. 13** Chemical 3′-end-labeling of oligodeoxynucleotides.

primers for enzymatic sequencing approaches, or standardized probes for screening routine diagnostic parameters.

Direct labeling procedures have been described for coupling the marker enzymes alkaline phosphatase with oligodeoxynucleotides (Jablonski *et al.*, 1986; Jablonski and Ruth, 1986) or HRP directly with DNA molecules (Renz and Kurz, 1983; Taub, 1986; Pollard-Knight *et al.*, 1990). Microperoxidase, a heme-carrying oligopeptide, was also used for covalently labeling DNA with an enzymatic active marker (Heller and Schneider, 1983). With alkaline phosphatase, detection is achieved either with dye substrates [BCIP/NBT; SNAP (small nuclear alkaline phosphatase) probes] or via phosphorylated phenyl dioxetane-mediated chemiluminescence [AMPPD; NICE (nonisotopic chemiluminescent enhanced) probes]. With HRP, chemiluminescent substrates like luminol combined with chemiluminescent enhancers (e.g., *p*-iodo phenol) are responsible for signal generation [ECL (enhanced chemiluminescence) system]. In the following section the only two most commonly applied systems, SNAP and ECL, will be described in more detail (see also Chapter 7).

## A. Labeling of Oligodeoxynucleotides with Alkaline Phosphatase (SNAP/NICE System)

Short synthetic oligodeoxynucleotides can be covalently cross-linked to the marker enzyme alkaline phosphatase using the homobifunctional reagent disuccinimyl suberate (Jablonski *et al.*, 1986). Oligodeoxynucleotides in the range of 21 to 26 bases can be labeled by this method.

The oligodeoxynucleotides are modified in a first reaction step with a linker arm, which carries a terminal reactive primary amine function. Modification is achieved by introducing a protected linker arm nucleoside 3′-phosphoramidite [5′-*O*-dimethoxy trityl-5-[N(7-trifluoracetylaminoheptyl)-3-acrylamido]-2′-deoxyuridine 3′-*O*-(methyl N,N-diisopropyl)-phosphoramidite] (Ruth *et al.*, 1985) directly during automated oligodeoxynucleotide synthesis (Ruth, 1984). The disuccinimidyl suberate cross-linker reacts quantitatively with the primary amine function of the activated oligodeoxynucleotide and provides a reactive succinimidyl group, approximately 25 Å apart from the nucleotide base. Enzyme coupling is obtained by acylation of enzyme amino groups via the active N-hydroxysuccinimidyl ester. The linking reaction results in defined enzyme : oligodeoxynucleotide complexes consisting of one enzyme label per oligodeoxynucleotide with a mass ratio of protein to DNA of about 20. Gel filtration chromatography and chromatography on DEAE cellulose result in pure oligomer–alkaline phosphatase (AP) conjugates. The marker enzyme retains full enzymatic activity throughout the conjugation and purification process.

Applying alkaline phosphatase : oligodeoxynucleotide conjugates as hybridization probes is as specific as with radioactively or hapten-labeled probes. However, the $T_m$ is decreased by approximately 10° C, indicating a slightly decreased binding energy between enzyme-modified probe and target DNA. In contrast to HRP, AP is stable at elevated hybridization temperatures; e.g., at 50° C there is only a negligible loss of activity during a 30-min hybridization period. Because AP is also stable against elevated concentrations of detergent, stringency can be controlled not only by temperature and salt concentration but also by the presence of sodium dodecyl sulfate (SDS) (0.5% PVP may also be present in the hybridization mixture). Signal generation can be obtained with a BCIP/NBT subtrate (colorimetric) (SNAP probes) or AMPPD as substrate (chemiluminescence) (NICE probes).

Because of higher concentrations of low-molecular-weight enzyme : oligomer complex as compared with high-molecular-weight enzyme : DNA complexes, short hybridization periods of 15 to 30 min and low nonspecific interactions can be realized. As a result, target DNA can be visualized in the pg range within 4 hr in the colorimetric dye-detection system. Overnight development further enhances visualization, but not sensitivity. With the AMPPD chemiluminescence substrate sub-pg sensitivities are observed within 1 hr on X-ray films after hybridization periods of only 20 min.

## B. Enhanced Chemiluminescence with Horseradish Peroxidase-Labeled Probes (ECL System)

Direct cross-linking of the reporter enzyme HRP to probe DNA is achieved by coupling a preformed enzyme-polyethyleneimine complex (Renz and Kurz, 1984) and the bifunctional cross-linking reagent glutaraldehyde. For this labeling reaction, denatured DNA is treated with a 50-fold excess (mass ratio) of polyethyleneimine-modified HRP in the presence of 1.25% glutaraldehyde (Pollard-Knight *et al.*, 1990). The labeled probe can be used for hybridization without any further purification. From high-performance liquid chromatography (HPLC) analysis it has been shown that the trimer HRP enzyme complex is preferentially bound to the DNA probes.

With this labeling method, DNA probes ranging from 50 to several thousand base pairs (e.g., pBR322 DNA) can be cross-linked with the marker enzyme. Applying these labeled probes in dot blots, the signal intensity is proportional to the probe length. It was shown that approximately one active perixodase molecule is linked every 50 to 100 bp. Therefore, probes smaller than 50 bp (e.g., 20mers) do not react with the labeling components. RNA chains several hundred bases in length can also be labeled with HRP using glutaraldehyde as complexing reagent.

The stability of the formed hybrid between the HRP-labeled probe and the target sequence is not changed because of $T_m$ measurements. This is in contrast to the labeling of oligodeoxynucleotides with AP, in which a decrease of about 10° C is observed; however, in this case the labeling density is significant higher (21–26 bases per enzyme) than that of peroxidase labeling.

During hybridization, high concentrations of urea (6 *M*) have to be used as denaturing agent; hybridization is performed at temperatures not higher than 42° C because of enzyme instability at higher hybridization temperatures. Thus, the stringency of the hybridization reaction has to be controlled by varying the salt concentration.

For detection of labeled hybrid complexes, HRP oxidizes the chemiluminescent substrate luminol in the presence of light-intensifying enhancer compounds like P-iodo phenol. Using this mode of detection as little as 1 pg DNA can be detected within less than 1 hr on blue-sensitive X-ray films.

Reprobing is possible with HRP-labeled probes for the same or a different sequence without removing the original bound probe; this may be the effect of displacement of the first probe by the second probe (Green and Tibbetts, 1981).

# V. OVERVIEW OF FACTORS INFLUENCING HYBRIDIZATION

Binding between analyte and modified probe leading to stable hybrid molecules is mediated by specific interaction between complementary purine and pyrimidine bases forming A : T and G : C base pairs. Whereas the specificity of the hybrid formation is influenced primarily by the geometry of the bases responsible for the hydrogen bonds between A and T (2 hydrogen bonds) and G and C (3 hydrogen bonds), the stability of the generated hybrids is a result of the generation of a variable number of hydrogen bonds and of the effect of electrostatic as well as hydrophobic forces. G : C base pairs are more stable than A : T base pairs because between G and C. In contrast, between A and T only two hydrogen bonds are possible. Therefore, G : C-rich sequences are more stable than A : T sequences and thus will have higher $T_m$.

Electrostatic forces are caused predominantly by the phosphate molecules of the nucleic acid backbone; thus, double-stranded sequences are stabilized by increasing ionic strength. The presence of a hydrophilic hydroxyl group at the 2′-position of the ribose also stabilizes the double-stranded structure of nucleic acids. Thus, the $T_m$ of DNA/RNA and RNA/RNA hybrids are significantly higher than those of the respective DNA/DNA hybrids. Hydrophobic interactions between the staggered bases also contribute to hybrid stability; this explains the destabilizing effect of organic solvents on hybrid formation. Because of the central role of hydrogen bonding in hybrid formation and hybrid stability, the most appropriate sites for base modifications are those not involved in hydrogen bonding: (1) C-5 of uracil and cytosine (Langer *et al.*, 1981; Ruth, 1984; Ruth *et al.*, 1985; Jablonski *et al.*, 1986; Jablonski and Ruth, 1986; Haralambidis *et al.*, 1987; Cook *et al.*, 1988). C-5 modification of uracil mimics thymine residues, and this makes these compounds useful as thymine analogs; (2) C-6 of cytidine (Verdlov *et al.*, 1974); and (3) C-8 of guanine and adenine (Reisfeld *et al.*, 1987; Huynh Dinh *et al.*, 1987; Keller *et al.*, 1988). The $N^4$-position of cytosine (Gillam and Tener, 1986; Urdea *et al.*, 1987; Gebeyehu *et al.*, 1987; Urdea *et al.*, 1988) and the $N^6$-position of adenine (Jablonski and Ruth, 1986) or guanine (Tchen *et al.*, 1984) are involved in hydrogen bonding and therefore less useful (Viscidi *et al.*, 1986; Gebeyehu *et al.*, 1987). Labeling of the probes at the backbone or at the 5′- or at the 3′-end is less critical; in these cases hydrogen bonding as well as base staggering is not directly hindered. The position of probe modification and the distance between the modified atom of the nucleotide and the reporter molecule are both important for the stability of the hybrid molecules. Therefore, both sites should be separated by mostly linear spacer molecules composed of a balanced mixture of hydrophilic and hydrophobic atoms connecting the nucleotides with the modifying group (Forster *et al.*, 1985; Mühlegger *et al.*, 1990).

## A. Rate of Reassociation

The rate of reassociation between analyte and probe nucleic acid depends on probe complexity, i.e., probe length and length and number of unique sequences, as well as probe concentration (Britten and Kohne, 1968). Reassociation rates are defined as $C_0t$ values, where $C_0$ is the initial DNA concentration [$M$] and t the incubation time [sec]. The reassociation rate is quantified in terms of $C_0t_{1/2}$ values; these values define when half of the nucleic acid molecules have reassociated. The reassociation rate is inversely proportional to probe complexity. At a given concentration the reassociation rates are therefore inversively proportional to sequence length and the length of the unique sequences within the probe (Davidson and Britten, 1979).

$C_0t$ analyses are performed in solution using equal concentrations of analyte and probe. Thus, in these studies second-order rates are applied for calculating reaction parameters. Because most hybridizations are performed with analytes immobilized on solid supports, such as nitrocellulose or nylon membranes, only first-order kinetics are used for the calculation of association parameters.

With excess probe concentrations—as with most hybridization experiments—the hybridization rate is primarily dependent on probe complexity, i.e., probe length and length of unique sequences, and probe concentration. Meinkoth and Wahl (1984) describe the first-order equations for hybridization with single-stranded probes as follows:

$$t_{1/2} = \frac{ln2}{k \times C}$$

$k$ = rate constant for hybrid formation [mol × 1/#nuc × sec]
$C$ = probe concentration [$M$]

Within short hybridization periods (< 4 hr), double-stranded probes can be handled similarly; with longer hybridization periods (16 hr), the actual probe concentration decreases because of probe reassociation: the rate constant $k$ is dependent on probe complexity ($L$ = probe length; $N$ = length of unique sequences) as well as reaction parameters like temperature, ionic strength, pH value, and viscosity of the incubation mixture.

$$k = \frac{k_n \times L^{0.5}}{N}$$

$$k_n = \text{nucleation constant}$$

The nucleation constant is $3.5 \times 10^5$ for standard hybridization conditions, i.e., $Na^+$ concentrations of 0.4 to 1.0 *M*, pH values between 5 and 9, and temperatures $T_m - 25°$ C (Marmur and Doty, 1959).

Under these standard hybridization conditions, the hybridization rate $t_{1/2}$ [sec] is calculated as follows:

$$t_{1/2} = \frac{N \times ln2}{3.5 \times 10^5 \times L^{0.5} \times C}$$

For a fragment probe of 500 bp, $t_{1/2}$ is calculated for a probe concentration of $6 \times 10^{-10}$ [*M*]:

$$t_{1/2} = \frac{500 \times 0.693 \text{ [sec]}}{3.5 \times 10^5 \times 22 \times 6 \times 10^{-10}} = 7.5 \times 10^3 \text{ sec} = 20 \text{ hr}$$

For an 25mer oligonucleotide probe, $t_{1/2}$ calculates to 1 hr. However, this value has to be considered a rough estimate because experimentally obtained $t_{1/2}$ values fit into the calculated linear relationships in a first approach only (Keller and Manak, 1989).

In addition to probe complexity and probe concentration, the reaction parameters temperature, salt concentration, base mismatches, and hybridization accelerators also influence the rate of reassociation (Hames and Higgins, 1985).

Reassociation is maximal at about 25° C below the $T_m$ of the double-stranded nucleic acid (Wetmur and Davidson, 1968; Bonner *et al.*, 1973; Britten *et al.*, 1974). Increasing salt enhances reassociation rate; at low NaCl concentrations (0.01–0.1 *M*) a twofold increase in concentration accelerates the rate by 5- to 10-fold or more; the increase continues up to 1.2 *M* NaCl (Britten *et al.*, 1974). For each 10% base mismatch, the reassociation rate is reduced for about twofold (Bonner *et al.*, 1973).

Inert polymers like dextran sulfate or polyethylene glycol (PEG) can be used as accelerators of the hybridization reaction with longer fragment probes (Wahl *et al.*, 1979). The accelerating rates are relatively low with single-stranded probes (threefold); values up to 100-fold are observed, however, with double-stranded probes. Nonpolymeric accelerators include phenol (Kohne *et al.*, 1977) and chaotropic salts, e.g., guanidinium isothiocyanate (Thompson and Gillespie, 1987). These chemicals act as water-exclusion reagents and thus lower the energy difference between single- and double-stranded nucleic acids.

## B. Factors Affecting Stability of Hybrids

The specificity of hybrid formation is determined by the stringency of the hybridization conditions and the stability of the formed hybrid complexes. Like the reassociation rate, the hybrid stability is directly related to the $T_m$, which itself is affected by base composition; salt concentration; presence of formamide; fragment length; nature of the nucleic acid within the hybrid (DNA : DNA, RNA : DNA, RNA : RNA); and mismatch formation.

For DNA : DNA hybrids, the influence of the first four parameters on $T_m$ is represented by the following equation:

$$T_m = \frac{[81.5°C + 16.6 \log M + 0.41 (\% G + C)] - 500}{n - 0.61 (\% \text{ formamide})}$$

$$M = C_{Na} + [M]$$
$$n = \text{length of hybridizing sequence}$$

With respect to the nature of the formed hybrid, $T_m$ of RNA : DNA hybrids is 10–15° C higher than that of DNA : DNA hybrids. $T_m$ of RNA : RNA hybrids is 20–25° C higher; to lower $T_m$ of these hybrids, formamide is normally applied.

Mismatches are less stable than normal base pairing; therefore, the presence of mismatches also reduce $T_m$; which decreases for about 1°C for every 1% mismatch (Hutton and Wetmur, 1973; Britten *et al.*, 1974).

The specificity of hybrid formation is predominantly influenced by the stringency of the final washing steps after hybridization. Stringency is mostly enhanced during the final washing steps by increasing the temperature to only 5–15° C below $T_m$ and lowering the salt concentration from 5 × SSC (0.75 $M$ $Na^+$) to 0.1 × SSC (0.015 $M$ $Na^+$). This holds true especially for oligonucleotides, for which the washing temperature is usually 5° C below $T_m$. However, the optimal washing temperature has to be determined experimentally, and compromise has to be made with respect to the washing conditions in which the probe binds strongly to the analyte and weakly to unspecific heterologous analyte components.

# VI. OVERVIEW OF DETECTION SYSTEMS

For nonradioactive detection of labeled nucleic acids, a large variety of detection systems are available. These detection approaches can be classified as follows.

- Optical systems; e.g., alkaline phosphatase/BCIP, NBT (Franci and Vidal, 1988) or FAST dyes (West *et al.*, 1990);

• Chemiluminescent systems; e.g., alkaline phosphatase/Lumiphos®, AMPPD (Schaap *et al.*, 1987; Bronstein *et al.*, 1989a, 1989b) or selective hydrolysis of acridinium esters (Arnold *et al.*, 1989);
• Bioluminescence systems; e.g., alkaline phosphatase/D-luciferin-*O*-phosphate/luciferase (Hauber and Geiger, 1987; Kricka, 1988; Geiger *et al.*, 1989);
• Fluorescence systems; e.g., alkaline phosphatase/Attophos™ (Donahue *et al.*, 1991) or fluorescein tags or lanthanide-directed time-resolved fluorescence (Agrawal *et al.*, 1986; Diamandis, 1988; Diamindis *et al.*, 1988, 1989);
• Metal precipitating systems; e.g. silver-enhanced immunogold (Tomlinson *et al.*, 1988; Guérin-Reverchon *et al.*, 1989); and
• Electrochemical systems; e.g., urease-catalyzed pH shift (McKnabb *et al.*, 1989), or electrochemiluminescence (Blackburn *et al.*, 1991).

An overview of the various detection systems is given in Table IV. Dependent on the nature of the reaction format, the various systems can be subdivided into matrix-based and soluble systems. In addition, various levels of signal enhancement can be coupled with signal generation:

• Cross-linking of probes; e.g., Christmas-tree structures (Urdea *et al.*, 1987; Fahrlander and Klausner, 1988);
• Polymeric binding components; e.g., poly-streptavidin, poly-haptens, PAP- or APAAP-complexes (Mason *et al.*, 1982; Mason 1985);
• Polyenzymes; e.g., polymeric AP (Ward *et al.*, 1987);
• Signal cascades; e.g., NAD, NADPH + $H^+$ cycles coupled to a redox color reaction (Self, 1985).

## A. Detection Reactions

In direct systems the most popular labels are marker enzymes like AP (Jablonski *et al.*, 1986; Jablonski and Ruth, 1986) or HRP (Renz and Kurz, 1984; Taub, 1986; Pollard-Knight, 1990), and fluorescent tags, like fluorescein ($\lambda_{max\ emission}$ = 527 nm), rhodamin ($\lambda_{max\ emission}$ = 622 nm) (Serke and Pachmann, 1988). Whereas the coupled enzymes are often used in blot or soluble formats, the fluorescent dyes are especially useful for *in situ* detection in tissue sections and for multiplex mapping on metaphase chromosomes (Lichter *et al.*, 1990). In the hybridization protection assay (HPA), a chemiluminescent acridinium dye is linked to the probe via an alkaline-labile linker (Arnold *et al.*, 1989). Hydrolysis of this linker takes place with single-stranded acridine-modified probes, whereas with the double-stranded hybrids the rate of hydrolysis is strongly reduced (see Chapter 11). This difference in hydrolysis is due to intercalation of the

**Table IV**
Nonradioactive Nucleic Acid Detection Systems

| Format | Mode of Detection: Optical | Reference(s) | Luminescence | Reference(s) |
|---|---|---|---|---|
| Blots | AP/BCIP,NBT | Lojda *et al*. (1973); Anderson and Deinard (1974); Franci and Vidal (1988) | AP,AMPPD<br>AP/Lumiphos® | Schaap *et al*. (1987); Bronstein *et al*. (1989a, 1989b) |
| | AP/FAST dyes | West *et al*. (1990) | β-Gal/AMPGD | Bronstein *et al*. (1989a) |
| | β-Gal/BCIG(X-gal),NBT | Lojda *et al*. (1973); Anderson and Deinard (1974); Franci and Vidal (1988) | | |
| | HRP/TMB | Bos *et al*. (1981) | | |
| | Immunogold | Tomlinson *et al*. (1988); Guérin-Reverchon *et al*. (1989) | | |
| Solution | AP/*p*-NPP | Garen and Levinthal (1960) | AP/AMPPD<br>AP/Lumiphos®<br>β-Gal/AMPGD | Schaap *et al*. (1987); Bronstein *et al*. (1989a; 1989b) |
| | β-Gal/CPRG | Wallenfels *et al*. (1960) | | |
| | POD/ABTS® | Galatti (1979); Porstmann *et al*. (1981) | POD/Luminol | Bronstein *et al*. (1989a) |
| | GOD : POD-pair/ABTS | Taub (1986) | Xanthine oxidase/cyclic dihydrazides | Pollard-Knight *et al*. (1990) |
| | Hexokinase : G-6-PDH-pair/ABTS | Albarella *et al*. (1985a, 1985b) | G-6-PDH/phenazinium salty | Baret and Fert (1989); Tsuji *et al*. (1987) |
| | | | Hydrolysis of acridinium esters | Arnold *et al*. (1989) |
| | | | Rhodamine: luminol-pair<br>AP/D-luciferin-*O*-phosphate: firefly luciferase/ATP/$O_2$ | Hauber and Geiger (1987); Kricka (1988); Geiger *et al*. (1989) |

*(continues)*

**Table IV** (*continued*)

| Format | Mode of Detection | | | |
|---|---|---|---|---|
| | Optical | Reference(s) | Luminescence | Reference(s) |
| *In situ* | AP/BCIP,NBT | Heiles *et al.* (1988); Tautz and Pfeifle (1989); Seibl *et al.* (1990); Lichter *et al.* (1990) | AP/AMPPD | Bronstein and Voyta (1989) |
| | POD/TMB | (Lichter *et al.* 1990) | | |
| | Immunogold | (Tomlinson *et al.* 1988; Guérin-Reverchon *et al.* 1989) | | |
| Blots | Fluorescein | Agrawal *et al.* (1986) | | |
| | Rhodamine | Agrawal *et al.* (1986) | | |
| | Hydroxy-coumarin | Agrawal *et al.* (1986) | | |
| | AP/AMPPD,fluorescein/rhodamine/hydroxy-coumarin | Voyta *et al.* (1988); Beck and Köster (1990) | | |
| | β-Gal/AMPPD,fluorescein/rhodamine/hydroxy-coumarin | Voyta *et al.* (1988); Beck and Köster (1990) | | |
| Solution | AP/4-MUF-P | Fernley and Walker (1965) | Urease/Urea | McKnabb and Tedesco (1989) |
| | β-Gal/4-MUF-β-Gal | Buonocore *et al.* (1980) | | |
| | POD/homovanillinic acid-*o*-dianisidine/$H_2O_2$ | Iwai *et al.* (1983) | | |
| | Aromatic peroxalate compounds/$H_2O_2$ | Arakawa *et al.* (1982) | | |
| | $Eu^{3+}$-micelles | Hemmilä *et al.* (1984); Lövgren *et al.* (1985) | | |
| | $Eu^{3+}$-chelates | Diamandis (1988); Diamandis *et al.* (1988; 1989) | | |
| *In situ* | Fluorescein | Agrawal *et al.* (1986) | | |
| | Rhodamine | Agrawal *et al.* (1986) | | |
| | Hydroxy-coumarin | Agrawal *et al.* (1986) | | |

acidinium dye specifically into the double-stranded helix only; this intercalation stabilizes the linker against alkaline hydrolysis.

In the indirect systems, the whole range of detection reactions can be utilized. The binding groups recognizing the modified probes are covalently bound to molecules, which in turn, generate the signal to be measured. In most cases, these conjugated molecules are enzymes that generate a stained, luminescent, or fluorescent reaction product through a coupled, catalytic substrate reaction (for review see Kessler, 1991).

The best known marker enzymes are AP from calf intestine (Ishikawa *et al.*, 1983); HRP from horseradish (Wilson and Nakane, 1978); and β-galactosidase from *E. coli* (β-Gal) (Inoue *et al.*, 1985). Urease (McKnabb *et al.*, 1989), glucose oxidase (GOD), and microperoxidase are also used but to a lesser extent.

Depending on the kind of substrate used, the enzymatically catalyzed substrate reactions yield either insoluble precipitates on blots and with *in situ* approaches, or soluble color products for quantitative measurements, e.g., in microtiter plates. Sensitive substrates resulting in insoluble precipitates include BCIP/nitroblue tetrazolium salt (NBT) for AP (Franci and Vidal, 1988), 3,3′,5,5′-tetramethylbenzidine (TMB) for peroxidase (Bos *et al.*, 1981), and 5-bromo-4-chloro-3-indolyl β-galactoside (BCIG or X-gal)/NBT for β-galactosidase (Lojda *et al.*, 1973); the respective soluble substrates are *p*-nitrophenyl phosphate (p-NPP), 2,2-azino-di-[3-ethylbenzthiazoline sulfate] (ABTS®) (Gallati, 1979; Porstmann *et al.*, 1981) and chlorophenolred-β-D-galactopyranoside (CPRG) (Wallenfels *et al.*, 1960).

Increase of sensitivity and speed of detection can be obtained in optical systems by coupling of an enzymatic cascade reaction using the following components (Self, 1985; Johannson *et al.*, 1985; Stanley *et al.*, 1985): alcohol dehydrogenase (ADH); diaphorase (DP); nicotinamide-dinucleotide, oxidated form (NAD); nicotinamide-dinucleotide, reduced form (NADH); and *p*-iodonitro-tetrazolium purple (INT violet).

In the first reaction step, a primary NADP substrate is dephosphorylated to NAD by hybrid-bound alkaline phosphatase. The NAD cofactor activates in a cyclic reaction a coupled secondary redox system that contains the two enzymes ADH and DP. In each cycle the generated NAD is reduced to NADPH + $H^+$ by linked oxydation of ethanol to acetaldehyde by ADH; this redox reaction is coupled to a second redox reaction in which NADPH + $H^+$ is oxidized to NAD, again linked with the reduction of INT violet to formazan. The intensively stained formazan can be photometrically quantified ($\lambda_{max\ emission}$ = 465 nm/EtOH,DMF).

Aside from the optical systems, a number of alternative detection systems have been developed based on luminescent or fluorescent substrates. Soluble chemiluminescent substrates are used with AP or β-Gal by exchanging the optical BCIP/NBT or BCIG/NBT substrates for the corresponding 1,2-dioxetane-luminescence substrates Lumiphos®/AMPPD or AMPGD (Schaap *et al.*, 1987; Voyta *et al.*, 1988; Bronstein and Kricka, 1989; Bronstein and Voyta, 1989; Thorpe and Kricka, 1989; Bronstein *et al.*, 1989a, 1989b; Beck and Köster, 1990). In the AMPPD and AMPGD chemiluminescent systems, enzymatic cleavage of a phosphate (AMPPD) or $\beta$-galactoside (AMPGD) residue forms the unstable AMPD-anion, which decomposes to produce light. The chemiluminescence intensity can be increased by inclusion of the 1,2-dioxetane derivatives in micelles or in covalently linked polymers which, in turn, contain fluorophores such as fluorescein that can be activated by the emitted luminescence light. By selecting suitable fluorophores, a secondary fluorescence signal can be generated as follows: hydroxy-coumarin ($\lambda_{max\ emission}$ = 467 nm); fluorescein ($\lambda_{max\ emission}$ = 527 nm); rhodamin ($\lambda_{max\ emission}$ = 622 nm).

Analogous micelles or chelates are used with time-resolved fluorescence (TRF) where the probes are coupled with divalent lanthanide ions (e.g., $Eu^{3+}$ or $Tb^{3+}$) complexed by the micelles (Hemmilä *et al.*, 1984; Lövgen *et al.*, 1985) or chelating agents (Soini and Kojola, 1983; Diamandis, 1988; Evangelista *et al.*, 1988; Oser and Valet, 1988; Diamandis *et al.*, 1988, 1989) (see Chapter 9).

In bioluminescence systems, a bioluminescence substrate is released by an initial enzymatic reaction mediated by an hybrid-bound marker enzyme (Miska and Geiger, 1987; Hauber and Geiger, 1987; Gould and Subramani, 1988; Kricka, 1988; Geiger *et al.*, 1989). Using D-luciferin-*O*-phosphate and alkaline phosphatase, D-luciferin is released by phosphate hydrolysis. D-Luciferin is converted into oxyluciferin, AMP, and $PP_i$ in a coupled reaction by the enzyme luciferase from *Photinus pyralis* in the presence of $O_2$ and ATP. This luciferase-catalyzed oxydation of D-luciferin into oxyluciferin is accompanied by the emission of light ($\lambda_{max\ emission}$ = 600 nm) (see Chapter 4).

## B. Detection Formats

Detection of nucleic acids by nonradioactively labeled probes can be performed in the solid phase with immobilized analytes or in solution. In the first case, the presence of absence of a particular sequence is recorded, whereas detection in solution allows quantitative measurements. For detection of extracted nucleic acids, a number of blot formats have been established: dot blot, slot blot, Southern blot (DNA analytes), Northern blot (RNA analytes), South-Western blot (protein-binding DNA sequences), genomic blot (analysis of whole genomes). Hybridization is also possible with isolated metaphase chromosomes. A variety of formats have been described for detection of nucleic acids *in situ:* colony hybridization (bacterial colonies), plaque hybridization (phage plaques), *in situ* hybridizations with tissue sections, fixed cells or whole organisms like *Drosophila* embryos (for review see Matthews and Kricka, 1988; Wilchek and Bayer, 1988; Kessler, 1991).

For quantitative detection of nucleic acids in diagnostic systems, reaction formats such as sandwich (Ranki *et al.*, 1983; Syvänen *et al.*, 1986a, 1986b; Yehle *et al.*, 1987; Jungell-Nortamo *et al.*, 1988; Nicholls and Malcolm, 1989; Newman *et al.*, 1989), replacement (Vary *et al.*, 1986; Ellwood *et al.*, 1986; Vary, 1987; Collins *et al.*, 1988), or energy-transfer formats (Heller *et al.*, 1982; Cardullo *et al.*, 1988) can be used. Compared with immunological antigen or antibody-detection systems, the nucleic acid-based systems have the following advantages (Zwadyk and Cooksey, 1987; Parsons, 1988; Pasternak, 1988; Kohne, 1990): high specificity of analyte recognition (base-pairing); sensitivity because of a prior amplification reaction (analyte amplification); and analyte detection before antigen expression (latency).

With the sandwich technique, two probes are used to hybridize with two different regions of the analyte. The first probe is used as *capture probe* for binding the sandwich complex to a solid support, whereas the second probe is used as a *detector probe* responsible for signal generation. After removal of free detector and capture probes, the signal generated by complexed detector probe is directly correlated with the amount of analyte. The replacement technique is based on the replacement of a prebound labeled oligonucleotide by the analyte from a binding complex; in this reaction format, the amount of released labeled oligonucleotide is a measure of the input analyte.

In energy-transfer formats, two probes are bound to the analyte. The first probe is labeled with an energy donor, and the second is labeled with an energy acceptor. Both probes are designed in such a way that when the two probes are bound to the analyte, the energy donor and acceptor moiety are located directly together. Thus energy transfer occurs only when both probes are bound to the analyte. In the unbound state no energy transfer is possible. An example of this format uses probes modified with the chemiluminescent POD-substrate luminol as energy-donor and rhodamine as energy acceptor (see Chapter 12).

In nucleic acid detection systems, unlike immunological detection systems analyte amplification steps can be integrated into the reaction format (Table V). A well-established analyte amplification reaction is the PCR; amplification factors of $10^6$ to $10^7$ can be obtained (Saiki *et al.*, 1985a, 1985b, 1988). By combination of these amplification reactions with nonradioactive labeling and detection systems, the quantitative analysis of single molecules is possible with isolated nucleic acids or in whole cells as analytes. This level of sensitivity is unique in bioanalytical indicator systems.

**Table V**
Nucleic Acid Amplification Systems

| Mode of amplification | Example(s) | Reference(s) |
|---|---|---|
| Target amplification | | |
| *In vitro* target amplification: Replication | | |
| Elongation temperature cycles | Polymerase chain reaction (PCR); oligonucleotide ligation assay (PCR/OLA) | Saiki *et al*. (1985a; 1985b; 1988); Li *et al*. (1988); Nickerson *et al*. (1990) |
| cDNA synthesis/elongation temperature cycles | Polymerase chain reaction or RNA basis | Murakawa *et al*. (1987) |
| Elongation isothermal replacement reactions | Strand displacement amplification (SDA) | Alexander *et al*. (1991) |
| *In vitro* target amplification: Transcription | | |
| cDNA synthesis/ds promoter-dependent transcription cycles | Nucleic acid sequence based amplification (NASBA), self-sustained sequence replication (3SR) | Davey and Malek (1988); Guatelli *et al*. (1990) |
| cDNA synthesis/ds promoter-dependent transcription | Transcription-based amplification system (TAS) | Gingeras *et al*. (1988); Joyce (1989); Kwoh *et al*. (1989) |
| *In vivo* target amplification: increased rRNA copy numbers | | |
| Bacterial rRNA detection | 16S/23S rRNA probes | Fox *et al*. (1980); Rossau *et al*. (1986); Yehle (1987); Stull (1988) |
| Target-specific signal amplification | | |
| *In vitro* indicator amplification: Replication | | |
| Replication cycles | $Q\beta$ replication ($Q\beta$) | Lizardi *et al*. (1988) |
| *In vitro* indicator amplification: Ligation | | |
| Ligation system | Ligase chain reaction (LCR) | Orgel (1989); Wu and Wallace (1989); Barringer *et al*. (1990) |

| | | |
|---|---|---|
| Ligation repair system | Repair chain reaction (RCR) | Segev, (1990); Segev *et al.* (1990) |
| *In vitro* Indicator amplification: Hydrolysis | | |
| RNA hydrolysis system | Target cycling amplification (TCA) | Duck and Bender (1989) |
| *In vivo* amplification of antibodies | Detection of antigen-specific antibodies | Malvano (1980); Avrameas *et al.* (1983) |
| Signal amplification | | |
| Tree structures | | |
| Network of binding partners | Probe network | Urdea *et al.* (1987) |
| Network of indicator molecules | Hybridization trees, sandwich complexes, primary/secondary antibody trees, peroxidase : antiperoxidase (PAP), alkaline phosphatase : antialkaline phosphatase (APAAP) | Fahrlander and Klausner (1988); Nicholls and Malcolm (1989); Oellerich (1983); Mason *et al.* (1982); Mason (1985) |
| Enzyme catalysis | | |
| Enzyme-catalyzed signal generation | Enzyme-linked immunosorbent assay (ELISA) | Vogt (1978); Maggio (1980); Ishikawa *et al.* (1981); Kemeny and Challacombe (1988) |
| Conjugates with precoupled marker enzymes (hedgehog conjugates) | ELISA with polymeric enzyme conjugates | Ward *et al.* (1987) |
| Coupled signal cascades | | |
| Cyclic NAD/NADH + $H^+$ redox reaction | NADH + $H^+$-coupled reduction of INT violet by ADH/DP (SELF) | Self (1985); Johannsson *et al.* (1985); Stanley *et al.* (1985) |

# REFERENCES


Adarichev, V. A., Dymshits, G. M., Kalachikov, S. M., Pozdnyakov, P. I., and Salganik, R. I. (1987). Introduction of aliphatic amino groups into DNA and their labeling with fluorochromes in preparation of molecular hybridization probes. *Bioorg. Khim.* **13,** 1066–1069.

Agrawal, S., Christodoulou, C., and Gait, M. J. (1986). Efficient methods for attaching nonradioactive labels to the 5′ ends of synthetic oligodeoxyribonucleotides. *Nucleic Acids Res.* **14,** 6227–6245.

Alexander, A., Fraiser, M., Little, M., Malinowski, D., Nadeau, J., Schram, J., Shank, D., and Walker, T. (1991). Isothermal, *in vitro* amplification of DNA by a novel restriction enzyme/DNA polymerase system—strand displacement amplification (SDA). *The San Diego Conference on Nucleic Acids: The Leading Edge,* San Diego, California, Abstract 17.

Al-Hakim, A. H., and Hull, R. (1986). Studies towards the development of chemically synthesized non-radioactive biotinylated nucleic acid hybridization probes. *Nucleic Acids Res.* **14,** 9965–9976.

Al-Hakeem, M., and Sommer, S. S. (1987). Terbium identifies double-stranded RNA on gels by quenching the fluorescence of intercalated ethidium bromide. *Anal. Biochem.* **163,** 433–439.

Albarella, J. P., and Anderson, I. H. (1985). Detection of polynucleotide sequence in medium and when single-stranded nucleic acids are present by using probe, intercalator, and antibody. *Eur. Patent Appl.* 0146815.

Albarella, J. P., and Anderson, L. H. (1985). Nucleic acid hybridization assay employing antibodies to intercalation complexes. *Eur. Patent Appl.* 146815.

Albarella, J. P., DeReimer, L. H. A., and Carrico, R. J. (1985). Hybridization assay employing labeled pairs of hybrid-binding reagents. *Eur. Patent Appl.* 0144914.

Albarella, J. P., DeReimer, L. H. A., and Carrico, R. J. (1985). Hybridization assay with immobilization of hybrids by antibody binding. *Eur. Patent Appl.* 146039.

Albarella, J. P., Minegar, R. L., Patterson, W. L., Dattagupta, N., and Carlson, E. (1989). Monoadduct forming photochemical reagents for labeling nucleic acids for hybridization. *Nucleic Acids Res.* **17,** 4293–4308.

Anderson, G. L., and Deinard, A. S. (1974). Nitroblue tetrazolium (NBT) test. Review. *Am. J. Med. Technol.* **40,** 345–353.

Ansorge, W., Sproat, B. S., Stegemann, J., and Schwager, C. (1986). A non-radioactive automated method for DNA sequence determination. *J. Biochem. Biophys. Methods* **13,** 315–323.

Ansorge, W., Sproat, B., Stegemann, J., Schwager, C., and Zenke, M. (1987). Automated DNA sequencing: Ultrasensitive detection of fluorescent bands during electrophoresis. *Nucleic Acids Res.* **15,** 4593–4603.

Arakawa, H., Maeda, M., and Tsuji, A. (1982). Chemiluminescence enzyme immunoassay of 17-hydroxyprogesterone using glucose oxidase and *bis*(2,4,6-trichlorophenyl)oxalate-fluorescent dye system. *Chem. Pharm. Bull.* **30,** 3036–3039.

Arnold, L. J., Hammond, P. W., Wiese, W. A., and Nelson, N. C. (1989). Assay formats involving acridinium-ester-labeled DNA probes. *Clin. Chem.* **35,** 1588–1594.

Asseline, U., Toulme, F., Thuong, N. T., Delarue, M., Montenay-Garestier, T., and Hélène, C. (1984). Oligodeoxynucleotides covalently linked to intercalating dyes as base sequence-specific ligands. Influence of dye attachment site. *EMBO J.* **3,** 795–800.

Ausubel, F. M., Brent, R., Kingston, R. E., Moore, D. D., Smith, J. A., Seidman, J. G., and Struhl, K. (1987). "Current Protocols in Molecular Biology." Greene Publishing Associates and Wiley-Interscience, New York.

Avrameas, S., Druet, P., Masseyeff, R., and Feldmann, G. (1983). "Immunoenzymatic Techniques." Elsevier Science Publishers, Amsterdam.

Baret, A., and Fert, V. (1989). $T_4$ and ultrasensitive TSH immunoassays using luminescent enhanced xanthine oxidase assay. *J. Biolumin. Chemilumin.* **4,** 149–153.

Barringer, K. J., Orgel, L., Wahl, G., and Gingeras, T. R. (1990). Blunt-end and single-strand ligations by *Escherichia coli* ligase: Influence on an *in vitro* amplification scheme. *Gene* **89,** 117–122.

Baumann, J. G., Wiegant, J., and van Duijin, P. (1983). The development, using poly(Hg-U) in a model system, of a new method to visualize cytochemical hybridization in fluorescence microscopy. *J. Histochem. Cytochem.* **31,** 571–578.

Bayer, E. A., and Wilchek, M. (1980). The use of the avidin-biotin complex as a tool in molecular biology. *Methods Biochem. Anal.* **26,** 1–45.

Bayer, E. A., and Wilchek, M. (1990). "Avidin-biotin technology" (M. Wilchek and E. A. Bayer, eds.). *In "Methods Enzymology"*, Vol. **184,** pp. 5–13. Academic Press, San Diego.

Beck, S., and Köster, H. (1990). Applications of dioxetane chemiluminescent probes to molecular biology. *Anal. Chem.* **62,** 2258–2270.

Ben-Hur, E., and Song, P. S. (1984). The photochemistry and photobiology of furocoumarins (psoralens), *Adv. Radiat. Biol.* **11,** 131–171.

Bergstrom, D. E., and Ruth, J. L. (1977). Preparation of carbon-5 mercurated pyrimidine nucleosides. *J. Carbohydr. (Nucleosides and Nucleotides)* **4,** 257–269.

Binder, M. (1987). *In situ* hybridization at the electron microscope level. *Scanning Microsc.* **1,** 331–338.

Blackburn, G. F., Shah, H. P., Kenten, J. H., Leland, J., Kamin, R. A., Link, J., Peterman, J., Powell, M. J., Shah, A., Talley, D. B., Tyagi, S. K., Williams, E., Wu, T.-G., and Massey, R. J. (1991). Electrochemiluminescence detection for development of immunoassays and DNA probe assays for clinical diagnosis. *Clin. Chem.* **37,** 1534–1539.

Bonner, T. I., Brenner, D. J., Neufeld, B. R., and Britten, R. J. (1973). Reduction in the rate of DNA reassociation by sequence divergence. *J. Mol. Biol.* **81,** 123–135.

Bos, E. S., van der Doelen, A. A., van Rooy, N., and Schuurs, A. H. (1981). 3,3′,5,5′-Tetramethylbenzidine as an Ames test negative chromogen for horseradish peroxidase in enzyme-immunoassay. *J. Immunoassay* **2,** 187–204.

Brakel, D. L., and Engelhardt, D. L. (1985). DNA hybridization method using biotin. *In* "Rapid Detection and Identification of Infectious Agents" (D. T. Kingsbury and S. Falcow, eds.), pp. 235–243. Academic Press, New York.

Britten, R. J., Graham, D. E., and Neufeld, B. R. (1974). Analysis of repeating DNA sequences by reassociation. *Methods Enzymol.* **29,** 363–418.

Britten, R. J., and Kohne, D. E. (1968). Repeated sequences in DNA. *Science* **161,** 529–540.

Bronstein, I., Edwards, B., and Voyta, J. C. (1989a). 1,2-Dioxetanes: Novel chemiluminescent enzyme substrates. Applications to immunoassay. *J. Biolumin. Chemilumin.* **4,** 99–111.

Bronstein, I., and Kricka, L. J. (1989). Clinical applications of luminescent assay for enzymes and enzyme labels. *J. Clin. Lab. Anal.* **3,** 316–322.

Bronstein, I., and Voyta, J. C. (1989). Chemiluminescent detection of herpes simplex virus I DNA in blot and *in situ* hybridization assay. *Clin. Chem.* **35,** 1856–1857.

Bronstein, I., Voyta, J. C., and Edwards, B. (1989b). A comparison of chemiluminescent and colorimetric substrates in a hepatitis B virus DNA hybridization assay. *Anal. Biochem.* **180,** 95–98.

Brosalina, E. B., and Grachev, S. A. (1986). The synthesis of 5′-biotin-labeled oligo- and polynucleotides and investigation of their complexes with avidin. *Bioorg. Khim.* **12,** 248–256.

Brown, D. M., Frampton, J., Goelet, P., and Karn, J. (1982). Sensitive detection of RNA using strand-specific M13 probes. *Gene* **20,** 139–144.

Bulow, S., and Link, G. (1986). A general and sensitive method for staining DNA and RNA blots. *Nucleic Acids Res.* **14,** 3973.

Buonocore, V., Sgambati, O., De Rosa, M., Esposito, E., and Gambacorta, A. (1980). A constitutive β-galactosidase from the extreme thermoacidophile archaebacterium *Caldariella acidophila:* Properties of the enzyme in the free state and in immobilized whole cells. *J. Appl. Biochem.* **2,** 390–397.

Butler, E. T., and Chamberlin, M. J. (1982). Bacteriophage SP6-specific RNA polymerase. *J. Biol. Chem.* **257,** 5772–5778.

Cardullo, R. A., Agrawal, S., Flores, C., Zamecnik, P. C., and Wolf, D. E. (1988). Detection of nucleic acid hybridization by nonradiative fluorescence resonance energy transfer. *Proc. Natl. Acad. Sci. U.S.A.* **85,** 8790–8794.

Chaiet, L., and Wolf, F. J. (1964). The properties of streptavidin, a biotin-binding protein produced by *Streptomycetes. Arch. Biochem. Biophys.* **106,** 1–5.

Chan, V. T.-W., Fleming, K. A., and McGee, J. O'D. (1985). Detection of subpicogram quantities of specific DNA sequences on blot hybridization with biotinylated probes. *Nucleic Acids Res.* **13,** 8083–8091.

Chan, V. T.-W., and McGee, J. O'D. (1990). Nonradioactive probes: Preparation, characterization, and detection. *In* "*In Situ* Hybridization. Principles and Practice" (J. M. Polak and J. O'D. McGee, eds.), pp. 59–70. Oxford University Press, Oxford.

Chollet, A., and Kawashima, E. H. (1985). Biotin-labeled synthetic oligodeoxyribonucleotides: Chemical synthesis and uses as hybridization probes. *Nucleic Acids Res.* **13,** 1529–1541.

Chu, B. C., and Orgel, L. E. (1985a). Detection of specific DNA sequences with short biotin-labeled probes. *DNA* **4,** 327–331.

Chu, B. C. F., and Orgel, L. E. (1985b). Nonenzymatic sequence-specific cleavage of single-stranded DNA. *Proc. Natl. Acad. Sci. U.S.A.* **82,** 963–967.

Chu, E. C. F., and Orgel, L. E. (1988). Ligation of oligonucleotides to nucleic acids or proteins via disulfide bonds. *Nucleic Acids Res.* **16,** 3671–3691.

Cimino, G. D., Gamper, H. B., Isaacs, S. T., and Hearst, J. E. (1985). Psoralens as photoactive probes of nucleic acid structure and function: Organic chemistry, photochemistry, and biochemistry. *Annu. Rev. Biochem.* **54,** 1151–1193.

Collins, M., Fritsch, E. F., Ellwood, M. S., Diamond, S. E., Williams, J. I., and Brewen, J. G. (1988). A novel diagnostic method based on strand displacement. *Mol. Cell. Probes* **2,** 15–30.

Connolly, B. A. (1987). The synthesis of oligonucleotides containing a primary amino group at the 5'-terminus. *Nucleic Acids Res.* **15,** 3131–3139.

Connolly, B. A., and Rider, P. (1985). Chemical synthesis of oligonucleotides containing a free sulphydryl group and subsequent attachment of thiol-specific probes. *Nucleic Acids Res.* **13,** 4485–4502.

Cook, A. F., Vuocolo, E., and Brakel, C. L. (1988). Synthesis and hybridization of a series of biotinylated oligonucleotides. *Nucleic Acids Res.* **16,** 4077–4095.

Cotton, F. A., and LaPrade, M. D. (1968). Stereochemically nonrigid organometallic molecules. XVI. The crystal and molecular structure of *p*-methyl-π-benzoyl-π-cyclopentadienyl-dicarbonylmolybdenum. *J. Am. Chem. Soc.* **90,** 5418–5422.

Coull, J. M., Weith, H. L., and Bischoff, R. (1986). A novel method for the introduction of an aliphatic primary amino group at the 5'-terminus of synthetic oligonucleotides. *Tetrahedron Lett.* **27,** 3991–3994.

Coutlee, F., Bobo, L., Mayur, K., Yolken, R. H., and Viscidi, R. P. (1989a). Immunodetection of DNA with biotinylated RNA probes: A study of reactivity of a monoclonal antibody to DNA–RNA hybrids. *Anal. Biochem.* **181,** 96–105.

Coutlee, F., Viscidi, P., and Yolken, H. (1989b). Comparison of colorimetric, fluorescent, and enzymatic amplification substrate systems in an enzyme immunoassay for detection of DNA–RNA hybrids. *J. Clin. Microbiol.* **27,** 1002–1007.

Coutlee, F., Yolken, R. H., and Viscidi, R. P. (1989c). Nonisotropic detection of RNA in an enzyme immunoassay using a monoclonal antibody against DNA–RNA hybrids. *Anal. Biochem.* **181,** 153–162.

Cremers, A. F., Jansen in de Wal, N., Wiegant, J., Dirks, R. W., Weisbeek, P., Van der Ploeg, M., and Landegent, J. E. (1987). Nonradioactive *in situ* hybridization. A comparison of several immunocytochemical detection systems using reflection-contrast and electron microscopy. *Histochemistry* **86,** 609–615.

Czichos, J., Koehler, M., Reckmann, B., and Renz, M. (1989). Protein–DNA conjugates produced by UV irradiation and their use as probes for hybridization. *Nucleic Acids Res.* **17,** 1563–1572.

Dahlen, P., Hurskainen, P., Lovgren, T., and Hyypia, T. (1988). Time-resolved fluorometry for the identification of viral DNA in clinical specimens. *J. Clin. Microbiol.* **26,** 2434–2436.

Dahlen, R., Syvänen, A. C., Hurskainen, P., Kwiatkowski, M., Sund, C., Ylikoski, J., Söderlund, H., and Lovgren, T. (1987). Sensitive detection of genes by sandwich hybridization and time-resolved fluorometry. *Mol. Cell. Probes.* **1,** 159–168.

Dale, R. M. K., Martin, E., Livingston, D. C., and Ward, D. C. (1975). Direct covalent mercuration of nucleotides and polynucleotides. *Biochemistry* **14,** 2447–2457.

Dattagupta, N., and Crothers, D. M. (1984). Labeled nucleic acid probes and adducts for their preparation. *Eur. Patent Appl.* 0131830.

Dattagupta, N., Knowles, W., Marchesi, V. T., and Crothers, D. M. (1984). Nucleic acid–protein conjugate. *Eur. Patent Appl.* 0154884.

Dattagupta, N., Rae, P. M. M., Huguenel, E. D., Carlson, E., Lygla, A., Shapiro, J. A., and Albarella, J. P. (1989). Rapid identification of microorganisms by nucleic acid hybridization after labeling the test sample. *Anal. Biochem.* **177,** 85–89.

Dattagupta, N., Rae, P. M. M., Knowles, W. J., and Crothers, D. M. (1985). Nucleic acid detection probe comprises hybridisable single stranded part of nucleic acid connected to non-hybridisable nucleic acid with specific recognition site. *Eur. Patent Appl.* 0147665.

Dattagupta, N., Rae, P. M. M., Knowles, W. J., and Crothers, D. M. (1988). Use of non-hybridizable nucleic acids for the detection of nucleic acid hybridization. *US Patent* 4724202.

Davey, C., and Malek, L. T. (1988). Nucleic acid amplification process. *Eur. Patent Appl.* 0329822.

Davidson, E. H., and Britten, R. J. (1979). Regulation of gene expression: Possible role of repetitive sequences. *Science* **204,** 1052–1059.

Diamandis, E. P. (1988). Immunoassays with time-resolved fluorescence spectroscopy: Principles and applications (Review). *Clin. Biochem.* **21,** 139–150.

Diamandis, E. P., Bhayana, V., Conway, K., Reichstein, E., and Papanastasiou-Diamandis, A. (1988). Time-resolved fluoroimmunoassay of cortisol in serum with a europium chelate as label. *Clin. Biochem.* **21,** 291–296.

Diamandis, E. P., Morton, R. C., Reichstein, E., and Khosravi, M. J. (1989). Multiple fluorescence labeling with europium chelators. Application to time-resolved fluoroimmunoassays. *Anal. Chem.* **61,** 48–53.

Donahue, C., Neece, V., Nycz, C., Weng, J. M. H., Walker, G. T., Vonk, G. P., Jurgensen, S. (1991). The San Diego Conference on Nucleic Acids: The Leading Edge, San Diego, California, Abstract 23.

Donovan, R. M., Bush, C. E., Peterson, W. R., Parker, L. H., Cohen, S. H., Jordan, G. W., Vanden Brink, K. M., and Goldstein, E. (1987). Comparison of nonradioactive DNA hybridization probes to detect human immunodeficiency virus nucleic acid. *Mol. Cell. Probes* **1,** 359–366.

Draper, D. E., and Gold, L. (1980). A method for linking fluorescent labels to polynucleotides: Application to studies of ribosome–ribonucleic acid interactions. *Biochemistry* **19,** 1774–1781.

Duck, P., and Bender, R. (1989). Methods for detecting nucleic acid sequences. *WO Patent Appl.* 8910415.
Ellwood, M. S., Collins, M., Fritsch, E. F., Williams, J. I., Diamond, S. E., and Brewen J. G. (1986). Strand displacement applied to assays with nucleic acid probes. *Clin. Chem.* **32,** 1631–1636.
Erlich, H. A., Gibbs, R., and Kazazian, H. H. (1989). "Polymerase Chain Reaction." Cold Spring Harbor Laboratory Press, New York.
Erlich, H. A. (1989). "PCR Technology. Principles and Applications for DNA Amplification." Stockton Press, New York.
Evangelista, R. A., Pollak, A., Allore, B., Templeton, E. F., Morton, R. C., and Diamandis, E. P. (1988). A new europium chelate for protein labeling and time-resolved fluorometric applications. *Clin. Biochem.* **21,** 173–178.
Fahrlander, P. D., and Klausner, A. (1988). Amplifying DNA probe signals: A "Christmas tree' approach. *BioTechnology* **6,** 1165–1168.
Feinberg, A. P., and Vogelstein, B. (1983). A technique for radiolabeling DNA restriction endonuclease fragments to high specific activity. *Anal. Biochem.* **132,** 6–13.
Feinberg, A. P., and Vogelstein, B. (1984). A technique for radiolabeling DNA restriction endonuclease fragments to high specific activity. (Addendum). *Anal. Biochem.* **137,** 266–267.
Fernley, H. N., and Walker, P. G. (1965). Kinetic behaviour of calf-intestinal alkaline phosphatase with 4-methylumbelliferyl phosphate. *Biochem. J.* **97,** 95–103.
Forster, A. C., McInnes, J. L., Skingle, D. C., and Symons, R. H. (1985). Nonradioactive hybridization probes prepared by the chemical labeling of DNA and RNA with a novel reagent, photobiotin. *Nucleic Acids Res.* **13,** 745–761.
Fox, G. E., Stackebrandt, E., Hespell, R. B., Gibson, J., Maniloff, J., Dyer, T. A., Wolfe, R. S., Balch, W. E., Tanner, R. S., Magrum, L. J., Zablen, L. B., Blakemore, R., Gupta, R., Bonen, L., Lewis, B. J., Stahl, D. A., Luehrsen, K. R., Chen, K. N., and Woese, C. R. (1980). The phylogeny of prokaryotes. *Science* **25,** 457–463.
Franci, C., and Vidal, J. (1988). Coupling redox and enzymic reactions improves the sensitivity of the ELISA-spot assay. *J. Immunol. Methods* **107,** 239–244.
Froehler, B., Ng, P., and Matteucci, M. (1988). Phosphoramidate analogues of DNA: Synthesis and thermal stability of heteroduplexes. *Nucleic Acids Res.* **16,** 4831–4839.
Gallati, H. (1979). Horseradish peroxidase: A study of the kinetics and the determination of optimal reaction conditions, using hydrogen peroxide and 2,2′-azinobis(3-ethylbenzthiazoline-6-sulfonic acid) (ABTS) as substrate. *J. Clin. Chem. Clin. Biochem.* **17,** 1–7.
Garen, A., and Levinthal, C. (1960). A fine-structure genetic and chemical study of the enzyme alkaline phosphatase of *E. coli.* I. Purification and characterization of alkaline phosphatase. *Biochim. Biophys. Acta* **38,** 470–483.
Gebeyehu, G., Rao, P. Y., SooChan, P., Simms, D. A., and Klevan, L. (1987). Novel biotinylated nucleotide analogs for labeling and colorimetric detection of DNA. *Nucleic Acids REs.* **15,** 4513–4534.
Geiger, R., Hauber, R., and Miska, N. (1989). New, bioluminescence-enhanced detection system for use in enzyme activity tests, enzyme immunoassays, protein blotting, and nucleic acid hybridization. *Mol. Cell. Probes* **3,** 309–328.
Gillam, I. C. (1987). Nonradioactive probes for specific DNA sequences. *Trends Biotech.* **5,** 332–334.
Gillam, I. C., and Tener, G. M. (1986). $N^4$-(6-aminohexyl)cytidine and -deoxycytidine nucleotides can be used to label DNA. *Anal. Biochem.* **157,** 199–207.
Gingeras, T. R., Merten, U., and Kwoh, D. Y. (1988). Transcription-based nucleic acid amplification/detection systems. *WO Patent Appl.* 8810315.
Gould, S. J., and Subramani, S. (1988). Review. Firefly luciferase as a tool in molecular and cell biology. *Anal. Biochem.* **175,** 5–13.

Graf, H., and Lenz, H. (1984). Derivatized nucleic acid sequence and its use in detection of nucleic acids. *Ger. Patent Appl.* 3431536

Green, C., and Tibbetts, C. (1981). Reassociation rate limited displacement of DNA strands by branch migration. *Nucleic Acids Res.* **9,** 1905–1918.

Greene, N. M. (1975). Avidin. *In* "Advances in Protein Chemistry" (C. B. Anfinsen and J. T. Edsall, eds.), Vol. 29, pp. 85–113. Academic Press, New York.

Gregersen, N., Koch, J., Koelvraa, S., Petersen, K. B., and Bolund, L. (1987). Improved methods for the detection of unique sequences in Southern blots of mammalian DNA by nonradioactive biotinylated DNA hybridization probes. *Clin. Chim. Acta* **169,** 267–280.

Gross, D. S., and Simpkins, H. (1981). Evidence for two-site binding in the terbium(III)–nucleic acid interaction. *J. Biol. Chem.* **256,** 9593–9598.

Guatelli, J. C., Whitfield, K. M., Kwoh, D. Y., Barringer, K. J., Richman, D. D., and Gingeras, T. R. (1990). Isothermal, *in vitro* amplification of nucleic acids by a multienzyme reaction modeled after retroviral replication. *Proc. Natl. Acad. Sci. U.S.A.* **87,** 1874–1878.

Guérin-Reverchon, I., Chardonnet, Y., Chignol, M. C., and Thivolet, J. (1989). A comparison of methods for the detection of human papillomavirus DNA by *in situ* hybridization with biotinylated probes on human carcinoma cell lines: Application to wart sections. *J. Immunol. Methods* **123,** 167–176.

Hames, B. D., and Higgins, S. J. (1985). "Nucleic Acid Hybridization." IRL Press, Oxford.

Haralambidis, J., Chai, M., and Tregear, G. W. (1987). Preparation of base-modified nucleosides suitable for nonradioactive label attachment and their incorporation into synthetic oligodeoxyribonucleotides. *Nucleic Acids Res.* **15,** 4857–4876.

Hauber, R., and Geiger, R. (1987). A new, very sensitive, bioluminescence-enhanced detection system for protein blotting. I. Ultrasensitive detection systems for protein blotting and DNA hybridization. *J. Clin. Chem. Clin. Biochem.* **25,** 511–514.

Hegnauer, R. (1971). Pflanzenstoffe und Pflanzensystematik. *Naturwissenschaften* **58,** 585–598.

Heiles, H. B. J., Genersch, E., Kessler, C., Neumann, R., and Eggers, H. J. (1988). *In situ* hybridization with digoxigenin-labeled DNA of human papillomaviruses (HPV 16/18) in HeLa and SiHa cells *BioTechniques* **6,** 978–981.

Heller, M. J., and Morrison, L. E. (1985). Chemiluminescent and fluorescent probes for DNA hybriditation systems. *In* "Rapid Detection and Identification of Infectious Agents" (D. T. Kingsbury and S. Falkow, eds.), pp. 245–256. Academic Press, New York.

Heller, M. J., Morrison, L. E., Prevatt, W. D., and Akin, C. (1982). Homogeneous nucleic acid hybridization diagnostics by nonradioactive energy transfer. *Eur. Patent Appl.* 0070685.

Heller, M. J., and Schneider, B. L. (1983). Chemiluminescent labeling of nucleic acids using azidophenyl glyoxal. *Fed. Proc.* **42,** 1954.

Hemmilä, I., Dakabu, S., Mukkala, V.-M., Siitari, H., and Lövgren, T. (1984). Europium as a label in time-resolved immunofluoremetric assays. *Anal. Biochem.* **137,** 335–343.

Herzberg, M. (1984). Molecular genetic probe, assay technique, and a kit using this molecular genetic probe. *Eur. Patent Appl.* 0128018.

Höfler, H. (1987). What's new in "*in situ*" hybridization." *Pathol. Res. Pract.* **182,** 421–430.

Höltke, H.-J., and Kessler, C. (1990). Nonradioactive labeling of RNA transcripts *in vitro* with the hapten digoxigenin (DIG); hybridization and ELISA-based detection. *Nucleic Acids Res.* **18,** 5843–5851.

Höltke, H. J., Sagner, G., Kessler, C., and Schmitz, G. (1991). Sensitive chemiluminescent detection of digoxigenin-labeled nucleic acids: A fast and simple protocol and its application. *BioTechniques,* in press.

Höltke, H.-J., Seibl, R., Burg, J., Mühlegger, K., and Kessler, C. (1990). Non-radioactive labeling and detection of nucleic acids: II. Optimization of the digoxigenin system. *Mol. Gen. Hoppe-Seyler* **371,** 929–938.

Hopman, A. H. N., Wiegant, J., Tesser, G. I., and Van Duijn, P. (1986a). A nonradioactive *in situ* hybridization method based on mercurated nucleic acid probes and sulfhydryl-hapten ligands. *Nucleic Acids Res.* **14,** 6471–6488.

Hopman, A. H. N., Wiegant, J., and van Duijn, P. (1986b). A new hybridocytochemical method based on mercurated nucleic acide probes and sulhydryl-hapten ligands. I. Stability of the mercurysulfhydryl bond and influence of the ligand structure on immunochemical detection of the hapten. *Histochemistry* **84,** 169–178.

Hutton, J. R., and Wetmur, J. G. (1973). Length dependence of the kinetic complexity of mouse satellite DNA. *Biochem. Biophys. Res. Commun.* **52,** 1148–1155.

Huynh Dinh, T., Sarfati, S., Igolen, J., and Guesdon, J. L. (1987). Markers for detecting nucleic acids are derived from 2′-desoxy adenosine derivatives. *Eur. Patent Appl.* 0254646.

Hyman, H. C., Yogev, D., and Razin, S. (1987). DNA probes for detection and identification of *Mycoplasma pneumoniae* and *Mycoplasma genitalium*. *J. Clin. Microbiol.* **25,** 726–728.

Innis, M. A., Gelfand, D. H., Sninsky, J. J., and White, T. J. (1990). "PCR Protocols. A Guide to Methods and Applications." Academic Press, New York.

Inoue, S., Hashida, S., Tanaka, K., Imigawa, M., and Ishikawa, E. (1985). Preparation of monomeric affinity-purified Fab′-β-D-galactosidase conjugate for immunoenzymometric assay. *Anal. Lett.* **18,** 1331–1344.

Ishikawa, E., Imagawa, M., Hashida, S., Yoshitake, S., Hamaguchi, Y., and Ueno, T. (1983). Enzyme labeling of antibodies and their fragments for enzyme immunoassay and immunohistochemical staining. *J. Immunoassay* **4,** 209–327.

Ishikawa, E., Kawai, T., and Miyai, K. (1981). "Enzyme Immunoassay." Igaku-Shoin, Tokyo.

Iwai, H., Ishihara, F., and Akihama, S. (1983). A fluorometric rate assay of peroxidase using the homovanillic acid-*o*-dianisidine-hydrogen peroxide system. *Chem. Pharm. Bull.* **31,** 3579–3582.

Jablonski, E., Moomaw, E. W., Tullis, R. H., and Ruth, J. L. (1986). Preparation of oligodeoxynucleotide–alkaline phosphatase conjugates and their use as hybridization probes. *Nucleic Acids Res.* **14,** 6115–6128.

Jablonski, E., and Ruth, J. L. (1986). Synthesis of oligonucleotide–enzyme conjugates and their use as hybridization probes. *DNA* **5,** 89.

Jäger, A., Levy, M. J., and Hecht, S. M. (1988). Oligonucleotide *N*-alkylphosphoramidates: Synthesis and binding to polynucleotides. *Biochemistry* **27,** 7237–7246.

Johannsson, A., Stanley, C. J., and Self, C. H. (1985). A fast highly sensitive colorimetric enzyme immunoassay system demonstrating benefits of enzyme amplification in clinical chemistry. *Clin. Chim. Acta* **148,** 119–124.

Joyce, G. F. (1989). Amplification, mutation, and selection of catalytic RNA. *Gene* **82,** 83–87.

Jungell-Nortamo, A., Syvänen, A. C., Luoma, P., and Söderlund, H. (1988). Nucleic acid sandwich hybridization: Enhanced reaction rate with magnetic microparticles as carriers. *Mol. Cell Probes* **2,** 281–288.

Kassavetis, G. A., Butler, E. G., Roulland, D., Chamberlin, M. J. (1982). Bacteriophage SP6-specific RNA polymerase. *J. Biol. Chem.* **257,** 5779–5788.

Keller, G. H., Cumming, C. U., Huang, D. P., Manak, M. M., and Ting, R. (1988). A chemical method for introducing haptens onto DNA probes. *Anal. Biochem.* **170,** 441–450.

Keller, G. H., Huang, D.-P., and Manak, M. M. (1989). Labeling or DNA probes with a photoactivatable hapten. *Anal. Biochem.* **177,** 392–395.

Keller, G. H., and Manak, M. M. (1989). "DNA Probes." Stockton Press, New York.

Kemeny, D. M., and Challacombe, S. J. (1988). "ELISA and other solid phase immunoassays. Theoretical and practical aspects." John Wiley and Sons, New York.

Kempe, T., Sundquist, W. I., Chow, F., and Hu, S. L. (1985). Chemical and enzymatic biotin-labeing of oligonucleotides. *Nucleic Acids Res.* **13,** 45–57.

Kessler, C. (1990). Detection of nucleic acids by enzyme-linked immuno-sorbent assay (ELISA) technique: An example for the development of a novel nonradioactive labeling and detection system with high sensitivity. *In:* "Advances in Mutagenesis Research" (G. Obe, ed.), Vol. 1, pp. 105–152. Springer-Verlag, Berlin/Heidelberg.

Kessler, C. (1991). The digoxigenin : antidigoxigenin (DIG) technology—a survey on the concept and realization of a novel bioanalytical indicator system. *Mol. Cell. Probes*, in press.

Kessler, C., Höltke, H.-J., Seibl, R., Burg, J., and Muhlegger, K. (1990). Nonradioactive labeling and detection of nucleic acids: I. A novel DNA labeling and detection system based on digoxigenin : antidigoxigenin ELISA principle (digoxigenin system). *Mol. Gen. Hoppe-Seyler* **371,** 917–927.

Kohne, D. E. (1990). The use of DNA probes to detect and identify microorganisms. *Adv. Exp. Med. Biol.* **263,** 11–35.

Kohne, D. E., Levinson, S. A., and Byers, M. J. (1977). Room temperature method for increasing the rate of DNA reassociation by many thousandfold: The phenol emulsion reassociation technique. *Biochemistry* **16,** 5329–5341.

Kricka, L. J. (1988). Review. Clinical and biochemical applications of luciferase and luciferins. *Anal. Biochem.* **175,** 14–21.

Krieg, P. A., and Melton, D. A. (1987). *In vitro* RNA synthesis with SP6 RNA polymerase. *Methods Enzymol.* **155,** 397–415.

Kumar, A., Tchen, P., Roullet, F., and Cohen, J. (1988). Nonradioactive labeling of synthetic oligonucleotide probes with terminal deoxynucleotidyl transferase. *Anal. Biochem.* **169,** 376–382.

Kwoh, D. Y., Davis, G. R., Whitfield, K. M., Chappelle, H. L., DiMichelle, L. J., and Gingeras, T. R. (1989). Transcription-based amplification system and detection of amplified human immunodeficiency virus type I with a bead-based sandwich hybridization format. *Proc. Natl. Acad. Sci. U.S.A.* **86,** 1173–1177.

Landegent, J. E., Jansen in de Wal, N., Baan, R. A., Hoeijmakers, J. H., and Van der Ploeg, M. (1984). 2-Acetylaminofluorene-modified probes for the indirect hybridocytochemical detection of specific nucleic acid sequences. *Exp. Cell. Res.* **153,** 61–72.

Landegent, J. E., Jansen in de Wal, N., Ploem, J. S., and Van der Ploeg, M. (1985). Sensitive detection of hybridocytochemical results by means of reflection-contrast microscopy. *J. Histochem. Cytochem.* **33,** 1241–1246.

Landes, G. M. (1985). Labeled DNA. *Eur. Patent Appl.* 0138357.

Langer, P. R., Waldrop, A. A., and Ward, D. C. (1981). Enzymatic synthesis of biotin-labeled polynucleotides: Novel nucleic acid affinity probes. *Proc. Natl. Acad. Sci. U.S.A.* **78,** 6633–6637.

Langer-Safer, P. R., Levine, M., and Ward, D. C. (1982). Immunological method for mapping genes on *Drosophila* polytene chromosomes. *Proc. Natl. Acad. Sci. U.S.A.* **79,** 4381–4385.

Lardy, H. A., and Peanasky, R. (1953). Metabolic functions of biotin. *Physiol. Rev.* **33,** 560–565.

Leary, J. J., Brigati, D. J., and Ward, D. C. (1983). Colorimetric method for visualizing biotin-labeled DNA probes hybridized to DNA or RNA immobilized on nitrocellulose: Bio-blots. *Proc. Natl. Acad. Sci. U.S.A.* **80,** 4045–4049.

Levinson, C., and Chang, C.-a. (1990). Nonisotopically labeled probes and primers. *In* "PCR Protocols. A Guide to Methods and Applications" (M. A. Innis, D. H. Gelfand, J. J. Sninsky, and T. J. White, eds.), pp. 99–112. Academic Press, New York.

Li, H., Gyllensten, U. B., Cui, X., Saiki, R. K., and Erlich, H. A. (1988). Amplification and analysis of DNA sequences in single human sperm and diploid cells. *Nature (London)* **335,** 414–417.

Li, P., Medon, P., Skingle, D. C., Lanser, J. A., and Symons, R. H. (1987). Enzyme-linked synthetic oligonucleotide probes: Nonradioactive detection of *Escherichia coli* in faecal specimens. *Nucleic Acids Res.* **15,** 5275–5287.

Lichter, P., Tang, C. J. C., Call, K., Hermanson, G., Evans, G. A., Housman, D., and Ward, D. C. (1990). High-resolution mapping of human chromosome 11 by *in situ* hybridization with cosmid clones. *Science* **247,** 64–69.

Lizardi, P. M., Guerra, C. E., Lomeli, H., Tussie-Luna, I., and Kramer, F. R. (1988). Exponential amplification of recombinant-RNA hybridization probes. *Biotechnology* **6,** 1197–1202.

Lo, Y.-M. D., Mehal, W. Z., and Fleming, K. A. (1988). Rapid production of vector-free biotinylated probes using the polymerase chain reaction. *Nucleic Acids Res.* **16,** 8719.

Lo, Y.-M. D., Mehal, W. Z., and Fleming, K. A. (1990). Incorporation of biotinylated dUTP. *In* "PCR Protocols. A Guide to Methods and Applications" (M. A. Innis, D. H. Gelfand, J. J. Sninsky, and T. J. White, eds.), pp. 113–118. Academic Press, New York.

Lojda, Z., Slaby, J., Kraml, J., and Kolinska, J. (1973). Synthetic substrates in the histochemical demonstration of intestinal disaccharidases. *Histochemie* **34,** 361–369.

Lovgren, T., Hemmilä, I., Pettersson, K., and Halonen, P. (1985). Time-resolved fluorometry in immunoassay. *In* "Alternative Immunoassays" (W. P. Collins, ed.), pp. 203–217. John Wiley and Sons, Chichester, England.

Maggio, E. T. (1980). "Enzyme-Immunoassay." CRC Press, Boca Raton, Florida.

Maitland, N. J., Cox, M. F., Lynas, C., Prime, S., Crane, I., and Scully, C. (1987). Nucleic acid probes in the study of latent viral disease. *J. Oral Pathol.* **16,** 199–211.

Malcolm, A. D. B., and Nicolas, J. L. (1984). Detecting a polynucleotide sequence and labelled polynucleotides useful in this method. *WO Patent Appl.* 8403520.

Malvano, R. (1980). "Immunoenzymatic Assay Techniques." Martinus Nujoff Publishers, The Hague, The Netherlands.

Marmur, J., and Doty, P. (1959). Heterogeneity in DNA. I. Dependence on composition of the configurational stability of deoxyribonucleic acids. *Nature* **183,** 1427–1428.

Mason, D. Y. (1985). Immunocytochemical labeling of monoclonal antibodies by the APAAP immunoalkaline phosphatase technique. *In* "Techniques of Immunocytochemistry" (G. R. Bullock and P. Petrusz, eds.), Vol. 3, pp. 25–42. Academic Press, London.

Mason, D. Y., Cordell, J. L., Abdulaziz, Z., Naiem, M., and Bordenave, G. (1982). Preparation of peroxidase-antiperoxidase (PAP) complexes for immunohistoligical labeling of monoclonal antibodies. *J. Histochem. Cytochem.* **30,** 1114–1122.

Matthews, J. A., and Kricka, L. J. (1988). Analytical strategies for the use of DNA probes. *Anal. Biochem.* **169,** 1–25.

McCracken, S. (1989). Preparation of RNA transcripts using SP6 RNA polymerase. *In* "DNA Probes" (G. H. Keller and M. M. Manak, eds.), pp. 119–120. Stockton Press, New York.

McKnabb, S., Rupp, R., and Tedesco, J. L. (1989). Measuring contamination DNA in bioreactor-derived monoclonals. *BioTechnology* **7,** 343–347.

Meinkoth, J., and Wahl, G. (1984). Hybridization of nucleic acids immobilized on solid supports. *Anal. Biochem.* **138,** 267–284.

Melton, D. A., Krieg, P. A., Rebagliati, M. R., Maniatis, T., Zinn, K., and Green, M. R. (1984). Efficient *in vitro* synthesis of biologically active RNA and RNA hybridization probes from plasmids containing a bacteriophage SP6 promoter. *Nucleic Acids Res.* **12,** 7035–7056.

Miska, W., and Geiger, R. (1987). I. Synthesis and characterization of luciferin derivatives for use in bioluminescence-enhanced enzyme immunoassays. New ultrasensitive detection systems for enzyme immunoassay. *J. Clin. Chem. Clin. Biochem.* **25,** 23–30.

Miska, W., and Geiger, R. (1989). Luciferin derivatives in bioluminescence-enhanced enzyme immunoassays. *J. Biolumin. Chemilumin.* **4,** 119–128.

Mock, G. A., Powell, M. J., and Septak, M. J. (1986). New (poly)-nucleotide labeled with acridine ester or lanthanide useful as analytical agents especially to detect organism gene in rapid diagnosis of infections. *Eur. Patent Appl.* 0212951.

Moench, T. R. (1987). *In situ* hybridization. *Mol. Cell. Probes* **1,** 195–205.

Morris, C. E., Klement, J. F., and McAllister, W. T. (1986). Cloning and expression of the bacteriophage T3 RNA polymerase gene. *Gene* **41,** 193–200.

Mühlegger, K., Huber, E., von der Eltz, H., Rüger, R., and Kessler, C. (1990). Nonradioactive labeling and detection of nucleic acids: IV. Synthesis and properties of the nucleotide compounds of the digoxigenin system and of photodigoxigenin. *Mol. Gen. Hoppe-Seyler* **371,** 939–951.

Murakawa, G. J., Wallace, B. R., Zaia, J. A., and Rossi, J. J. (1987). Method for amplification and detection of RNA sequences. *Eur. Patent Appl.* 0272098.

Nelson, P. S., Frye, R. A., and Liu, E. (1989a). Bifunctional oligonucleotide probes synthetized using a novel CPG support are able to detect single base pair mutations. *Nucleic Acids Res.* **17,** 7187–7194.

Nelson, P. S., Sherman-Gold, R., and Leon, R. (1989b). A new and versatile reagent for incorporating multiple primary aliphatic amines into synthetic oligonucleotides. *Nucleic Acids Res.* **17,** 7179–7186.

Newman, C. L., Modlin, J., Yolken, R. H., and Viscidi, R. P. (1989). Solution hybridization and enzyme immunoassay for biotinylated DNA–RNA hybrids to detect enteroviral RNA in cell culture. *Mol. Cell. Probes* **3,** 375–382.

Nicholls, P. J., and Malcolm, A. D. B. (1989). Nucleic acid analysis by sandwich hybridization. *J. Clin. Lab. Anal.* **3,** 122–135.

Nickersen, D. A., Kaiser, R., Lappin, S., Steward, J., Hood, L., and Landgren, U. (1990). Automated DNA diagnostics using an ELISA-based oligonucleotide ligation assay. *Proc. Natl. Acad. Sci. U.S.A.* **87,** 8923–8927.

Nur, I., Reinhartz, A., Hyman, H. C., Razin, S., and Herzberg, M. (1989). Chemiprobe, a nonradioactive system for labeling nucleic acid. *Ann. Biol. Clin.* **47,** 601–606.

Oellerich, M. (1983). Principles of enzyme-immunoassays. *In* "Methods of Enzymatic Analysis" (H. U. Bergmeyer, J. Bergmeyer, and M. Grassl, eds.), Vol. 1, pp. 233–260. Verlag Chemie, Weinheim, Germany.

Orgel, L. E. (1989). Ligase-based amplification method. *WO Patent Appl.* 8909835.

Oser, A., Roth, W. K., and Valet, G. (1988). Sensitive nonradioactive dot-blot hybridization using DNA probes labeled with chelate group substituted psoralen and quantitative detection by europium ion fluorescence. *Nucleic Acids Res.* **16,** 1181–1196.

Oser, A., and Valet, G. (1988). Improved detection by time-resolved fluorometry of specific DNA immobilized in microtiter wells with europium/metal-chelator labeled DNA probes. *Nucleic Acids Res.* **16,** 8178.

Paau, A., Platt, S. G., and Sequeiro, L. (1983). Assay method and probe for polynucleotide sequence. *UK Patent* 2125964.

Parsons, G. (1988). Development of DNA probe-based commercial assay. *J. Clin. Immunoassay* **11,** 152–160.

Pasternak, J. J. (1988). Microbial DNA diagnostic technology. *Biotech. Adv.* **6,** 683–695.

Pataki, S., Meyer, K., and Reichstein, T. (1953). Die Konfiguration des Digoxigenins (Teilsynthese des 3$\beta$-, 12$\alpha$- und des 3$\beta$, 12$\alpha$-Dioxyatiansäure-methylesters). Glykoside und Aglycone, 116. Mitteling. *Helv. Chim. Acta* **36,** 1295–1308.

Pereira, H. G. (1986). Nonradioactive nucleic acid probes for the diagnosis of virus infections. *BioEssays* **4,** 110–113.

Pezzella, M., Pezzella, F., Galli, C., Macchi, B., Verani, P., Sorice, F., and Baroni, C. D. (1987). *In situ* hybridization of human immunodeficiency virus (HTLV-III) in cryostat sections of lymph nodes of lymphadenopathy syndrome patients. *J. Med. Virol.* **22,** 135–142.

Pfeiffer, F., and Gilbert, W. (1988). Vecbase,a cloning vector sequence data base. *Protein Seq. Data Analysis* **1,** 269–280.

Pitcher, D. G., Owen, R. J., Dyal, P., and Beck, A. (1987). Synthesis of a biotinylated DNA probe to detect ribosomal RNA cistrons in *Providencia stuartii*. *FEMS Microbiol. Lett.* **48,** 283–287.

Pollard-Knight, D. (1990). Current methods in nonradioactive nucleic acid labeling and detection. *Technique* **2,** 113–132.

Pollard-Knight, D., Read, C. A., Downes, M. J., Howard, L. A., Leadbetter, M. R., Pheby, S. A., McNaughton, E., Syms, A., and Brady, M. A. W. (1990). Nonradioactive nucleic acid detection by enhanced chemiluminescence using probes directly labeled with horseradish peroxidase. *Anal. Biochem.* **185,** 84–89.

Porstmann, B., Porstmann, T., and Nugel, E. (1981). Comparison of chromogenes for the determination of horseradish peroxidase as a marker in enzyme immunoassay. *J. Clin. Chem. Clin. Biochem.* **19,** 435–439.

Porstmann, T., Ternynck, T., and Avrameas, S. (1985). Quantitation of 5-bromo-2-deoxyuridine incorporation into DNA: An enzyme immunoassay for the assessment of the lymphoid cell proliferative response. *J. Immunol. Methods* **82,** 169–179.

Proverenny, A. M., Podgorodnichenko, V. K., Bryksina, L. E., Monastyrskaya, G. S., and Sverdlov, E. D. (1979). Immunochemical approaches to DNA structure investigation—I. *Mol. Immunol.* **16,** 313–316.

Rabin, B. R., Taylorson, C. J., and Hollaway, M. R. (1985). Assay method using enzyme fragments as labels and new enzyme substrates producing coenzymes or prostetic groups. *Eur. Patent Appl.* 0156641.

Ranki, M., Palva, A., Virtanen, M., Laaksonen, M., and Söderlund, H. (1983). Sandwich hybridization as convenient method for the detection of nucleic acids in crude samples. *Gene* **21,** 77–85.

Rashtchian, A., Eldredge, J., Ottaviani, M., Abbott, M., Mock, G., Lovern, D., Klinger, J., and Parsons, G. (1987). Immunological capture of nucleic acid hybrids and application to nonradioactive DNA probe assay. *Clin. Chem.* **33,** 1526–1530.

Reckmann, B., and Rieke, E. (1987). Verfahren und Mittel zur Bestimmung von Nucleinsäuren. *Eur. Patent* 0286958.

Reichstein, T., and Weiss, E. (1962). The sugars of the cardiac glycosides. *Adv. Carbohydr. Chem.* **17,** 65–120.

Reisfeld, A., Rothenberg, J. M., Bayer, E. A., and Wilchek, M. (1987). Nonradioactive hybridization probes prepared by the reaction of biotin hydrazide with DNA. *Biochem. Biophys. Res. Commun.* **142,** 519–526.

Renz, M. (1983). Polynucleotide-histone H1 complexes as probes for blot hybridization. *EMBO J.* **2,** 817–822.

Renz, M., and Kurz, C. (1984). A colorimetric method for DNA hybridization. *Nucleic Acids Res.* **12,** 3435–3444.

Rigby, P. W. J., Dieckmann, M., Rhodes, C., and Berg, P. (1977). Labeling deoxyribonucleic acid to high specific activity *in vitro* by nick translation with DNA polymerase I. *J. Mol. Biol.* **113,** 237–251.

Riley, L. K., Marshall, M. E., and Coleman, M. S. (1986). A method for biotinylating oligonucleotide probes for use in molecular hybridizations. *DNA* **5,** 333–337.

Rossau, R., Heyndrickx, L., and Van Heuverswyn, H. (1988). Nucleotide sequence of a 16S ribosomal RNA gene from *Neisseria gonorrhoeae*. *Nucleic Acids Res.* **16,** 6227.

Rossau, R., van Landschoot, A., Mannheim, W., and De Ley, J. (1986). Intergeneric and intrageneric similarities of ribosomal RNA cistrons of the Neisseriaceae. *Int. J. Syst. Bacteriol.* **36,** 323–332.

Rothenberg, J. M., and Wilchek, M. (1988). *p*-Diazobenzoyl-biocytin: A new biotinylating reagent for DNA. *Nucleic Acids Res.* **16,** 7197.

Roychoudhury, R., Tu, C.-P. D., and Wu, R. (1979). Influence of nucleotide sequence adjacent to duplex DNA termini on 3′-terminal labeling by terminal transferase. *Nucleic Acids Res.* **6,** 1323–1333.

Ruth, J. L. (1984). Chemical synthesis of nonradioactively-labeled DNA hybridization probes. *DNA* **3,** 123.

Ruth, J. L., Morgan, C., and Pasko, A. (1985). Linker arm nucleotide analogs useful in oligonucleotide synthesis. *DNA* **4,** 93.

Saiki, R. K., Arnheim, N., and Erlich, H. A. (1985a). A novel method for the detection of polymorphic restriction sites by cleavage of oligonucleotide probes: Application to a sickle-cell anemia. *BioTechnology* **3,** 1008–1012.

Saiki, R. K., Gelfand, D. H., Stoffel, S., Scharf, S. J., Higuchi, R., Horn, G. T., Mullis, K. B., and Erlich, H. A. (1988). Primer-directed enzymatic amplification of DNA with a thermostable DNA polymerase. *Science* **239,** 487–491.

Saiki, R. K., Scharf, S., Faloona, F., Mullis, K. B., Horn, G. T., Erlich, H. A., and Arnheim, N. (1985b). Enzymatic amplification of beta-globin genomic sequences and restriction site analysis for diagnosis of sickle cell anemia. *Science* **230,** 1350–1354.

Sakamoto, H., Traincard, F., Vo-Quang, T., Ternynck, T., Guesdon, J. L., and Avrameas, S. (1987). 5-Bromodeoxyuridine *in vivo* labeling of M13 DNA, and its use as a nonradioactive probe for hybridization experiments. *Mol. Cell. Probes* **1,** 109–120.

Sambrook, J., Fritsch, E. F., and Maniatis, T. (1989a). "Molecular Cloning. A Laboratory Manual" 2nd Ed. Cold Spring Harbor Laboratory Press, New York.

Schaap, A. P., Sandison, M. D., and Handley, R. S. (1987). Chemical and enzymatic tiggering of 1,2-dioxetanes. Alkaline phosphatase-catalyzed chemiluminescence from an aryl phosphate-substituted dioxetane. *Tetrahedron Lett.* **28,** 1159–1162.

Schmitz, G. G., Walter, T., and Kessler, C. (1991). Nonradioactive labeling of oligonucleotides *in vitro* with the hapten digoxigenin (DIG) by tailing with terminal transferase. *Anal. Biochem.* **192,** 222–231.

Segev, D. (1990). Amplification and detection of target nucleic acid sequences—for *in vitro* diagnosis of infectious disease, genetic disorders, and cellular disorders, e.g., cancer. *WO Patent Appl.* 9001069.

Segev, D., Zehr, S., Lin, P., and Park-Turkel, H. S. (1990). Amplification of nucleic acid sequences by the repair chain reaction (RCR). *5th San Diego Conference on Nucleic Acids,* AACC, Abstract Poster 44.

Seibl, R., Höltke, H.-J., Rüger, R., Meindl, A., Zachau, H.-G., RaBhofer, G., Roggendorf, M., Wolf, H., Arnold, N., Weinberg, J., and Kessler, C. (1990). Nonradioactive labeling and detection of nucleic acids: III. Applications of the digoxigenin system. *Mol. Gen. Hoppe-Seyler* **371,** 939–951.

Self, C. H. (1985). Enzyme amplification—a general method applied to provide an immunoassisted assay for placental alkaline phosphatase. *J. Immunol. Methods* **76,** 389–393.

Serke, S., and Pachmann, K. (1988). An immunocytochemical method for the detection of fluorochrome-labeled DNA probes hybridized *in situ* with cellular RNA. *J. Immunol. Methods* **112,** 207–211.

Sheldon, E., Kellog, D. E., Levenson, C., Bloch, W., Aldwin, L., Birch, D., Goodson, R., Sheridan, P., Horn, G., Watson, R., and Erlich, H. (1987). Nonisotopic M13 probes for detecting the $\beta$-globin gene: Application to diagnosis of sickle cell anemia. *Clin. Chem.* **33,** 1368–1371.

Sheldon, E. L., Kellogg, D. E., Watson, R. E., Levinson, C. H., and Erlich, H. A. (1986). Use of nonisotopic M13 probes for genetic analysis: Application to class II loci. *Proc. Natl. Acad. Sci. U.S.A.* **83,** 9085–9089.

Sheldon, E. L., III, Levenson, C. H., Mullis, D. B., and Rapoport, H. (1985). Process for labeling nucleic acids using psoralen derivates. *Eur. Patent Appl.* 0156287.

Smith, L. M., Fung, S., Hunkapiller, M. W., Hunkapiller, T. J., and Hood, L. E. (1985). The synthesis of oligonucleotides containing an aliphatic amino group at the 5′ terminus: Synthesis of fluorescent DNA primers for use in DNA sequence analysis. *Nucleic Acids Res.* **13,** 2399–2412.

Smith, L. M., Kaiser, R. J., Sanders, J. Z., and Hood, L. E. (1987). The synthesis and use of fluorescent oligonucleotides in DNA sequence analysis. *Methods Enzymol.* **155,** 260–301.

Smith, L. M., Sanders, J. Z., Kaiser, R. J., Hughes, P., Dodd, C., Connell, C. R., Heiner, C., Kent, S. B. H., Hood, L. E. (1986). Fluorescence detection in automated DNA sequence analysis. *Nature* **321,** 674–679.

Sodja, A., and Davidson, N. (1978). Gene mapping and gene enrichment by the avidin-biotin interaction: Use of cytochrome-c as a polyamine bridge. *Nucleic Acids. Res.* **5,** 385–401.

Soini, E., and Kojola, H. (1983). Time-resolved fluorometer for lanthianide chelates—a new generation of nonisotopic immunoassay. *Clin Chem.* **29,** 65–68.

Sproat, B. S., Beijer, B., and Rider, P. (1987). The synthesis of protected 5′-amino-2′,5′-dideoxyribonucleoside-3′-*O*-phosphoramidites: Applications of 5′-amino-oligodeoxyribonucleotides. *Nucleic Acids Res.* **15,** 6181–6196.

Stanley, C. J., Johannsson, A., and Self, C. H. (1985). Enzyme amplification can enhance both the speed and the sensitivity of immunoassays. *J. Immunol. Methods* **83,** 89–95.

Stollar, B. D., and Rashtchian, A. (1987). Immunochemical approaches to gene probe assays. *Anal. Biochem.* **161,** 387–394.

Stull, T. L., LiPuma, J. J., and Edling, T. D. (1988). A broad spectrum probe for molecular epidemiology of bacteria: Ribosomal RNA. *J. Infect. Dis.* **157,** 280–286.

Stürzl, M., and Roth, W. K. (1990). "Run-off" synthesis and application of defined single-stranded DNA hybridization probes. *Anal. Biochem.* **185,** 164–169.

Syvänen, A. C., Alanen, M., and Söderlund, H. (1985). A complex of single-strand binding protein and M13 DNA as hybridization probe. *Nucleic Acids Res.* **13,** 2789–2802.

Syvänen, A. C., Laaksonen, M., and Söderlund, H. (1986a). Fast quantification of nucleic acid hybrids by affinity-based hybrid collection. *Nucleic Acids Res.* **14,** 5037–5048.

Syvänen, A. C., Tchen, P., Ranki, M., and Söderlund, H. (1986b). Time-resolved fluorometry: A sensitive method to quantify DNA-hybrids. *Nucleic Acids Res.* **14,** 1017–1028.

Takahashi, Y., Arakawa, H., Maeda, M., and Tsuiji, A. (1989). A new biotinylating system for DNA using biotin aminocaproyl hydrazide and gluataraldehyde. *Nucleic Acids Res.* **17,** 4899–4900.

Taub, F. (1986). An assay for nuclic acid sequences, particularly genetic lesions. *WO Patent Appl.* 8603227.

Tautz, D., and Pfeifle, C. (1989). A nonradioactive *in situ* hybridization method for the localization of specific RNAs in *Drosophila* embryos reveals translational control of the segmentation gene hunchback. *Chromosoma* **98,** 81–85.

Tchen, P., Fuchs, R. P. P., Sage, E., and Leng, M. (1984). Chemically modified nucleic acids as immunodetectable probes in hybridization experiments. *Proc. Natl. Acad. Sci. U.S.A.* **81,** 3466–3470.

Theissen, G., Richter, A., and Lukacs, N. (1989). Degree of biotinylation in nuleic acids estimated by a gel retardation assay. *Anal. Biochem.* **179,** 98–105.

Thompson, J., and Gillespie, D. (1987). Molecular hybridization with RNA probes in concentrated solutions of guanidine thiocynate. *Anal. Biochem.* **163,** 281–291.

Thorpe, G. H., and Kricka, L. J. (1989). Incorporation of enhanced chemiluminescent reactions into fully automated enzyme immunoassays. *J. Biolumin. Chemilumin.* **3,** 97–100.

Thuong, N. T., Asseline, U., Roig, V., Takasugi, M., and Hélène, C. (1987). Oligo($\alpha$-deoxynucleotide)s covalently linked to intercalating agents: Differential binding to ribo- and deoxyribopolynucleotides and stability towards nuclease digestion. *Proc. Natl. Acad. Sci. U.S.A.* **84,** 5129–5133.

Tomlinson, S., Lyga, A., Huguenel, E., and Dattagupta, N. (1988). Detection of biotinylated nucleic acid hybrids by antibody-coated gold colloid. *Anal. Biochem.* **171,** 217–222.

Tous, G., Fausnaugh, J., Vieira, P., and Stein, S. (1988). Preparation and chromatographic use of 5′-fluorescent-labeled DNA probes. *J. Chromatogr.* **444,** 67–77.

Traincard, F., Ternynck, T., Danchin, A., and Avrameas, S. (1983). An immunoenzymic procedure for the demonstration of nucleic acid molecular hybridization. *Ann Immunol.* **134,** 339–405.

Trainor, G. L., and Jensen, M. A. (1988). A procedure for the preparation of fluorescence-labeled DNA with terminal deoxynucleotidyl transferase. *Nucleic Acids Res.* **16,** 11846.

Tsuji, A., Maeda, M., Arakawa, H., Shimizu, S., Tanabe, K., and Sudo, Y. (1987). Chemiluminescence enzyme immunoassay using invertase, glucose-6-phosphate dehydrogenase and β-D-galactosidase as label. *In* "Bioluminescence and Chemiluminescence" (J. Scholmerich, R. Anderson, A. Kapp, M. Ernst, and W. G. Woods, eds.), pp. 233–235. Wiley Interscience, Chichester, England.

Tu, C.-P. D., and Cohen, S. (1980). 3′-End labeling of DNA with [$\alpha$-$^{32}$P]cordycepin-5′-triphosphate. *Gene* **10,** 177–183.

Urdea, M. S., Running, J. A., Horn, T., Clyne, J., Ku, L., and Warner, B. D. (1987). A novel method for the rapid detection of specific nucleotide sequences in crude biological samples without blotting or radioactivity; application to the analysis of hepatitis B virus in human serum. *Gene* **61,** 253–264.

Urdea, M. S., Warner, B. D., Running, J. A., Stempien, M., Clyne, J., and Horn, T. (1988). A comparison of nonradioisotopic hybridization assay methods using fluorescent, chemiluminescent, and enzyme-labeled synthetic oligodeoxyribonucleotide probes. *Nucleic Acids Res.* **16,** 4937–4956.

Van Prooijen-Knegt, A. C., Van Hoek, J. F., Bauman, J. G., Van Duijin, P., Wool, I. G., and Van der Ploeg, M. (1982). *In situ* hybridization of DNA sequences in human metaphase chromosomes visualized by an indirect fluorescent immunocytochemical procedure. *Exp. Cell Res.* **141,** 397–407.

Vary, C. P. H. (1987). A homogeneous nucleic acid hybridization assay based on strand displacement. *Nucleic Acids Res.* **15,** 6883–6897.

Vary, C. P. H., McMahon, F. J., Barbone, F. P., and Diamond, S. E. (1986). Nonisotopic detection methods for strand displacement assays of nucleic acids. *Clin. Chem.* **32,** 1696–1701.

Verdlov, E. D., Monastyrskaya, G. S., Guskova, L. I., Levitan, T. L., Sheichenko, V. I., and Budowsky, E. I. (1974). Modification of cytidine residues with a bisulfite-*O*-methylhydroxylamine mixture. *Biochem. Biophys. Acta* **340,** 153–165.

Viscidi, R. P., Connelly, C. J., and Yolken, R. H. (1986). Novel chemical method for the preparation of nucleic acids for nonisotopic hybridization. *J. Clin. Microbiol.* **23,** 311–317.

Viscidi, R. P., and Yolken, R. G. (1987). Molecular diagnosis of infectious diseases by nucleic acid hybridization. *Mol. Cell. Probes* **1,** 3–14.

Vogt, W. (1978). "Enzymimmunoassay." Georg Thieme Verlag, Stuttgart, Germany.

Voyta, J. C., Edwards, B., and Bronstein, I. (1988). Unltrasensitive chemiluminescent detection of alkaline phosphatase activity. *Clin. Chem.* **34,** 1157.

Wachter, L., Jablonski, J.-A., and Ramachandran, K. L. (1986). A simple and efficient procedure for the synthesis of 5′-aminoalkyl oligodeoxynucleotides. *Nucleic Acids Res.* **14,** 7985–7994.

Wahl, G. M., Stern, M., and Stark, G. R. (1979). Efficient transfer of large DNA fragments from agarose gels to diazobenzyloxymethyl-paper and rapid hybridization by using dextran sulfate. *Proc. Natl. Acad. Sci. U.S.A.* **76,** 3683–3687.

Wallenfels, K., Lehmann, J., and Malhotra, O. P. (1960). Untersuchungen über milchzuckerspaltende Enzyme—Die Spezifität der β-Galactosidase von *E. coli* ML309. *Biochem. Z.* **333,** 209–225.

Ward, D. C., Leary, E. H., and Brigati, D. J. (1987). Visualization polymers and their application to diagnostic medicine. *U.S. Patent* 4687732.

Ward, D. C., Waldrop, A. A., and Langer, P. R. (1982). Modified nucleotides and their use. *Eur. Patent* 0063879.

West, S., Schröder, J., and Kunz, W. (1990). A multiple-staining procedure for the detection of different DNA fragments on a single blot. *Anal. Biochem.* **190.** 254–258.

Wetmur, J. G., and Davidson, N. (1968). Kinetics of renaturation of DNA. *J. Mol. Biol.* **31,** 349–370.

Wilchek, M., and Bayer, E. (1984). The avidin-biotin complex in immunology. *Immunol. Today* **5,** 39–43.

Wilchek, M., and Bayer, E. A. (1988). The avidin-biotin complex in bioanalytical applications. *Anal. Biochem.* **171,** 1–32.

Wilson, M. B., and Nakane, P. K. (1978). Recent developments in the periodate method of conjugating horseradish peroxidase (HRPO) to antibodies. *In* "Immunofluorescence and Related Staining Techniques" (W. Knapp, K. Holubar, and G. Wick, eds.), pp. 215–224. Elsevier/North Holland Biomedical Press, New York.

Woodhead, J. L., and Malcolm, A. D. B. (1984). Nonradioactive gene-specific probes. *Biochem. Soc. Trans.* **12,** 279–280.

Wu, D. Y., and Wallace, R. B. (1989). The ligation and amplification reaction (LAR)—amplification of specific DNA sequences using sequential rounds of template-dependent ligation. *Genomics* **4,** 560–569.

Yamakawa, K., Oymada, H., and Nakagomi, O. (1989). Identification of rotavirus by dot-blot hybridization using an alkaline phosphatase conjugated synthetic oligonucleotide probe. *Mol. Cell. Probes* **3,** 397–401.

Yehle, C. O., Patterson, W. L., Boguslawski, S. J., Albarella, J. P., Yip, K. F., and Carrico, R. J. (1987). A solution hybridization assay for ribosomal RNA form bacteria using biotinylated DNA probes and enzyme-labeled antibody to DNA : RNA. *Mol. Cell. Probes* **1,** 277–193.

Yolken, R. H. (1988). Nucleic acids or immunoglobulins: Which are the molecular probes of the future? *Mol. Cell. Probes.* **2,** 87–96.

Zwadyk, P., and Coodsey, R. C. (1987). Nucleic acid probes in clinical microbiology. *CRC Crit. Rev. Clin. Lab. Sci.* **25,** 71–103.

# PART TWO
# Detection Methods

# 3 Detection of Alkaline Phosphatase by Time-Resolved Fluorescence

Eva Gudgin Templeton, Hector E. Wong, and Alfred Pollak
Kronem Systems, Inc.,
Mississauga, Ontario, Canada



## I. INTRODUCTION

Replacement of radioactive labeling systems with alternative highly sensitive nonisotopic detection systems has become recognized as a priority if nucleic acid hybridization assays are to become a routine laboratory and diagnostic tool. Detection systems that combine the use of an enzyme label and a substrate that is converted to a light-emitting product have the highest potential sensitivities, where such a light-emitting product is typically either fluorescent, bioluminescent, or chemiluminescent (Barnard and Collins, 1987; Gosling, 1990; Jackson and Ekins, 1986; Tijssen, 1985).

Fluorescence-based detection systems in general must compensate for the problem of nonspecific background fluorescence, which is present in all biological samples and reduces the practical sensitivity of the system. This problem can be overcome with time-resolved fluorescence detection using lanthanide chelates as the light-emitting species (Diamandis, 1988; Soini, 1984). The lanthanide chelate consists of a lanthanide metal ion (typically $Tb^{3+}$, $Eu^{3+}$, or $Sm^{3+}$), which is bound to an organic chelator. The chelator contains a number of nucleophilic groups that bind strongly to the lanthanide ion, resulting in complex formation. If, in addition, the

*Nonisotopic DNA Probe Techniques*

chelator contains a UV-absorbing chromophore of the correct structure, optimized for efficient light absorption and subsequent energy transfer to the lanthanide ion, the chelate may be highly luminescent (Soini and Lovgren, 1987). Lanthanide ions themselves, in the absence of a suitable chelator, have only very weak absorption bands in the ultraviolet, and their emission is inefficient and highly quenched by the presence of water molecules in their coordinating sphere.

The desirable features of lanthanide chelate compounds that permit efficient elimination of background signals have been extensively described (Diamandis and Christopoulos, 1990; Soini and Lovgren, 1987). Luminescent lanthanide chelates, particularly those involving terbium and europium, possess a number of spectroscopic features making them ideal labels for biological assays. They have strong absorption bands in the UV region, characteristic of the organic chelator; subsequent to UV light absorption, the excitation energy can be efficiently transferred to the lanthanide ion. This process will be efficient only when the excited-state energy levels (most probably the first excited triplet level) of the chelator are properly matched to the excited-state levels of the lanthanide ion. The lanthanide ion may then emit the transferred excitation energy as characteristic lanthanide ion luminescence or fluorescence. $Tb^{3+}$ and $Eu^{3+}$ emission bands have narrow halfwidths (in the range of 1 to 20 nm) and are highly Stokes shifted, located in the green to red regions of the spectrum. This results in an absence of self-quenching, and easy discrimination between the specific lanthanide ion luminescence, which can be selected with the use of appropriate filters, and the excitation light and the normal nonspecific backgrounds, which have smaller Stokes shifts. In addition, owing to the nature of the electronic transitions involved, europium and terbium chelates have long lifetimes, typically on the order of 100 $\mu$sec to 2 msec, permitting the facile use of pulsed excitation time-resolved fluorescence (TRF) detection in order to eliminate both scattered light and normal short-lived fluorescence background emission. Finally, their luminescence yields and lifetimes are largely insensitive to minor variations in their environment, resulting in robust emission properties.

Two different types of time-resolved luminescence detection schemes, in which a lanthanide chelate is used directly as label, are currently available. In one type, a luminescent chelate can itself be used as a labeling agent, usually on a protein such as avidin or an antibody (Canfi *et al.*, 1989; Evangelista *et al.*, 1988). In the second type, the chelate used as the label is not itself luminescent, and the lanthanide ion must be released into a micellar *enhancing solution*, which contains a number of reagents necessary to observe the luminescence (Hemmila, 1988; Jackson and Ekins, 1986). These methods have some limitations for DNA hybridization applications, in particular with regard to their applicability to membrane-based formats.

Therefore, in order to achieve an alternative lanthanide-based detection system for this application, enzyme amplification, using alkaline phosphatase as label, has been combined with TRF detection of a lanthanide chelate reagent, giving the method termed enzyme amplified lanthanide luminescence (EALL) (Evangelista *et al.*, 1991).

In EALL, a substrate that is incapable of forming a luminescent lanthanide chelate is converted enzymatically into a product that does form such a chelate. In this way, the excellent background rejection allowed by the principle of TRF detection is combined with the signal amplification provided by the use of an enzyme as label. This is performed by designing an enzymatic reaction that modifies either, or preferably both, the spectroscopic properties and chelating abilities of the substrate by the enzymatic reaction. The alteration of these properties on transformation of substrate to product can result in any combination of the following changes: increase in excitation light absorption efficiency, increase in energy transfer efficiency, increase in formation constant of the chelate with the lanthanide ion, or reduction in quenching of the lanthanide ion luminescence by bound water molecules. The net effect of such changes is to ultimately promote formation of a luminescent chelate between the product molecule and the lanthanide ion.

With an optimized design of substrate/product pairs, the method is applicable to a wide variety of assay formats, including standard heterogeneous DNA hybridization assays on nylon membranes or polystyrene microwells with the use of alkaline phosphatase (AP) labels either directly or indirectly conjugated to the probe. The same or slightly modified EALL substrates can also be used in microwell immunoassays or other heterogeneous bioassays. In addition, they may have potential uses in *in situ* hybridization and other applications that can use AP-labeled probes. Preliminary EALL substrates giving good detectability have also been developed for a variety of other oxidative and hydrolytic enzymes, including $\beta$-galactosidase, xanthine oxidase, glucose oxidase and esterase, showing the versatility of the method for enzyme-based detection formats (Evangelista *et al.*, 1991).

# II. MATERIALS

## A. Reagents

The chemical reaction for EALL-based detection of AP labels is illustrated in Fig. 1. The substrates consist of substituted salicyl phosphates; the AP label present on the probe promotes the dephosphorylation of the salicyl phosphate substrate. The product salicylic acid forms a luminescent ternary chelate with $Tb^{3+}$ and ethylenediaminetetraacetic acid (EDTA)

**Fig. 1** EALL detection of alkaline phosphatase using substituted salicyl phosphate substrates. After dephosphorylation by alkaline phosphatase, the luminescent ternary complex is formed between the product salicylic acid, terbium ion, and EDTA. For membrane applications, the substituent is a branched alkyl group in the 5-position, while for solution applications, the substituent is 5-fluoro.

(Hemmila, 1985; Poluektov *et al.*, 1973) at pH 12–13. The substituent has been tailored to provide maximal luminescence of the product chelate while giving the desired solubility of the product molecule on dephosphorylation.

For solution-based assays, 5-fluorosalicyl phosphate (FSAP, REALL-S substrate, Kronem Systems Inc., Mississauga, Ontario, Canada) provides good solubility of the product and high luminescence yield of the product chelate. The values for $K_M$ and $v_{max}$ for 5-fluorosalicyl phosphate have been determined as $1.2 \times 10^{-4}$ *M* and 0.19 $\mu$mol $min^{-1}$ $U^{-1}$, respectively, in the standard substrate buffer used for DNA hybridization assays (0.1 *M* Tris, 0.1 *M* NaCl, $10^{-3}$ *M* $MgCl_2$, pH 9.0) (Evangelista *et al.*, 1991). Alternatively, for membrane-based assays such as dot blot or Southern blot assays, a branched alkyl substituted salicyl phosphate (REALL-M substrate, Kronem Systems Inc., Mississauga, Ontario, Canada) gives a less soluble product in order to provide good localization of signal in the vicinity of the label over a few hours of incubation with enzyme. The detectability of AP using FSAP in solution-based microwell assays with quantitative instrumentation has been found to be 1 amol for 1 to 2 hr incubation and 0.3 amol for overnight incubation with substrate (Evangelista *et al.*, 1991). Substrates with different substituents are under development in order to improve both the detectability and diffusive properties of solution- and membrane-based substrates. In addition, substrates that can be detected with $Eu^{3+}$ giving a maximum emission band in the red at 615 nm are under development for two-color applications.

The substituted salicyl phosphate substrate (XSAP) molecules absorb light only below 300 nm, and do not form luminescent Tb:EDTA complexes under the conditions of the measurement. The product of the enzymatic reaction, XSA, in the presence of Tb:EDTA forms a complex whose excitation maximum is in the vicinity of 320 to 330 nm (Hemmila, 1985, Evangelista *et al.*, 1991). The emission is characteristic of $Tb^{3+}$ luminescence, namely a number of narrow bands in the visible (green to red regions), with a main band at 547 nm whose half-bandwidth is approximately 10 nm; the emission lifetime is approximately 1 msec (Evangelista *et al.*, 1991).

The optimum conditions for the enzymatic reaction are as follows: $10^{-3}$ *M* substrate at pH 9, in 0.1 *M* Tris or other amine-containing buffer, 1 m*M* $Mg^{2+}$. A lower pH can cause spontaneous hydrolysis of the substrate (Bromilow and Kirby, 1972) over prolonged periods, resulting in elevated backgrounds. The enzymatic reaction is typically performed for 2 to 4 hr. Solution-based assays can be left longer; however, membrane-based assays if left for $>$ 4 hr may show loss of resolution of bands or spots due to excessive diffusion of the product of the enzymatic reaction during the incubation with enzyme. The luminescence signal development is performed by adding an equal or slight excess volume of developing solution (Kronem Systems Inc., Mississauga Ontario, Canada; 5X REALL Developing Reagent, 5X REALL Developing Buffer, deionized distilled water 1:1:3 v/v). The luminescence of the product complex is reasonably stable and can be repeatedly measured up to 2 hr after signal development. Always ensure that sufficient developing solution is prepared to develop the entire membrane or microwell assay at once; different preparations may not perform identically.

## B. Instrumentation

The recommended detection instrumentation for use with EALL is TRF based. The principle of TRF detection and instrumentation is illustrated in Fig. 2. Typical TRF instrumentation uses pulsed UV excitation (either from a laser or flashlamp source), visible wavelength emission filters (chosen to transmit selectively the luminescence of the lanthanide ion used as opposed to the background fluorescence and scattered light signals), and time-gated detection (which measures only that signal emitted in a window starting several hundred microseconds after the excitation pulse).

For use in a wide variety of hybridization assay formats, both membrane and microwell based, the TRP100 time-resolved photographic camera system (Kronem Systems Inc., Mississauga, Ontario, Canada) has been developed, as outlined in Fig. 3. The camera is designed to provide rapid photographic results on high-speed instant film (Polaroid ISO 3000 or 20,000). The pulsed UV source consists of two xenon flashlamps (pulsewidth $< 50$ $\mu$sec). Selection of desired excitation and emission wavelengths is provided by the use of different filter pairs. The filters provided with the instrument give excitation over the range of 320 to 400 nm, and emission detection at wavelengths $>$515 nm, suitable for EALL detection of alkaline phosphatase using REALL reagents. Uniform sample illumination results from the use of a reflective tunnel between the

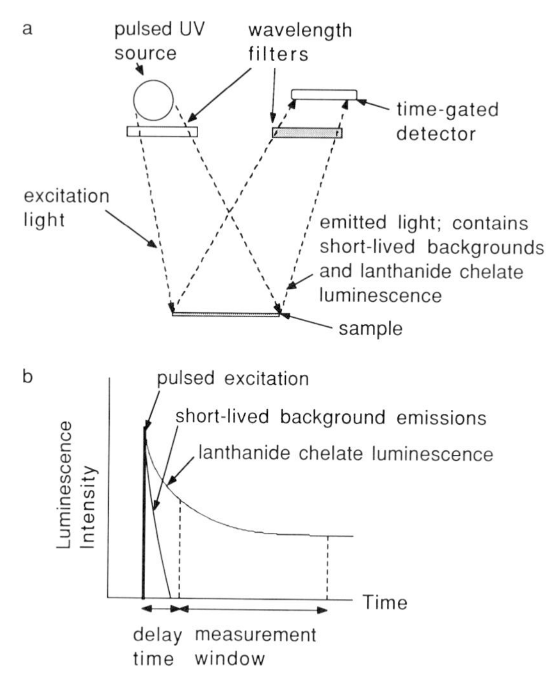


**Fig. 2** (a) General representation of time-resolved fluorescence instrumentation used for measurement of lanthanide chelate luminescence. Pulsed ultraviolet light excites the sample; emission from the sample is filtered and time-gated to select for lanthanide luminescence, and then detected electronically or photographically. (b) Principle of time-gated detection of lanthanide luminescence. Gating can be accomplished by electronic or mechanical means.

lamps and the sample. Time-gating of both the excitation pulse and the emission is provided by the use of a rotating chopper wheel; additional time-gating of the excitation lamps is needed owing to the presence of a long-lived thermal emission in the visible, which could contribute to background signals by reflection off the sample. The lamps are triggered by the rotation of the chopper wheel, and both the delay between excitation and emission, and the emission time window are determined by the selected rotation speed of the chopper, as given in Table 1. The time-gated luminescence signal is then repeatedly recorded on the film each time the chopper rotates through one cycle until sufficient signal has been reached. Typical exposure times are in the range of 10 sec to 3 min, allowing for easy reexposure to permit optimization of signal levels on the film. The height is adjusted as necessary so that the top of the sample is located in the focal plane of the instrument. The unit is designed for use with either microplate or membrane formats, with sample sizes of up to $18 \times 13$ cm$^2$ accommodated.

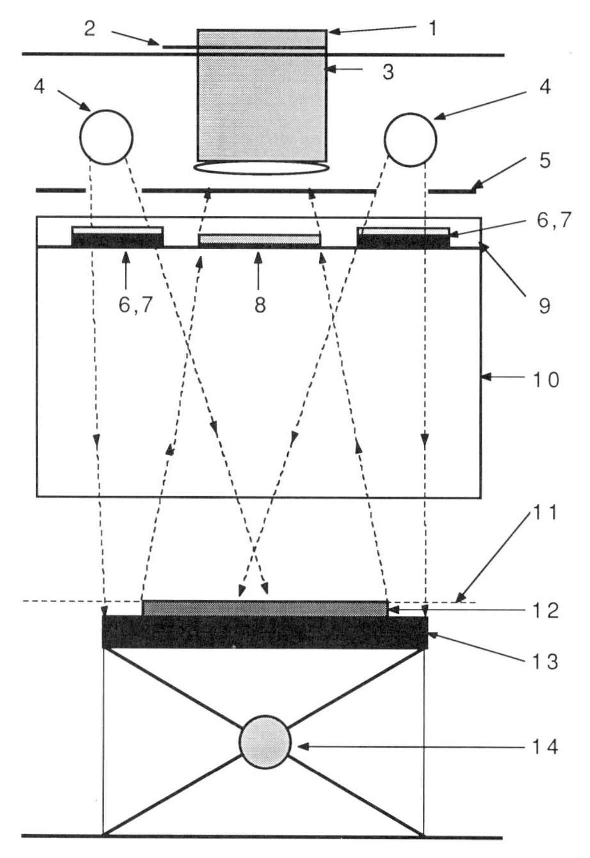


**Fig. 3** Schematic of TRP100 time-resolved photographic instrument. (1) Polaroid film back; (2) shutter; (3) camera; (4) flashlamps; (5) rotating chopper wheel; (6,7) excitation filters; (8) emission filter; (9) filter plate; (10) mirror tunnel; (11) focal plane; (12) sample, e.g., membrane or microplate; (13) sample tray; (14) focusing adjustment.

For microplate solution-based formats, quantitative results can be obtained using either of two commercial TRF microplate readers available (both originally designed for use with direct lanthanide labeling systems in immunoassay applications (Diamandis *et al.*, 1989; Soini and Kojola, 1983), namely the CyberFluor 615 Immunoanalyzer (CyberFluor Inc., Toronto, Canada) and the Arcus (Delfia) 1234 Plate Reader (Pharmacia LKB, Uppsala, Sweden). The CF615 unit is designed for measurement of $Eu^{3+}$ luminescence in opaque white polystyrene microwells, exciting from above and reading signal from the surface of the microwell or solution contained therein. It provides no wavelength choice or adjustable time gate, having a nitrogen laser for excitation at 337 nm, and emission detec-

**Table I**
Exposure Delay Time and Exposure Window Settings on TRP100 Instrument

| Chopper speed setting | Speed of rotation | Excitation rate | Exposure delay time | Exposure window |
|---|---|---|---|---|
| Slow | 6000 rpm | 300 Hz | 440 μsec | 4.1 msec |
| Fast | 9000 rpm | 450 Hz | 290 μsec | 2.8 msec |

tion at 615 ± 5 nm, with a time gate of approximately 200–600 μsec, pulse repetition rate of 20 Hz, and data collection time/well of 1 sec. Note that terbium emission in the EALL system is equally well quantitated using these conditions, as $Tb^{3+}$ has a minor emission band in the region of 615 nm, in addition to its main emission at 545 nm. The Arcus unit is designed for use with transparent polystyrene microwells, and measures emission in a *straight through* geometry with excitation filters and three emission filters; the filter combination that provides excitation at >320 nm (excitation filter #1) and emission detection at 615 nm (emission filter # 1) gives the best signal to background discrimination and dynamic range. The best choice of time window is approximately 200–400 μsec with a data collection time/well of about 1 sec. The two instruments have comparable detection sensitivities with the EALL detection system using these parameters. The use of quantitative instrumentation improves the detectability of AP by about three to five times over that obtained with the TRP100.

A variety of other normal fluorescence instrumentation can be used also without time-resolved detection, by utilizing narrow-band emission detection, which can eliminate a large fraction of the potentially interfering emissions. For example, for photography of membrane-based assays, the Spectroline UV CC-80 lightbox system (Spectronics, Westbury, New York) equipped with 312 nm (UV-B) lamps, and Fotodyne UV 300-nm transilluminator (Fotodyne Inc., New Berlin, Wisconsin) have been used, with a Polaroid camera equipped with a 515-nm cutoff or 545-nm (bandpass 10 nm) interference filter. The sensitivity of the normal instrumentation compared to TRF instrumentation is similar in the best case; however, artifacts resulting from green-emitting backgrounds may appear. Extreme care in the handling of the membranes during the assay may reduce the likelihood of such artifacts.

## III. PROCEDURES

The progress of the enzymatic dephosphorylation reaction can be followed in all cases by viewing the sample in subdued lighting with a hand-held 300 to 320-nm or UV-B lamp; a 300-nm transilluminator may be used also for membranes. Blue fluorescence from the dephosphorylated product will be visible. Samples may be degraded by excessive exposure to UV light, so viewing should be kept brief.

Recommended solid supports for use with photography are Nytran nylon membrane (0.45 $\mu$m, Schleicher and Schuell, Keene, New Hampshire) or Microlite 1 Removawell opaque white strips (Dynatech, Chantilly, Virgina). Other supports may give less satisfactory results. Alternatively, with TRF plate readers, use microwell strips or plates as recommended by the manufacturer of the instrumentation. Solution formulations are given at the end of this section.

## A. Membrane-Based Assays

Southern and dot blotting should be performed following standard protocols (Maniatis *et al.*, 1982) for nylon membranes, with appropriate modifications for the use of alkaline phosphatase-labeled probes. The quantities of solutions given in the following protocols are for 100 $cm^2$ of membrane, and volumes should be scaled accordingly for different membrane areas. Protocols are given for biotinylated, long DNA probes, detected with alkaline phosphatase-labeled streptavidin, or digoxigenin-labeled probes detected with AP-labeled antidigoxigenin, and can be modified for oligonucleotide probes or direct or other indirect labeling systems and conjugates. For photography using the TRP100 camera, membrane formats of a size up to 9 $\times$ 13 $cm^2$ are recommended, as they are most conveniently handled and photographed; samples of up to 18 $\times$ 13 $cm^2$ can be photographed in two exposures.

### 1. DNA Southern Blot Hybridization Assay

The protocol for agarose gel electrophoresis and Southern transfer generally follows standard techniques. The loading buffer described below is recommended; the tracking dye should not be run in lanes containing the samples of interest, as the dye may interfere with uniform illumination of the samples during the final photography.

*Solution Formulations*

| | |
|---|---|
| Loading buffer | 15% Ficoll type 400 in deionized water |
| Denaturation solution | 1.5 *M* NaCl, 0.5 *M* NaOH |
| Neutralization solution | 1.0 *M* Tris pH 8.0, 1.5 *M* NaCl |
| Transfer buffer | 10× SSC (0.15 *M* sodium citrate, 1.5 *M* NaCl, pH 7.0) |
| 6× SSC | 0.09 *M* sodium citrate, 0.9 *M* NaCl, pH 7.0 |

| | |
|---|---|
| DNA dilution buffer | Phosphate buffered saline (1.5 m$M$ $NaH_2PO_4$, 8.0 m$M$ $K_2HPO_4$, 137 m$M$ NaCl, 2.5 m$M$ KCl, pH 7.2) containing 2 $\mu$g/ml sheared salmon sperm DNA |
| 20× SSC | 0.3 $M$ sodium citrate, 3.0 $M$ NaCl, pH 7.0 |
| Prehybridization buffer | 6× SSC, 50% formamide, 0.1% Ficoll, 0.1% polyvinylpyrrolidone, 0.5% powdered skim milk, 0.2 mg/ml freshly denatured sheared salmon sperm DNA, 5% polyethyleneglycol |
| Hybridization solution | 50–200 ng/ml of labeled DNA probe in prehybridization buffer |
| Membrane wash solution 1 | 5× SSC, 0.5% SDS |
| Membrane wash solution 2 | 0.1× SSC, 1% SDS |
| Membrane wash solution 3 | 2× SSC |
| TBS-T20 | 0.1 $M$ Tris pH 7.5, 0.15 $M$ NaCl, 0.05% Tween 20 |
| Blocking solution | 1% BSA in TBS-T20 |
| Alkaline phosphatase-labeled streptavidin solution | Working dilution of conjugate in TBS-T20, for example, 1 : 6000 dilution of ExtrAvidin streptavidin–alkaline phosphatase conjugate (Sigma), approx. 500 mU/ml |
| Substrate buffer | 0.1 $M$ Tris pH 9.0, 0.1 $M$ NaCl, 1 m$M$ $MgCl_2$ |
| Substrate stock solution | $10^{-2}$ $M$ REALL-M in 0.1 $M$ NaOH |
| Substrate solution | 10× dilution of substrate stock solution in substrate buffer |
| Developing solution | 1× REALL Developing Reagent, 1× REALL Developing Buffer in distilled, deionized water |

1. The transfer of the DNA from the agarose gel to nylon membrane is performed as follows. Denature the DNA by gently shaking the gel in denaturation solution (2–3 gel volumes) for 30 min at room temperature; repeat this once. Neutralize the gel by gentle shaking in neutralization solution (2–3 gel volumes) for 30 min at room temperature. Check the pH of the gel with pH paper and repeat neutralization step if necessary. Perform the Southern transfer to nylon membrane cut to precisely the size of the gel and prewetted in transfer buffer. Perform the transfer in transfer buffer for 18 hr. Wash the membrane in 6X SSC for 5 min at room temperature,

and allow it to dry for 30 min on a sheet of clean blotting paper. Irradiate the membrane with 254 nm UV light for 3 min, or alternately place in a vacuum oven at 80°C for 1 to 2 hr. The membrane can be stored dry at this point.

2. Prehybridize the membrane in a sealed plastic bag for 1 to 2 hr at 42°C in 10 ml prehybridization buffer. Remove the prehybridization buffer and add 5 ml hybridization solution containing 50–200 ng/ml biotinylated long probe. Seal the membrane in a plastic bag and hybridize at 42°C overnight with shaking. Wash the membrane twice in 100 ml membrane wash solution 1 for 5 min at 65°C, once in 100 ml membrane wash solution 2 for 30 min at 65°C (this wash solution temperature can be adjusted for desired level of stringency), and once in 100 ml in membrane wash solution 3 for 5 min at room temperature.
3. Soak the membrane for 5 min in 100 ml TBS-T20 and then block with 100 ml of blocking solution at 65°C for 1 hr. Incubate the membrane with 50 ml of the alkaline phosphatase-labeled streptavidin solution for 10 min. Remove nonspecifically bound alkaline phosphatase conjugate, by washing twice with 100 ml of TBS-T20 for 15 min and once with 100 ml substrate buffer for 1 hr.
4. Before adding the substrate solution, lay the membrane (DNA side up) on heavy blotting paper until the membrane is uniformly damp but not wet, to remove excess liquid. Place the membrane inside a development bag (consisting of a 0.4-mm thick transparent polyethylene plastic bag that has been cut open on three sides) leaving a gap of about 1 cm around the edge of the membrane on all four sides. With the top of the bag pulled away, add 1.0 ml of REALL-M substrate solution in drops over the surface of the membrane. Close the top of the bag gently over the surface of the membrane in order to exclude air bubbles and spread the solution. Using a 10 ml disposable pipet, roll over the top of the bag gently in several directions to ensure even distribution of the substrate. Incubate for 1 to 4 hr in subdued lighting (longer incubations will reduce sharpness of bands without substantially increasing sensitivity). Do not handle the bad during the incubation period, and at no time handle the membrane other than as described below, in order to prevent smearing of the signal.
5. After the desired incubation time has elapsed, turn the development bag containing the membrane face down and gently open the back side of the bag to one side. Cut a piece of heavy blotting paper to a size larger than the membrane and apply it to the back side of the membrane. Remove excess substrate solution and then remove the blotting paper. Add 1.5 ml of developing solution in

drops to the back of the membrane around all four sides. Close the bag and gently roll with a pipet. After a few seconds, blot the excess solution from behind the membrane as described above. At this point, seal the bag to prevent leakage of luminescent solution and degradation of the luminescent signal. The membrane is now ready for photography. Photograph the membrane within 2 hr of development. To photograph the membrane in the TRP100, place the membrane in the plastic bag in the sample tray of the TRP100 and clamp in place, and then adjust height of the sample tray as needed to obtain correct focus. Select the correct operating parameters for the TRP100 for use with REALL reagents. Photograph the sample for an exposure time in the range of about 30 sec to 3 min. Typical results of a Southern blotting analysis are presented in Fig. 4.

## 2. DNA Dot Blot Hybridization Assay Protocol

A dot blot DNA hybridization assay can be performed on nylon membranes in a manner similar to that described for Southern blotting above, omitting the Southern transfer procedure. All solution volumes should be scaled as described previously, except for sample DNA. Typical results of a dot blot analysis are presented in Fig. 5.

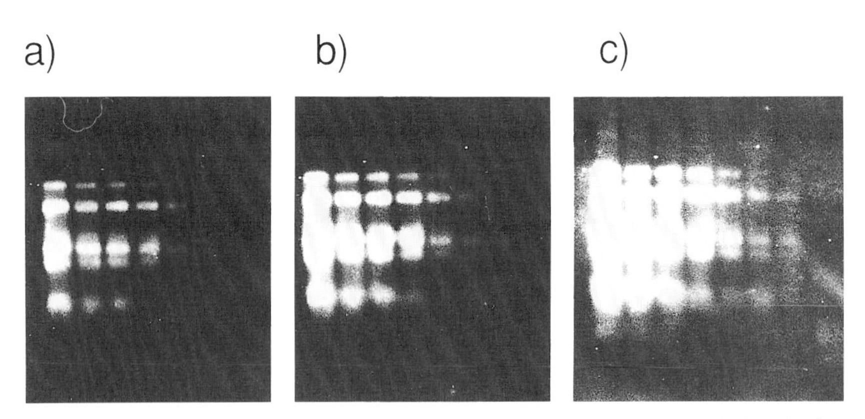


**Fig. 4** Southern blot of BstN I digest of pBR322 DNA probed with biotinylated, nick-translated pBR322 probe. REALL-M substrate was incubated for 2 hr. Lanes contain (from left) 20, 10, 5, 2.5, 1.3, 0.63, 0.31 ng pBR322, and 543 ng *Hind* III digest λ-DNA (not visible). Photographic exposure (a) 20 sec; (b) 40 sec; (c) 80 sec.

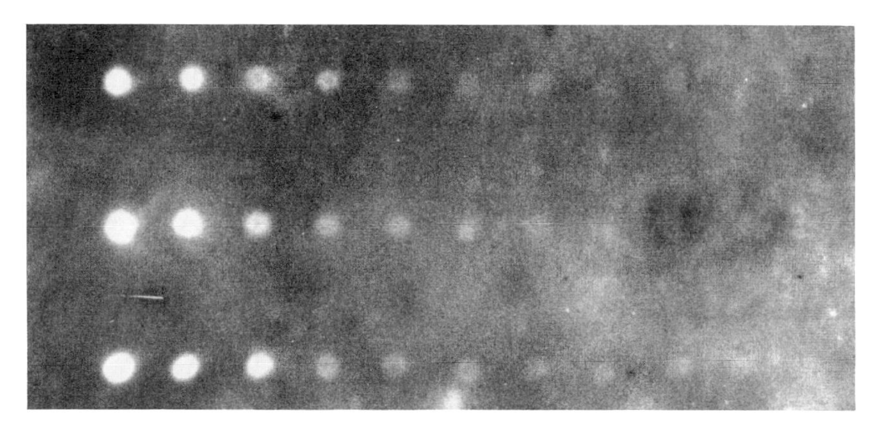

**Fig. 5** Dot blot *Hind* III-digested pBR322 DNA. Samples contain (from left) 250, 125, 63, 31, 16, 8, 4, 2, 1, 0 pg pBR322. REALL-M substrate was incubated for 2 hr.

---

1. Prepare dilutions of sample DNA in DNA dilution buffer. Denature the sample at 100°C for 10 min and place on ice. Spot samples in 2–5 μl volumes on Nytran nylon membrane (0.45 μm) and allow to dry. Irradiate the membrane with 254 nm UV light for 3 min or alternately place in a vacuum oven at 80°C for 1 to 2 hr.

2–5. These are performed exactly as described for Southern blot protocol.

---

### 3. Detection Using Digoxigenin-Labeled Probes

The procedures given above can be modified as follows for the use of digoxigenin-labeled probes and AP-labeled antidigoxigenin antibody (Genius, Boehringer Mannheim).

---

1–2. These are identical to steps 1 and 2 of the DNA Dot Blot Hybridization assay except that the hybridization solution contains 50–200 ng/ml denatured digoxigenin-labeled probe prepared by random primed incorporation of digoxigenin-labeled deoxyuridine triphosphate.

Wash the membrane twice in 100 ml 2X SSC, 0.1% SDS for

5 min at room temperature, then twice in 100 ml 0.1X SSC, 0.1% SDS for 15 min at 65 to 68°C (this wash solution temperature can be adjusted for desired level of stringency).

Soak the membrane for 5 min in 100 ml TBS and then block with 100 ml blocking solution (0.5% blocking reagent, a purified subfraction of dry milk, included in the Genius kit) for 30 min. After blocking, incubate in a 1:5000 dilution of antidigoxigenin Fab fragments conjugated to alkaline phosphatase for 30 min. Remove unbound antibody conjugate by washing the membrane twice for 15 min with 100 ml TBS. Afterwards, incubate the membrane with substrate buffer.

## B. Microwell-Based Assays

### 1. DNA Microwell Hybridization Assay Protocol

A number of DNA hybridization assays for use with polystyrene microwells have been reported (Dahlen, 1987; Morrissey and Collins, 1989; Nagata *et al.*, 1985; Dahlen *et al.*, 1987). One recommended protocol for use with REALL detection reagents, which uses biotinylated long probes and alkaline phosphatase-labeled avidin, is given below.

*Solution Formulations*

| | |
|---|---|
| Phosphate-buffered saline | 1.5 m*M* $NaH_2PO_4$, 8.0 m*M* $K_2HPO_4$, 137 m*M* NaCl, 2.5 m*M* KCl, pH 7.2 |
| DNA dilution buffer | phosphate buffered saline containing 2 μg/ml sheared salmon sperm DNA |
| 2× immobilization buffer | 0.2 *M* $MgCl_2$ in phosphate buffered saline |
| Well wash solution 1 | 2× SSC, 0.1% SDS |
| Well wash solution 2 | 0.1× SSC, 0.1% SDS |
| TBS | 0.1 *M* Tris pH 7.5, 0.15 *M* NaCl |
| Blocking solution | 0.5% skim milk powder in TBS |
| Alkaline phosphatase-labeled avidin solution | working dilution of conjugate in TBS, for example, 1 : 5000 dilution of avidin-alkaline phosphatase (Sigma) |

1. Denature the sample containing the target DNA sequence in a volume of 25 μl DNA dilution buffer at 100°C for 10 min and place on ice. Add this volume to the polystyrene microwell followed by an equal volume of 2× immobilization buffer.

Thoroughly agitate the samples, seal, and leave to immobilize overnight at room temperature. Aspirate the wells, allow to dry, and irradiate using a 254-nm lamp for 6 min at 1.6 kJ/m$^2$ to bind the DNA. Samples can be stored dry at this point.

2. Prehybridize the wells for 1 hr at 42°C in 200 μl prehybridization buffer and then aspirate. For hybridization add 100 μl hybridization solution containing 50–200 ng/ml freshly denatured biotinylated probe, seal the wells, and shake overnight at 42°C. Wash the wells twice with 200 μl well wash solution 1 for 5 min at room temperature, and twice with 200 μl well wash solution 2 for 20 min at 42°C to remove nonspecifically bound probe.
3. Add 200 μl of TBS to the wells for 5 min at room temperature. Block the wells with 200 μl/well blocking solution for 30 min at room temperature. Wash the wells once with 200 μl of TBS for 5 min and then incubate with 100 μl alkaline phosphatase-labeled avidin solution for 30 min. Remove nonspecifically bound avidin-AP by washing 3 times with 200 μl TBS solution for 10 min at room temperature with thorough agitation, or 6 times using an automatic plate washer.

4–5. Wash the wells with 200 μl substrate buffer for 30 min at room temperature. Detect the bound alkaline phosphatase by adding 150 μl REALL-S substrate solution. Incubate for the desired length of time (30 min to overnight) and terminate the reaction by adding 150 μl of developing solution followed by gentle

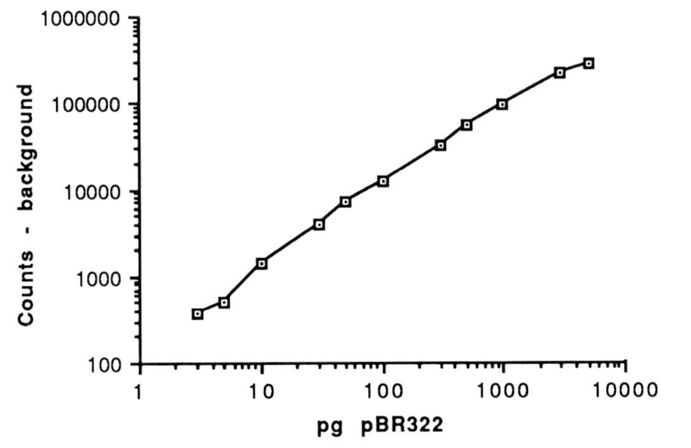


**Fig. 6** Microwell dot blot assay for *Hind* III-digested pBR322 DNA. REALL-S substrate was incubated overnight, and results were read on CF615 plate reader.

agitation. The wells are then ready for photography. To photograph in the TRP100, place the holder containing the samples in the sample compartment, and adjust the height of the sample tray as needed to obtain correct focus. Select the correct instrument parameters for use with REALL reagents. Photograph the sample for an exposure time in the range of about 5 to 30 sec. For quantitative results, wells can be read on TRF plate readers. Typical results of an assay are illustrated in Fig. 6.

# REFERENCES

Barnard, G. J. R., and Collins, W. P. (1987). The development of luminescence immunoassays. *Med. Lab. Sci.* **44,** 249–266.

Bromilow, R. H., and Kirby, A. J. (1972). Intramolecular general acid catalysis of phosphate monoester hydrolysis. The hydrolysis of salicyl phosphate. *J. Chem. Soc. Perkin II* 149–155.

Canfi, A., Bailey, M. P., and Rocks, B. F. (1989). Multiple labeling of immunoglobulin G, albumin, and testosterone with a fluorescent terbium chelate for fluorescence immunoassays. *Analyst* **114,** 1407–1411.

Dahlen, P. (1987). Detection of biotinylated DNA probes by using Eu-labeled streptavidin and time-resolved fluorometry. *Anal. Biochem.* **164,** 78–83.

Dahlen, P., Syvanen, A. C., Hurskainen, P., Kwiatkowski, M., Sund, C., Ylikoski, J., Soderlund, H., and Lovgren, T. (1987). Sensitive detection of genes by sandwich hybridization and time-resolved fluorometry. *Mol. Cell. Probes* **1,** 159–168.

Diamandis, E. P. (1988). Immunoassays with time-resolved fluorescence spectroscopy: Principles and applications. *Clin. Biochem.* **21,** 139–150.

Diamandis, E., and Christopoulos, T. K. (1990). Europium chelate labels in time-resolved fluorescence immunoassays and DNA hybridization assays. *Anal. Chem.* **62,** 1149A–1157A.

Diamandis, E. P., Evangelista, R. A., Pollak, A., Templeton, E. F., and Lowden, J. A. (1989). Time-resolved fluoroimmunoassays with europium chelates as labels. *Am. Clin. Lab.* **Feb.,** 26–28.

Evangelista, R. A., Pollak, A., Allore, B., Templeton, E. F., Morton, R. C., and Diamandis, E. P. (1988). A new europium chelate for protein labeling and time-resolved fluorometric applications. *Clin. Biochem.* **21,** 173–178.

Evangelista, R. A., Pollak, A., and Templeton, E. F. (1991). *Anal. Biochem.* **197,** 213–224.

Gosling, J. P. (1990). A decade of development in immunoassay methodology. *Clin. Chem.* **36,** 1408–1427.

Hemmila, I. (1985). Time-resolved fluorometric determination of terbium in aqueous solution. *Anal. Chem.* **57,** 1676–1681.

Hemmila, I. (1988). Lanthanides as probes for time-resolved fluorometric immunoassays. *Scand. J. Clin. Lab. Invest.* **48,** 389–400.

Jackson, T. M., and Ekins, R. P. (1986). Theoretical limitations on immunoassay sensitivity. Current practice and potential advantages of fluorescent $Eu^{3+}$ chelates as nonradioisotopic tracers. *J. Immunol. Methods* **87,** 13–20.

Maniatis, T., Fritsch, E. F., and Sambrook, J., (1982). "Molecular Cloning. A Laboratory Manual." Cold Spring Harbor Laboratory, Cold Spring Harbor, New York.

Morrissey, D. V., and Collins, M. L. (1989). Nucleic acid hybridization assays employing dA-tailed capture probes. Single capture methods. *Mol. Cell. Probes* **3,** 189–207.

Nagata, Y., Yokota, H., Kosuda, O., Yokoo, K., Takemura, K., and Kikuchi, T. (1985). Quantification of picogram levels of specific DNA immobilized in microtiter wells. *FEBS Lett.* **183,** 379–382.

Poluektov, N. S., Alakaeva, L. A., and Tishchenko, M. A. (1973). Fluorometric determination of terbium as its complex with salicylic acid and ethylenediaminetetraacetic acid. *Zh. Anal. Khim.* **28,** 1621–1623.

Soini, E. (1984). Pulsed light, time-resolved fluorometric immunoassay. *In* "Monoclonal Antibodies and New Trends in Immunoassays" (C.A. Bizollon, ed.), pp. 197–208. Elsevier Science Publishers, New York.

Soini, E., and Kojola, H. (1983). Time-resolved fluorometer for lanthanide chelates—a new generation of nonisotopic immunoassays. *Clin. Chem.* **29,** 65–68.

Soini, E., and Lovgren, T. (1987). Time-resolved fluorescence of lanthanide probes and applications in biotechnology. *CRC Crit. Rev. Anal. Chem.* **18,** 105–154.

Tijssen, P. (1985). "Practice and Theory of Enzyme Immunoassays. Laboratory Techniques in Biochemistry and Molecular Biology, Vol. 15." Elsevier, New York.

# 4 Detection of Alkaline Phosphatase by Bioluminescence

Reinhard Erich Geiger
University of Munich
Munich, Germany



## I. INTRODUCTION

Bioluminescence is a natural phenomenon found in many lower forms of life (DeLuca, 1978; DeLuca and McElroy, 1986; Herring, 1987). Naturally occurring bioluminescent systems differ with regard to the structure and function of enzymes and cofactors as well as in the mechanism of the light-emitting reactions (Burr, 1985; Schölmerich *et al.*, 1987). Because of its high sensitivity, firefly (*Photinus pyralis*) bioluminescence had been used for many years for the sensitive determination of adenosine triphosphate (ATP) (Lundin *et al.*, 1976). More recently, further highly sensitive bioluminescent and chemiluminescent methods have become available for many different analytes (Kricka *et al.*, 1984; Wood, 1984; Gould and Subramani, 1988; Kricka, 1988).

*Nonisotopic DNA Probe Techniques*

A new type of highly sensitive enzyme substrate based on luciferin derivates (Miska and Geiger, 1987) can be used for unmodified enzymes and for enzyme conjugates, applied in enzymatic activity test systems, reporter gene tests, enzyme immunoassays, protein blot analysis, and nucleic acid hybridization tests (Miska and Geiger, 1988; Berger *et al.*, 1988; Geiger, 1990; Hauber *et al.*,1988a, Hauber *et al.*, 1988b, Hauber *et al.*, 1989a). The test principle of these new substrates is the release of D-luciferin from D-luciferin derivatives by the action of hydrolytic enzymes (Fig. 1). Released D-luciferin can be quantified by a luminometric detection system (Fig. 2). The substrate, D-luciferin-*O*-phosphate, is not a substrate for the firefly luciferin reaction, and no light emission is observed for concentrations up to 1 m*M*. The high sensitivity of these bioluminogenic substrates is owing to the combination of the amplification that occurs in the releasing step (e.g., one molecule of alkaline phosphatase can convert 1000 molecules of D-luciferin-*O*-phosphate to D-luciferin per second) and the very sensitive firefly bioluminescence system (concentrations of $5 \times 10^{-12}$ *M* of D-luciferin can be detected) (Fig. 3) (Miska and Geiger, 1987). Structures and kinetic data of some further luciferin derivatives are summarized in Fig. 4 and Table I.

## II. MATERIALS

### A. Reagents

Synthetic D-luciferin (*Photinus pyralis*) and D-luciferin-*O*-phosphate and other D-luciferin derivatives can be purchased from Medor GmbH, W-8036 Herrsching, Germany or Novabiochem AG, CH-4448 Läufelfingen,

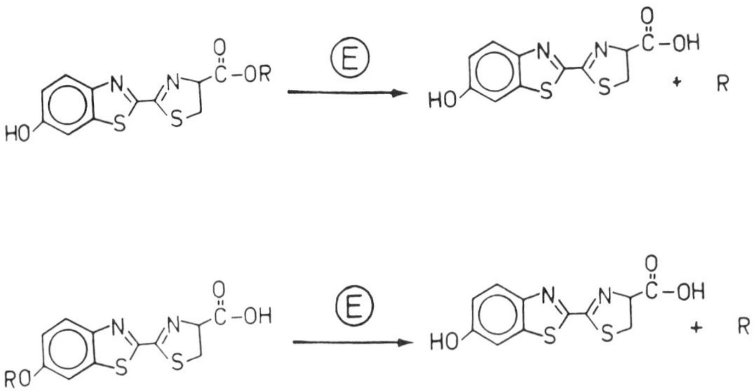

**Fig. 1** Scheme and principle of the test systems. E, enzymes; R, residues (e.g., carboxypeptidase, phosphatases, and others).

1) D - Luciferin - *O* - phosphate $\xrightarrow{\text{alk. phosphatase}}$ Luciferin + P

2) Luciferin + $O_2$ + ATP $\xrightarrow[(\ Mg^{2+})]{\text{Luciferase}}$ Oxyluciferin + $\underline{h \cdot v}$ + AMP + PP

**Fig. 2** Scheme of bioluminescence reaction using firefly (*Photinus pyralis*) luciferase.

Switzerland. Nitrocellulose filters (BA 85, 0–45 μm) are products of Schleicher and Schüll (Dassel, Germany). Nitrocellulose purchased from different distributors may contain substances that interfere with firefly luciferase. A reduction in luciferase activity was obtained by adding buffer to the test system in which nitrocellulose was soaked or stored for a short time. The luciferase may be inhibited or may be denaturated by compounds existing in the nitrocellulose sheets, and washing in high-quality water is recommended.

In earlier stages of synthesis of D-luciferin-*O*-phosphate, very small amounts of free D-luciferin could be detected in the preparations by the very sensitive bioluminescence reaction after purification of D-luciferin-*O*-phosphate. These traces of D-luciferin sometimes influence the blank values. By improving the purification methods, D-luciferin-*O*-phosphate is now available in a highly purified grade (Miska and Geiger, 1987). After

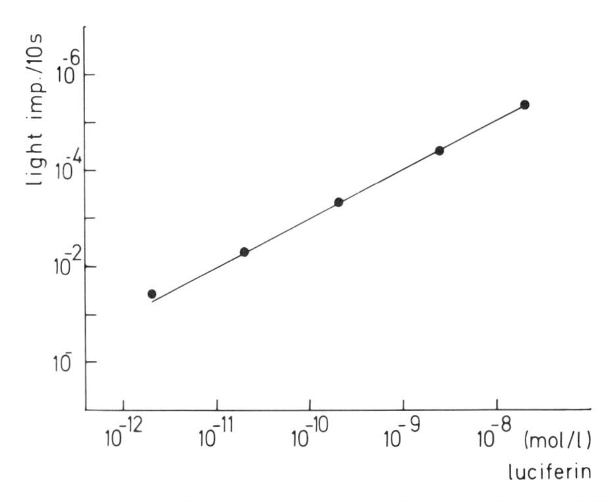


**Fig. 3** Determination of D-luciferin in the luminometric assay. Test conditions (Biolumat, Fa. Berthold Typ 9500 T, Wildbad, FRG; 25°C; volume. 0.5 ml, Lundin *et al.*, 1976): 0.4 ml test buffer (30 mM HEPES, 6.6 mM $MgCl_2$, 0.66 mM EDTA, 0.1 mM DTT, 5 mM ATP, 1 μg luciferase, pH 7.75) were preincubated at 25°C for 5 min. Then 0.1 ml luciferin solution (standard or incubation mixtures of the enzymatic tests, was added and light impulses measured for 10 sec (peak mode) (Miska and Geiger, 1987).

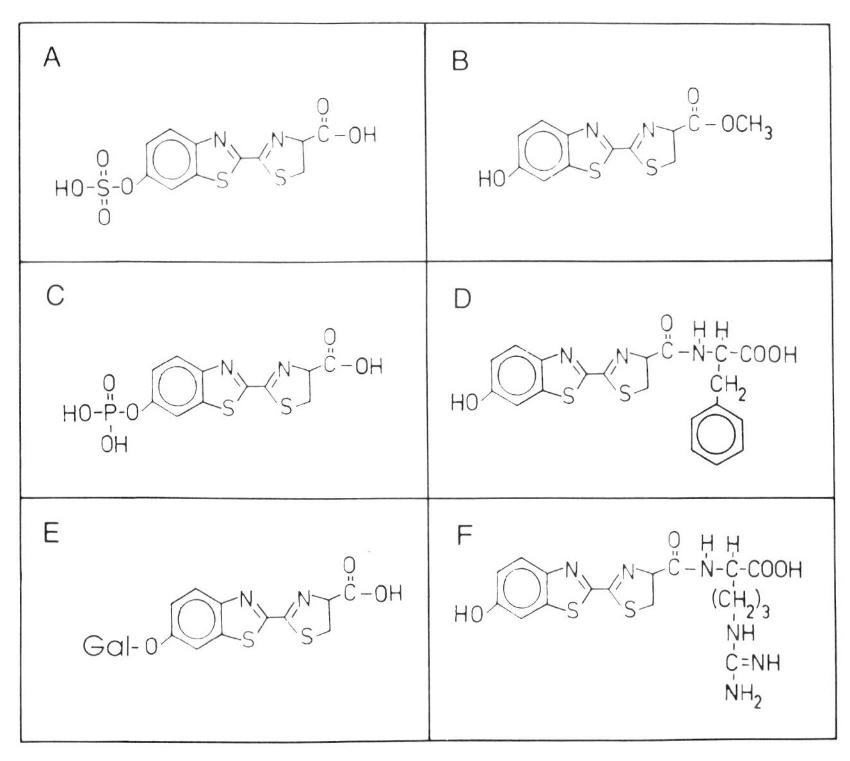


**Fig. 4** Structures, molecular weights, and molar absorbtion coefficients of some properties of D-luciferin derivatives. (A) D-Luciferin-*O*-sulphate [$M_r$, 360.39; $E_{290}$, 5690 (1/mol × cm) in water]; (B) D-luciferin methyl ester [$M_r$, 294.35; $E_{335}$, 8520 (1/mol × cm)]; (C) D-luciferin-*O*-phosphate [$M_r$, 359.39; $E_{315}$, 8360 (1/mol × cm) in water]; (D) D-Luciferyl-L-phenylalanine [$M_r$, 402.49; $E_{335}$, 5770 (1/mol × cm) in water]; (E)D-luciferin-*O*-β-galactoside [$M_r$, 442.46; $E_{260}$, 5520 (1/mol × cm) in 0.05 *M* Tris/HC1, pH 7.5]; (F)D-luciferyl-L-*N*α-arginine [$M_r$, 436.52; $E_{335}$, 5710 (1/mol × cm) in water].

dissolving D-luciferin-*O*-phosphate in water or 0.05 *M* ammonium acetate, pH 6.5, aliquots are stored at −80°C or −30°C until use. For each experiment, a fresh aliquot is used.

## B. Instrumentation

Bioluminescence-enhanced detection systems require convenient and reliable light-measuring instruments (luminometers); a wide range of lumenometers suitable for measuring light emission from tubes or microtite

**Table I**
Kinetic Data for the Hydrolysis of D-Luciferin Derivatives by Carboxypeptidase *N*, β-Galactosidase and Alkaline Phosphatase

| Enzymes | Substrates | $K_m$ [mol/l] | $k_{cat}$ [l/s] | $k_{cat}/K_m$ [$l \times mmol^{-1} \times s^{-1}$] |
|---|---|---|---|---|
| β-Galactosidase | D-luciferin-*o*-β-galctoside | $2.9 \times 10^{-6}$ | 256 | 90.000 |
| Alkaline phosphatase | D-luciferin-*o*-phosphate | $4.3 \times 10^{-5}$ | 1010 | 23.500 |
| Carboxypeptidase | D-Luciferyl-*N*-arginine | $0.5 \times 10^{-4}$ | 11.8 | 236 |

plates are now available. Recently, Hamamatsu Photonics Deutschland GmbH (W-8036 Herrsching, Germany) has produced a very convenient microtiter plate reader for bioluminescent measurements with a parallel measuring mode for rapid and precise measurements. For electronic measurement of emitted photons produced on nitrocellulose sheets a photon-counting camera system (e.g., Argus-100, Hamamatsu Photonics) is recommended. This innovative and precise technique can detect photons within a very short time at an extreme sensitivity (1 photon $\times$ $cm^3$ $\times$ $sec^{-1}$). Photographic film (e.g., Tri-X pan, 380 ASA or Polaroid films for light detection were purchased from Kodak AG and Polaroid GmbH and developed using procedures given by the manufacturers.

# III. PROCEDURES

## A. Dot–Slot Blots—Protein Blots

Our new bioluminescent method was developed first with protein blots; the procedures are described here for completeness.Protein blot analyses are now extensively used for research and for diagnosis of e.g., infectious diseases (Gershoni, 1985). The first practical methods for transfer of proteins from gel electropherograms to cellulose or nitrocellulose were published in 1979 (Towbin *et al.,* 1979; Renart *et al.,* 1979) and since that time there has been an explosive increase in applications of these techniques (Towbin and Gordon, 1984).

The principle of two new, ultrasensitive bioluminescence-enhanced detection systems for protein blotting based on firefly D-luciferin-*O*-phosphate and an alkaline phosphatase label is shown in Fig. 5.

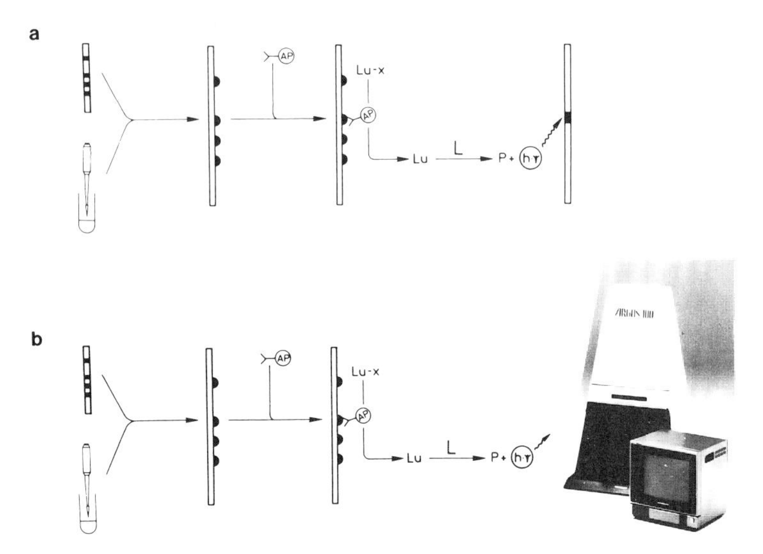


**Fig. 5** Scheme of the bioluminescent-enhanced detection system of protein blotting using (a) photographic films or (b) a photon-counting camera system.>—AP, antibody alkaline phosphatase conjugate; Lu-X, D-luciferin-*O*-phosphate; Lu, D-luciferin; P, oxyluciferin.

## 1. Bioluminescence-Enhanced Detection of Immunoblotted Proteins Using a Photographic Film

1. Transfer protein to nitrocellulose soaked with phosphate buffer (6.5 m*M* $Na_2HPO_4$, 1.5 m*M* $KH_2PO_4$, 137 m*M* NaCl, 2.7 m*M* KCl, pH 8.0) (Bjerrum and Schafer-Nielsen, 1986; Lin and Kasamatsu, 1983).
2. Block the nitrocellulose with 5% (w/v) bovine serum albumin in phosphate buffer for 2 hr at 25°C. Incubate the filter in antirabbit IgG alkaline phosphatase conjugate (1:500 dilution) for 5 min at 25°C, and then wash it three times in the phosphate buffer. Transfer the membrane to a transparent plastic dish containing approximately 6 ml of the detection solution (40 m*M* HEPES buffer, 1.6 m*M* diethanolamine, 6 m*M* $MgCl_2$, 0.54 m*M* ethylenediaminetetraacetic acid (EDTA), 3.4 m*M* DTT, 2.6 m*M* ATP, 2 m*M* luciferin-*O*-phosphate, 0.15 mg luciferase, pH 8.0).
3. Place the dish on a photographic film (in the dark) and expose it for 2 hr. After development (5 min at 25°C in a solution of Kodak HC-110) and fixation of the film, protein-bound antibody can be visualized as dark spots (Fig. 6).

### 2. Bioluminescence-Enhanced Detection of Proteins Using a Photon-Counting Camera System (Argus-100)

1. Transfer protein to nitrocellulose as described in the previous section.
2. Block the nitrocellulose with 5% (w/v) bovine serum albumin in phosphate buffer for 2 hr at 25°C. Incubate the filter for 5 min at 25°C in phosphate buffer containing antirabbit IgG alkaline phosphate conjugate (1:1000 dilution of conjugate in phosphate buffer), wash it three times with phosphate buffer without conjugate. Dip it into detection solution (40 m*M* HEPES buffer, 1.6 m*M* diethanolamine, 6 m*M* $MgCl_2$, 0.54 m*M* EDTA, 3.4 m*M* DTT, 2.6 m*M* ATP, 2 m*M* luciferin-*O*-phosphate, 0.08 mg luciferase/3 ml, pH 8.0) for a few seconds.
3. Place the nitrocellulose membrane in a plastic dish and image bound label using a photon-counting camera system (Argus-100) (count and integrate for 5 min). Protein-bound antibodies can be visualized as bright spots using a computer program developed by Hamamatsu Photonics (Fig. 7).

   The bioluminescent blotting procedure has excellent sensitivity, and spots containing 5 pg protein (corresponding to $30 \times 10^{-18}$ mol of rabbit immunoglobin G) can be detected on photographic film. The detection limit can be lowered into the femtogram range using a photon-counting camera system to detect light emission.

## B. Dot–Slot Blots— Nucleic Acid Hybridization

### 1. Bioluminescence-Enhanced Detection System for DNA Hybridization

Nucleic acid hybridization techniques are now extensively used for the prenatal diagnosis of hereditary disorders, infectious diseases, and in forensic medicine (Blasi, 1987; Viscidi and Yolken, 1987). DNA probes are most commonly labelled with $^{32}P$. However, the label presents a hazard to the operator and has a short shelf-life. A number of nonradioactive methods have now become available based on enzymatically or chemically modified nucleotide probes, which can be recognized either by enzyme-labeled antibodies or by biotinylated enzyme avidin/streptavidin com-

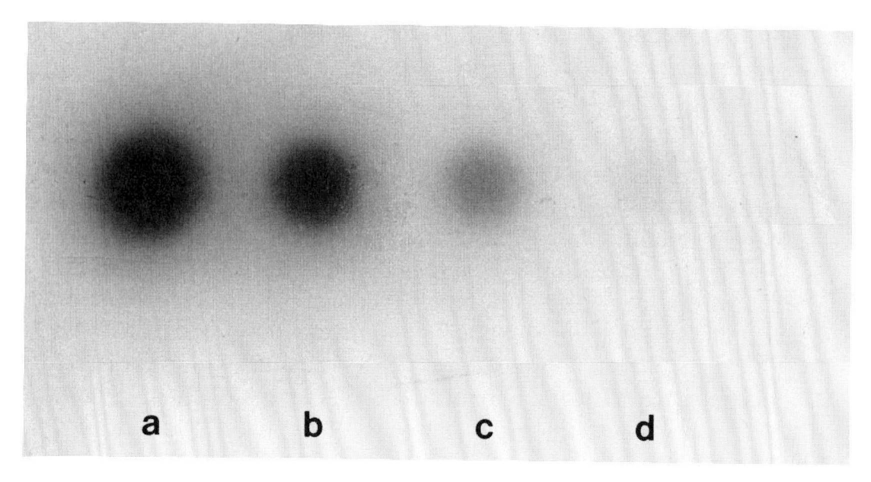


**Fig. 6** Photographic detection of a protein blot on nitrocellulose filter (Hauber and Geiger, 1987). (a) 5 ng; (b) 500 pg; (c) 50 pg; (d) 5 pg rabbit immunoglobulin G.

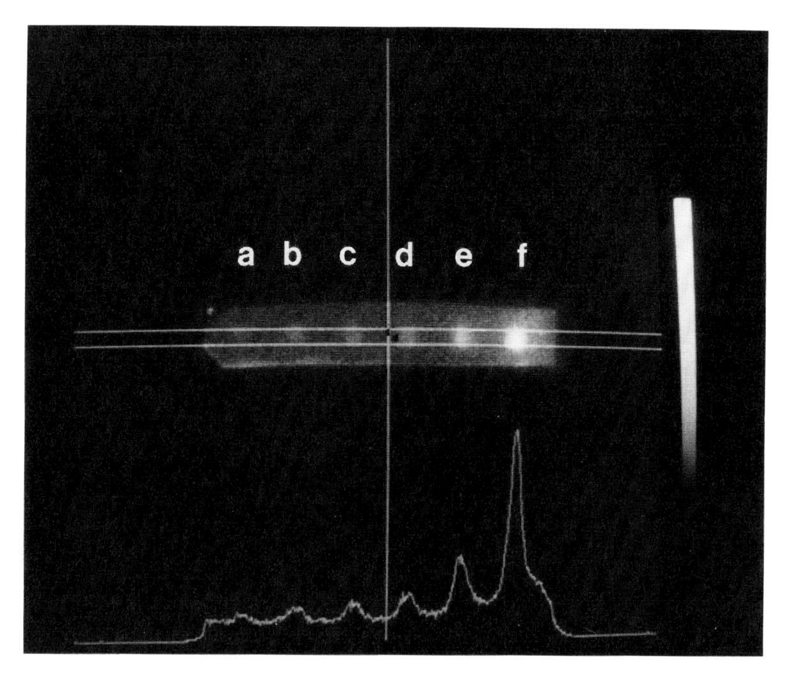


**Fig. 7** Detection of a protein blot by the use of a photon-counting camera (Argus-100; Hauber *et al.*, 1988a). (a) control, no immunoglobulin G applied; (b) 5 fg; (c) 50 fg; (d) 500 fg; (e) 5 pg; (f) 50 pg of rabbit immunoglobulin G.

plexes (Bayer and Wilchek, 1980; Matthews *et al.*, 1985). Direct labeling of DNA with enzyme labels has also been described (Li *et al.*, 1987; Renz and Kurz, 1984). Bound enzyme activity can be detected by either chromogenic, fluorogenic, and/or chemiluminogenic substrates (Leary *et al.*, 1983; Matthews and Kricka, 1988). We have developed a sensitive bioluminescent assay system for alkaline phosphatase labels in nucleic acid hybridization (Geiger and Miska, 1987; Hauber and Geiger, 1988b; Hauber and Geiger, 1989, Miska and Geiger, 1987) based on firefly luciferin-*O*-phosphate (Fig. 8). The bioluminescence is detected either photographically or using a photon-counting camera system (Argus-100).

## 2. Bioluminescence-Enhanced Detection of Nucleic Acid Hybridization Using a Photographic Film

1. Apply pBR322 to nitrocellulose (Southern, 1975), and hybridize the pBR322 with biotinylated pBR322 according to a standard procedure (Schnell *et al.*, 1980). Block the membrane with bovine serum albumin (5%).
2. Incubate the nitrocellulose filter with a streptavidin biotinylated alkaline phosphatase complex, and then wash the filter.
3. Transfer the filter to a transparent plastic dish containing 6 ml of

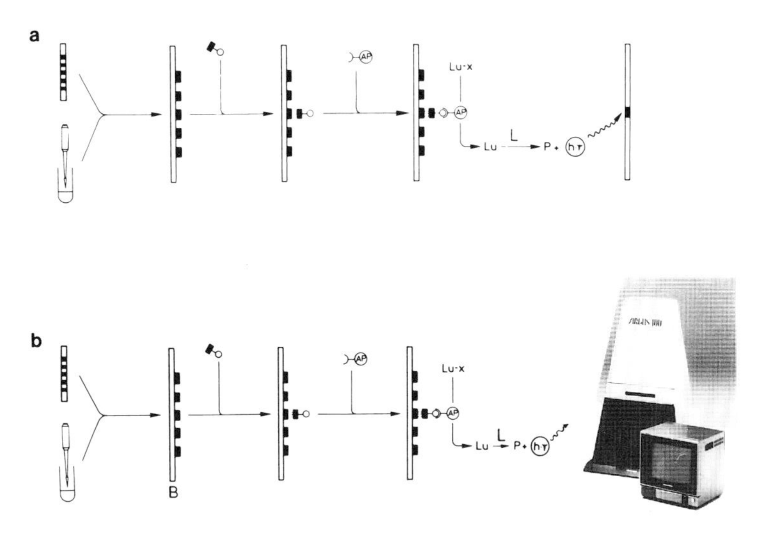


**Fig. 8** Scheme of the bioluminescence-enhanced detection system of DNA hybridization using (a) photographic films or (b) a photon-counting camera system. >—AP, antibody alkaline phosphatase conjugate; -o, sulphonylated probe; Lu-x, luciferin; P, oxyluciferin.

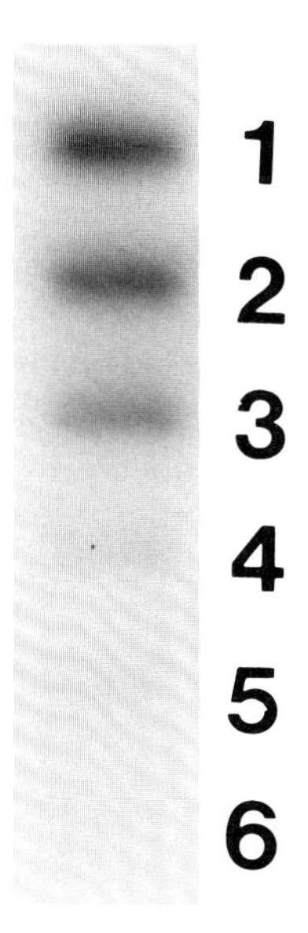


**Fig. 9** Detection of nucleic acid hybridization by the use of a photographic film (Hauber and Geiger, 1988b). (1) 600 ng; (2) 300 ng; (3) 150 ng; (4) 75 ng; (5) 38 ng; (6) 19 ng; of pBR322. Hybridization was performed using a biotinylated probe.

the detection solution (40 m*M* HEPES buffer, 1.6 m*M* diethanolamine, 6 m*M* $MgCl^2$, 0.54 m*M* EDTA, 3.4 m*M* DTT, 2.6 m*M* ATP, 2 m*M* luciferin-*O*-phosphate, 0.15 mg luciferase, pH 8.0).

Place the dish on a photographic film (in the dark) and expose the film for 2 hr. After development (5 min at 25°C in a solution of Kodak HC-110) and fixation of the film, the hybridized probe can be visualized as dark spots (Fig. 9).

### 3. Bioluminescence-Enhanced Detection of Nucleic Acid Hybridization Using a Photon-Counting Camera System

1. Hybridize pBR322 bound to nitrocellulose with biotinylated pBR322 as described in the previous section.
2. Incubate the nitrocellulose filter with a streptavidin biotinylated alkaline phosphatase complex, and then wash the filter.
3. Transfer the filter to a plastic dish containing 2–5 ml of the detection solution (41 m*M* HEPES, 7.8 m*M* diethanolamine, 5m*M* $MgCl_2$, 3.5 m*M* DTT, 2.6 m*M* ATP, 0.33 $\mu M$ luciferin-*O*-phosphate, 0.008 mg luciferase /ml, pH 8.0) for a few seconds, and

then transfer it to a clean plastic dish. Image the bound label using a photon-counting camera system (Argus-100).

Hybridized DNA can be visualized as bright spots using a computer program developed by Hamamatsu Photonics (Fig. 10). Similar results can be obtained using a sulfonated pBR322 probe and an alkaline phosphatase-labeled antibody directed against the sulfonated nucleotide (Fig. 11). The sulfonated pBR322 was prepared using a Chemiprobe kit (Biozym, Hameln, Germany) following the manufacturer's instructions.

## IV. CONCLUSIONS

The bioluminescent assay for alkaline phosphatase labels based on firefly D-luciferin-*O*-phosphate is sensitive and versatile. In conjunction with a photon-counting camera (Argus-100), the detection limits can be improved

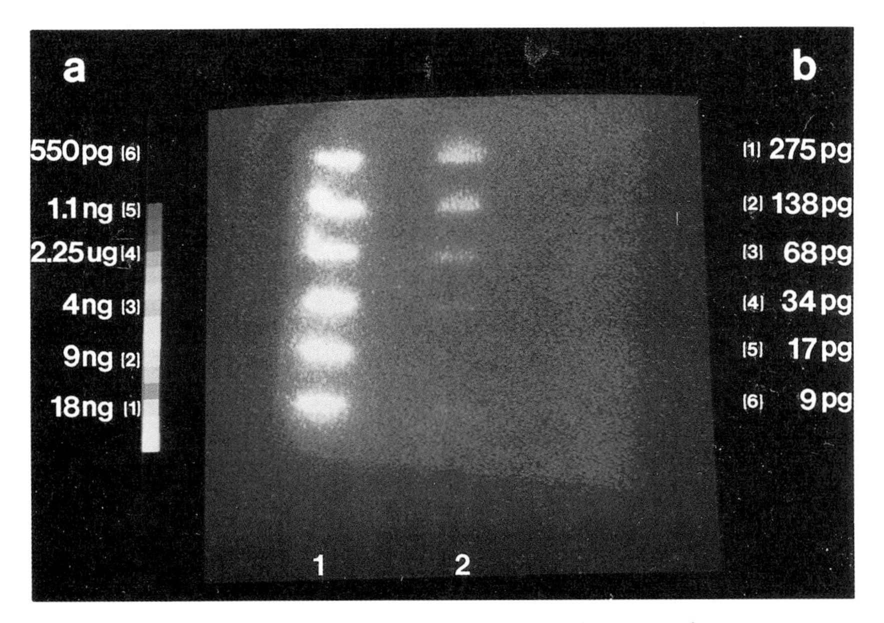


**Fig. 10** Detection of nucleic acid hybridization using a photon-counting camera system (Argus-100; Hauber and Geiger 1989b). (a): (1) 18 ng; (2) 9ng; (3) 4 ng; (4) 2.25 ng; (5) 1.1 ng; (6) 550 pg. (b): (1) 275 pg; (2) 138 pg; (3) 68 pg; (4) 34 pg; (5) 17 pg, (6) 9 pg of pBR322. (Hybridization was performed using a biotinylated pBR322 probe).

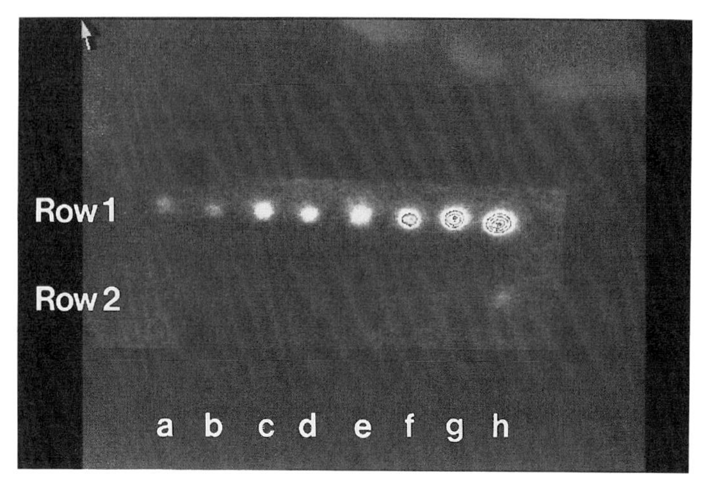


**Fig. 11** Detection of nucleic acid hybridization using a photon-counting camera system (Argus-100). Row 1: (a) 100 fg; (b) 200 fg; (c) 400 fg; (d) 800 fg; (e) 1.6 pg; (f) 3.2 pg; (g) 6.4 pg; (h) 15 pg of pBR322. Row 2: (h) control (hybridization was performed using a sulfonylated pBR322 probe).

by several orders of magnitude (5 to 10 min counting period) compared to detection using photographic film.

# REFERENCES

Bayer, E. A., and Wilchek, M. (1980). The use of avidin-biotin complex as a tool in molecular biology. *Methods Biochem. Anal.* **26,** 1–45.

Berger, J., Hauber, J., Hauber, R., Geiger, R., and Cullen, B. R. (1988). Secreted placental alkaline phosphatase: A powerful new qualitative indicator of gene expression in eukaryotic cells. *Gene* **66,** 1–10.

Bjierrum, O. J., and Schafer-Nielsen, C. (1986). Buffer systems and transfer parameters for semidry electroblotting with a horizontal apparatus. *In* "Electrophoresis 1986" M. J. Dunn, ed.) pp. VCH Weinheim Bergstrasse, Germany.

Blasi, F. (1987). "Human Genes and Diseases." John Wiley, Chichester, England.

Burr, G. J. (1985). "Chemi- and Bioluminescence." Marcel Dekker, New York.

DeLuca, M. A. (1978). Bioluminescence and chemiluminescence. Methods Enzymology **57,** 653 pp., Academic Press, Orlando.

DeLuca, M., and McElroy, W.D. (1986). Bioluminescence and chemiluminescence, Part B. Methods Enzymology, Vol. **133,** 649 pp., Academic Press, Orlando.

Geiger, R., and Miska, W. (1987). The bioluminescence-enhanced immunoassay. New ultrasensitive detections for enzyme immunoassays. *J. Clin. Chem. Clin. Biochem.* **25,** 31–38.

Geiger, E. R. (1990). Bioluminescence enzyme immunoassays. *In* "Luminescence Immunoassays and Molecular Application" (K van Dyke and R. van Dyke, eds.), CRC Press, Boca Raton, Florida.

Gershoni, J. M. (1985). Protein blotting: Developments and perspectives. *TIBS* **10,** 103–106.

Gould, S. J. and Subramani, S. (1988). Firefly luciferase as a tool in molecular and cell biology. *Anal. Biochem.* **175,** 5–13.

Hauber, R., and Geiger, R. (1987). A new, very sensitive detection system for protein blotting. Ultrasensitive detection systems for blotting and nucleic acid hybridization, I. *J. Clin Chem. Clin. Biochem.* **25,** 511–514.

Hauber, R., Miska, W., Schleinkofer, L., and Geiger, R. (1988a). The application of a photon-counting camera in very sensitive, bioluminescence-enhanced detection system for protein blotting. ultrasensitive detection systems for protein blotting and nucleic acid hybridization, II. *J. Clin. Chem. Clin. Biochem.* **26,** 147–148.

Hauber, R., Miska, W., Schleinkofer, L., and Geiger, R. (1989a). New, sensitive, radioactive-free bioluminescence-enhanced detection system in protein blotting and nucleic acid hybridization. *J. Biolumin. Chemilumin.* **4,** 367–372.

Hauber, R., and Geiger, R. (1988a). A sensitive, bioluminescence-enhanced detection method for DNA dot-hybridization. *Nucleic Acids Res.* **16,** 1213.

Hauber, R., and Geiger, R. (1988b). The application of a photon-counting camera in a sensitive, bioluminescence-enhanced detection system for nucleic acid hybridization. Ultrasensitive detection systems for protein blotting and nucleic acid hybridization, III. *J. Clin. Chem. Clin. Biochem.* in press.

Herring, P. J. (1987). Systematic distribution of bioluminescence in living organisms. *J. Biolumin. Chemilumin.* **1,** 146–163.

Kricka, L. J., Stanley, P. E., Thorpe, G. H. G., and Whitehead, T. P. (1984). "Analytical Applications of Bioluminescence and Chemiluminescence." Academic Press, New York.

Kricka, L. J. (1988). Clinical and biochemical applications of luciferase and luciferins. *Anal. Biochem.* **175,** 14–21.

Leary, J. J., Brigati, D. J., and Ward, D. C. (1983). Rapid and sensitive colorimetric method for visualizing biotin-labeled DNA probes hybridized to DNA or RNA immobilized on nitrocellulose: Bio-blots. *Proc. Natl. Acad. Sci. U.S.A.* **80,** 4045–4049.

Li, P., Medon, P. P., Skingle, D. C., Lanser, J. A., and Symons, R. H. (1987). Enzyme-linked synthetic oligonucleotide probes: Non-radioactive detection of enterotoxogenic *Escherichia coli* in faecal specimens. *Nucleic Acids Res.* **15,** 5275–5287.

Lin, W., and Kasamatsu, H. (1983). On the electrotransfer of polypeptides from gel to nitrocellulose membranes. *Anal. Biochem.* **128,** 302–311.

Lundin, A., Richardsson, A., and Thorpe, A. (1976). Continuous monitoring of ATP-converting reactions of purified firefly luciferase. *Anal. Biochem.* **75,** 611–620.

Matthews, J. A., Batki, A., Hynds, C., and Kricka, L. J. (1985). Enhanced chemiluminescent method for the detection of DNA dot-hybridization assays. *Anal. Biochem.* **151,** 205–209.

Matthews, J. A., and Kricka, L. J. (1988). Analytical strategies for the use of DNA probes. *Anal. Biochem.* **169,** 1–25.

Miska, W., and Geiger, R. (1987). I. Synthesis and characterization of luciferin derivatives for use in bioluminescence-enhanced enzyme immunoassays. New ultrasensitive detection systems for enzyme immunoassays. *J. Clin. Chem. Clin. Biochem.* **25,** 23–30.

Miska, W., and Geiger, R. (1988). A new type of ultrasensitive bioluminogenic enzyme substrates. I. Enzyme substrates with D-Luciferin as leaving group. *Biol. Chem. Hoppe Seyler* **369,** 407–411.

Renart, J., Reiser, J., and Stark, G. R. (1979). Transfer of proteins from gels to diazobenzyloxymethyl paper and detection with antisera: A method for studying antibody specificity and antigen structure. *Proc. Natl. Acad. Sci. U.S.A.* **76,** 3116–3120.

Renz, M., and Kurz, C. (1984). A colorimetric method for DNA hybridizations. *Nucleic Acids Res.* **12,** 3435–3444.

Schnell, H., Steinmetz, M., Zachau, H. G., and Schechter, I. (1980). An unusual translocation of immunoglobulin gene segments in variants of the mouse myeloma MPC11. *Nature* **268,** 170–173.

Schölmerich, J., Andreesen, R., Kapp, A., Ernest, M., and Woods, W. G. (1987). "Bioluminescence and Chemiluminescence, New Perspectives." John Wiley, Chichester, England.

Southern, E. M. (1975). Detection of specific sequences among DNA fragments separated by gene electrophoresis. *J. Mol. Biol.* **98,** 503–517.

Towbin, H., Staehelin, T., and Gordon, J. (1979). Electrophoretic transfer of proteins from polyacrylamide gels to nitrocellulose sheets: Procedure and some applications. *Proc. Natl. Acad. Sci. U.S.A.* **76,** 4350–4435.

Towbin, H., and Gordon, J. (1984). Immunoblotting and dot immunoblotting—current status and outlook. *J. Immunol. Methods.* **72,** 313–340.

Viscidi, R. P., and Yolken, R. G. (1987). Molecular diagnosis of infectious diseases by nucleic acid hybridization. *Mol. Cell. Probes* **1,** 3–14.

Wood, W. G. (1984). Luminescence immunoassays: Problems and possibilities. *J. Clin. Chem. Clin. Biochem.* **22,** 905–918.

# 5 Detection of DNA on Membranes with Alkaline Phosphatase-Labeled Probes and Chemiluminescent AMPPD Substrate

Annette Tumolo, Quan Nguyen, and Frank Witney, Genetic Systems Division, Bio-Rad Laboratories, Richmond, California

Owen J. Murphy, John C. Voyta, and Irena Bronstein[1], Tropix, Inc., Bedford, Massachusetts



[1] To whom correspondence should be addressed.

# I. INTRODUCTION

Southern blotting is used for the routine detection and quantitation of nucleic acids in various samples (Southern, 1975). This membrane-based methodology requires a transfer of DNA fragments from a gel followed by hybridization with a labeled probe. Although the most frequently used labels for DNA/RNA detection are radioisotopes such as $^{32}P$ and $^{35}S$, various nonisotopic labels have been described, including fluorophores, metal atoms, enzymes, and others (see Chapter 1). Lately, enzymes have become the label of choice for nucleic acid detection. The inherent catalytic amplification afforded by the enzyme label, when coupled with luminescent substrates, permits extremely sensitive detection of target DNA-RNA (Bronstein *et al.*, 1990).

Alkaline phosphatase has been widely used as a durable label in nucleic acid detection. In particular, the thermal stability of the enzyme precludes degradation of the label during high-temperature hybridizations. The recent development of novel chemiluminescent substrates for alkaline phosphatase simplifies nonisotopic nucleic acid analysis while increasing detection sensitivity (Bronstein *et al.*, 1989a; Bronstein and Voyta, 1989; Tizard *et al.*, 1990). One such substrate, disodium 3-(4-methoxyspiro[1,2-dioxetane-3,2′–tricyclo[3.3.1.1$^{3,7}$]-decan]4-yl)phenyl phosphate (AMPPD), functions as a direct substrate for alkaline phosphatase, requiring no additional reagents for chemiluminescent activation (Bronstein *et al.*, 1989b). Upon dephosphorylation by the enzyme, AMPPD decomposes with concurrent light emission at 477 nm. For AMPPD detection of target DNA, a complementary nucleic acid probe is labeled with alkaline phosphatase by one of the techniques described in related chapters. The principles of AMPPD chemiluminescent detection of alkaline phosphatase-labeled DNA probes are described in this chapter along with detailed protocols for the chemiluminescent detection of alkaline phosphatase avidin:biotin, and covalently labeled nucleic acid probe systems. These methods can be applied to the detection of DNA in Southern blots, dot blots, colony hybridizations, plaque lifts, and other hybridization assays.

# II. GENERAL SOUTHERN BLOTTING PROCEDURE WITH CHEMILUMINESCENCE

DNA can be detected in Southern blots by AMPPD chemiluminescence using biotinylated DNA probes coupled with streptavidin-labeled alkaline phosphatase. The following sections of this paper describe the optimized conditions for the detection of hybridized probes that have been biotinylated by various methods.

## A. Selection and Preparation of Membranes

Neutral nylon membrane is recommended when using biotinylated probes and avidin or streptavidin–alkaline phosphatase conjugates. Tropilon-45 membrane (available from Tropix, Inc., Bedford, Massachusetts) provides optimal results and is highly recommended.

Best results are obtained when DNA is fixed to Tropilon membrane by UV irradiation. This may be performed with a shortwave UV (254 nm) mineralight lamp, which delivers 1200 to 1500 microwatts per $cm^2$ at a distance of 10 to 15 cm from the membrane for up to 5 min (optimal distances and times may vary). Baking ($<30$ min, 80°C) and base transfer may also be used.

## B. Materials

### 1. Reagents

AMPPD substrate, supplied as a 10 mg/ml stock solution (Tropix, Inc., Bedford, Massachusetts). Store at 2 to 8°C.

I-Block reagent supplied as a dry powder (Tropix, Inc., Bedford, Massachusetts). Store at room temperature.

AVIDx-AP, alkaline phosphatase conjugated to streptavidin (Tropix, Inc., Bedford, Massachusetts). Supplied as a concentrated stock solution. Store at 2 to 8°C.

DEA, 99% pure diethanolamine (Tropix, Inc., Bedford, Massachusetts). Store at room temperature.

Other reagents, such as Tween 20, SDS, PVP, and others should be of the highest grade available.

Water used to prepare buffers should be of the highest possible grade.

We recommend using freshly deionized high performance liquid chromatography (HPLC) grade water (Milli-Q, Millipore, Bedford, Massachusetts).

### 2. Instrumentation

Water bath or oven for hybridizations, UV light source, bag sealer. Optimal results are obtained when sealed bags are used for hybridizations, washes, and blocking.

**Note:** In order to assure low nonspecific backgrounds, all reagents must be kept free from bacterial contamination by alkaline phosphatase.

## C. Hybridization Conditions

### 1. Biotinylated Oligonucleotide Probes

Oligonucleotide probes can be biotinylated using several methods. Oligonucleotides may be synthesized with an amine group attached to the 5′ end of the probe and subsequently reacted with NHS-biotin. Alternatively, terminal deoxynucleotidyl transferase can be used to attach one or more biotinylated nucleotides to the 3′ end of an existing oligonucleotide probe. The recommended hybridization buffers for oligonucleotide probes are described below.

#### *a. Stock Solutions*

20% Sodium dodecyl sulfate (SDS)
0.5*M* disodium phosphate, pH 7.2
- 0.5 *M* $Na_2 HPO_4 \cdot 7H_20$
- adjust pH with 85% $H_3P0_4$

25× SSC*, pH 7.0
- 3.75 *M* sodium chloride
- 0.375 *M* sodium citrate dihydrate

0.2 *M* Ethylenediaminetetraacetic acid
- 0.2 *M* EDTA, disodium dihydrate[1], pH 8.0

#### *b. Preparation of Solutions for Hybridization and Stringency Washes*

Blot wetting buffer
- 0.25 *M* disodium phosphate, dilute 0.5 *M* disodium phosphate 1:2

Hybridization buffer

[1] Adjust pH with HCI or NaOH and sterile filter (0.45 $\mu$) before use.

- add 0.2 *M* EDTA to a final concentration of 1 m*M*
- add 20% SDS to a final concentration of 7%
- add 0.5 *M* disodium phosphate to a final concentration of 0.25 *M*
- add powdered I-Block Reagent to a final concentration of 1% w/v

Dissolve I-Block Reagent in disodium phosphate buffer (heat to 65°C with stirring or microwave). Subsequently add EDTA and SDS. It may be necessary to heat this mixture to 65°C to keep it in solution. Adjust the volume to 50 ml with deionized $H_2O$.

### c. *Stringency Wash Buffers*

2× SSC, 1.0% SDS

- add 25× SSC to a final concentration of 2×
- add 20% SDS to a final concentration of 1%

1× SSC, 1.0% SDS

- add 0.25× SSC to a final concentration of 1×
- add 20% SDS to a final concentration of 1%

1× SSC

- add 25× SSC to a final concentration of 1×

Adjust the volume of each buffer with deionized $H_2O$.

### d. *Hybridization and Stringency Wash Conditions*

1. Wet membrane in blot wetting buffer.
2. Prehybridize in hybridization buffer (10–100 μl per cm$^2$) for 1 hr at 55°C or an appropriate hybridization temperature. Drain buffer.
3. Dilute heat-denatured biotinylated probe (0.1–5.0 pmol/ml) in fresh hybridization buffer and add to membrane (10–100 μl per cm$^2$). Incubate 2 hr at the appropriate temperature.
4. Wash membrane 2 × 5 min at room temperature (RT) in 2× SSC, 1.0% SDS, then 2 × 15 min at the hybridization temperature in 1× SSC, 1.0% SDS then 2 × 5 min at room temperature in 1× SSC (1 ml per cm$^2$).
5. Follow the procedure for chemiluminescent detection of biotinylated DNA. Membranes may be stored for up to 3 days in blocking buffer before continuing with the detection protocol.

## 2. Biotinylated DNA Probes Prepared by Nick Translation or Random Hexamer Labeling

The following protocol has been developed for the hybridization of biotin-labeled long DNA probes. These probes may be prepared by nick translation, random primer labeling or photoreactive biotinylation procedures to incorporate biotin. Kits that contain all the necessary reagents to produce biotin-labeled probes are available from Tropix.

### *a. Stock Solutions*

5% Polyvinylpyrrolidone (PVP)
20% Sodium dodecyl sulfate (SDS)
50% Dextran sulfate
1 *M* Tris buffer[2], pH 7.5
- 1 *M* tris base
- adjust pH with HCI

25 × SSC[2], pH 7.0
- 3.75 *M* sodium chloride
- 0.375 *M* sodium citrate dihydrate
- adjust pH with HCI

0.2 *M* Ethylenediaminetetraacetic acid
- 0.2 *M* EDTA, disodium dihydrate[2], pH 8.0

### *b. Preparation of Solutions for Hybridization and Stringency Washes*

Blot Wetting Buffer
5× SSC
- add 25× SSC to a final concentration of 5×

Hybridization Buffer
- add 5% PVP to a final concentration of 0.5%
- add 0.2 *M* EDTA to a final concentration of 1 m*M*
- add NaCl to a final concentration of 1 *M*
- add 1 *M* Tris to a final concentration of 50 m*M*
- add 50% dextran sulfate to a final concentration of 5%
- add heparin (Sigma grade 1-A) to a final concentration of 0.2%
- add 20% SDS to a final concentration of 4%

Dissolve NaCl in deionized $H_2O$ and subsequently add PVP, EDTA, Tris, heparin, and SDS. It may be necessary to heat this mixture to 65°C to keep it in solution. Add dextran sulfate and adjust volume with deionized $H_2O$ and filter (using a 0.45 $\mu$m pore size filter) before use. Aliquots of this buffer may be frozen for later use.

[2] Adjust pH with HCI or NaOH and sterile filter (0.45 $\mu$) before use.

Posthybridization Wash Buffer
- add 5% PVP to a final concentration of 0.5% PVP
- add NaCl to a final concentration of 1 *M*
- add 1 *M* Tris to a final concentration of 50 m*M*
- add 20% SDS to a final concentration of 2%

Dissolve NaCl in deionized $H_2O$ and then add PVP, Tris, and SDS. It may be necessary to heat this mixture to 65°C to keep it in solution. Adjust the volume with deionized $H_2O$ and filter (using a 0.45 μ pore size filter) before use. Aliquots of this buffer may be stored frozen for later use.

***c. Stringency Wash Buffers***

2× SSC. 1% SDS
- add 25× SSC to a final concentration of 2×
- add 20% SDS to a final concentration of 1%

0.1× SSC, 1% SDS
- add 25× SSC to a final concentration of 0.1×
- add 20% SDS to a final concentration of 1%

1× SSC
- add 25× SSC to a final concentration of 1×

Adjust the volume of each buffer with deionized $H_2O$.

***d. Hybridization and Stringency Wash Conditions***

1. Wet membrane in blot wetting buffer.
2. Prehybridize in hybridization buffer (10–100 μl per $cm^2$) for 1 hr at 55°C or an appropriate hybridization temperature. Drain buffer.
3. Dilute heat-denatured biotinylated probe (0.1–5.0 pmol/ml) in fresh hybridization buffer and add to membrane (10–100 μl per $cm^2$). Incubate 2 hr at the appropriate temperature.
4. Wash membrane 1 × 10 min at 55°C with posthybridization wash buffer, 2 × 5 min at RT in 2× SSC, 1.0% SDS, then 2 × 15 min at the hybridization temperature in 0.1× SSC, 1.0% SDS, then 2 × 5 min at RT in 1× SSC (1 ml per $cm^2$).
5. Follow the procedure for chemiluminescent detection of biotinylated DNA described below. Membranes may be stored for up to 3 days in blocking buffer before continuing with the detection protocol.

**Note:** Biotinylated probes should not be subjected to alkaline denaturation at elevated temperature since biotin may be cleaved from the probe.

## D. Chemiluminescent Detection of Biotinylated DNA

The following protocol has been developed for the detection of immobilized biotinylated or hybridized biotinylated DNA on neutral nylon membrane. All recommendations regarding volumes in the following protocol apply to a single blot (100 $cm^2$). If larger sections of membrane or multiple blots are being processed simultaneously, the volumes must be increased accordingly. AVIDx-Alk Phos (Tropix, Inc., Bedford, Massachusetts), a streptavidin alkaline phosphatase that has been screened for low backgrounds in membrane-based chemiluminescence assays, should be used in the following protocol. All procedures should be performed at room temperature unless stated otherwise.

### *a. Solutions*

Prepare all solutions with filter sterilized, freshly deionized $H_2O$.

10× Phosphate buffered saline (PBS)

- add $Na_2HPO_4$ to a final concentration of 0.58 *M*
- add $NaH_2PO_4 \cdot H_2O$ to a final concentration of 0.1 *M*
- add NaCl to a final concentration of 0.68 *M*
- adjust volume using deionized $H_2O$. A 1× PBS solution should have a pH of 7.3 to 7.4.

Conjugate Buffer

- add I-Block Reagent to a final concentration of 0.2%
- add 10× PBS to a final concentration of 1×
- add sodium azide to a final concentration of 0.02%

Suspend I-Block Reagent in deionized water, heat gently and stir. Immediately add 10× PBS and stir at 70°C for 30 min or until solution is clear of particles. Alternatively, the solution may be microwaved (high) for 30 to 60 sec per 100 ml and stirred to dissolve. Adjust volume with deionized $H_2O$. The solution may remain cloudy. Cool to room temperature before use.

Blocking buffer

- 1× conjugate buffer
- add Tween 20 to a final concentration of 0.1%

Wash buffer

- add 10× PBS to a final concentration of 1×
- add Tween 20 to a final concentration of 0.3%
- add sodium azide to a final concentration of 0.02%
- adjust volume with deionized $H_2O$

Assay buffer

- add DEA to a final concentration of 0.1 *M*

- add $MgCl_2$ to a final concentration of 1 m*M*
- add sodium azide to a final concentration of 0.02%
- Dissolve above components in deionized $H_2O$ and adjust pH to 10.0 with HCI. Adjust to final volume.

AMPPD substrate solution

- dilute AMPPD Stock Solution 1:100 in assay buffer. Final concentration is 0.24 m*M*

### *b. Membrane Blocking and Chemiluminescent Detection*

1. Wash membrane 2 × 5 min in blocking buffer (0.5 ml per $cm^2$).
2. Incubate for 30 min in 100 ml of blocking buffer (1 ml per $cm^2$).
3. Microcentrifuge the tube of AVIDx-AP conjugate for 4 min at RT. With a sterile pipet tip, remove 1.0 $\mu$l per 100 $cm^2$ membrane. Dilute AVIDx-AP conjugate 1:10,000 in conjugate buffer (use 1.0 $\mu$l of AVIDx-AP conjugate in 10 ml conjugate buffer per 100 $cm^2$).
4. Incubate membrane for 10 min with constant agitation in diluted conjugate solution.
5. Wash 1 × 5 min in blocking buffer (0.5 ml per $cm^2$).
6. Wash 4 × 5 min in wash buffer (1 ml per $cm^2$).
7. Wash filter 2 × 5 min in assay buffer (0.5 ml per $cm^2$).
8. Add AMPPD substrate solution to the blot (5 ml per 100 $cm^2$).
9. Slowly shake or agitate the immersed membrane for 5 min.
10. Drain off excess substrate from the membrane. Do not blot the membrane or allow it to dry at all. The membrane must remain wet.
11. Seal the membrane in a plastic bag (e.g., hybridization bag or Saran wrap, etc.). Do not fold or bend the membrane; it should remain flat.

**Note:** Careful handling of blots is very critical. Mishandled blots exhibit high nonspecific background signals. Never touch membranes with ungloved hands.

## E. Film Exposure and Development

Blots that are sealed in plastic wrap or a hybridization bag may be imaged by placing them in direct contact with standard Kodak XAR X-ray film or Polaroid Instant Photographic Black and White film. Initial exposures of 5 to 60 min are recommended to assess the optimal exposure time for your application. Owing to the kinetics of light emission on nylon membranes, incubation of the plastic wrapped membrane at room temperature for 1 hr to overnight will permit shorter film exposures. Very short film exposures are possible after overnight incubations. Imaging of a small blot up to 7.3 cm × 9.5 cm in size may be performed within minutes on Polaroid instant film in a handheld camera luminometer (Tropix Cat. No. ICL-901). Kodak XAR X-ray film may also be used to detect the chemiluminescent signal with similar sensitivity.

## F. Reprobing Membranes

Blots that have remained wet following hybridization may be stripped of hybridized DNA and used for subsequent reprobing utilizing the following procedure:

1. Heat buffer containing 1× SSC, 1% SDS to 95°C.
2. Pour this solution over the membrane and agitate rapidly for 5 min at RT, and drain buffer.
3. Repeat above two steps, if necessary.
4. Wash 2 × 5 min in 1× SSC at RT and store air dried or wet, wrapped in Saran wrap at 4°C.

Successful removal of biotinylated probes and AVIDx-AP conjugate may be confirmed by reincubating the membrane in the presence of AMPPD substrate solution as described above. If a detectable light signal is observed, the biotin-labeled probe was not fully stripped, and more rigorous treatment is necessary.

## G. Troubleshooting

Since AMPPD substrate is the most sensitive detector of alkaline phosphatase activity available, it is recommended that only ultrapure water and other reagents free of AP contamination be used.

For the detection of DNA with biotinylated probes labeled by nick translation, random primer labeling, or 3′ end labeling, Tropix has optimized the above protocols using Tropilon membrane, I-Block reagent, AVIDx-AP conjugate, and AMPPD substrate. If other membranes or enzyme conjugates are used, results may vary.

If the expected sensitivity is not attained, try the following:

1. For best results, the assay buffers should be prepared daily.
2. To detect lower levels of DNA, increase the film exposure time as much as possible or until background signal obscures the image.

3. Increase the incubation time during the hybridization step to overnight and/or conjugate incubation to 60 min if background signal is sufficiently low.

4. Increase the concentration of labeled DNA and/or AP conjugate in order to increase the signal. This may, however, contribute to increased nonspecific binding.

5. Check that the probe is sufficiently labeled and denatured before use.

If the nonspecific background signal is too high (e.g., the film appears overexposed or the image is uneven or spotty) try the following:

1. Splotchy images may result from bacterial contamination of the membrane. Make sure that all buffers are free of contamination before use and that the membrane, blotting paper, and hybridization bags are clean and fingerprint free.

2. Decrease the film exposure time until appropriate resolution is achieved. If the expected sensitivity is not attained, try step 3 below.

3. If the background signal appears evenly across the membrane but obscures the specific signal, try incubating the membrane in the blocking buffer overnight at 4°C or increasing the number of wash steps after conjugate incubation.

4. In order to reduce nonspecific binding of the AVIDx-AP conjugate, increase the dilution of the conjugate to 1:15,000 and spin down any particulate material before use.

5. In order to reduce nonspecific binding of DNA to the membrane, reduce the biotin-labeled probe concentration in the hybridization buffer, or increase the duration of the final two stringency washes.

6. If the background signal is spotty, try precipitating the probe with EtOH before the hybridization.

## III. TWO-STEP HYBRIDIZATION SOUTHERN BLOTTING PROCEDURE—DETECTION OF SINGLE-COPY GENES

Nonisotopic methods employing biotinylated or digoxogenin-labeled probes work very well for plaque and colony-screening applications or for dot blots and Southern blots containing nanogram amounts of target DNA. (Bronstein *et al.*, 1990; Kincaid and Nightingale, 1988; Martin *et al.*, 1990). However, two problems associated with these methods make their application unreliable where picogram levels of detection are required. First, high-sensitivity applications such as northern blots or genomic Southern blots require that the probe DNA have a high efficiency of label incorporation. Since there is no easy method for monitoring the efficiency of the biotin or digoxigenin incorporation into the probe, users often proceed with a hybridization without knowing whether or not the probe DNA is adequately labeled. Second, the high backgrounds associated with the avidin or antidigoxigenin–alkaline phosphatase conjugates often reduce the signal to noise to an extent that obscures the results.

In response to the need for a better nonisotopic methodology for Northern and genomic Southern hybridizations, we developed a two-step hybridization method that eliminates the problems outlined. The procedure, a variation of the sandwich hybridization method first proposed by Dunn and Hassell (Dunn and Hassell, 1977) involves a primary hybridization of single-stranded phagemid or M13 DNA containing the probe DNA to a

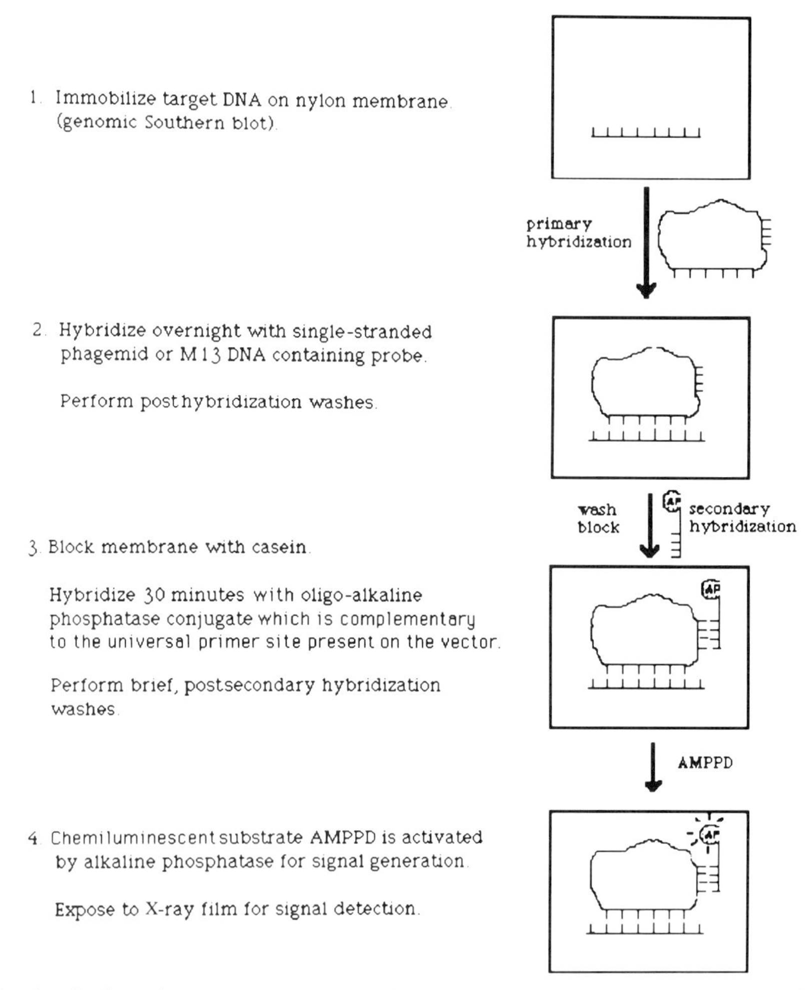


**Fig. 1** Outline of the two-step hybridization method of chemiluminescent detection for Northerns and genomic Southerns.

Northern or Southern blot, followed by a secondary hybridization with an oligonucleotide–alkaline phosphatase conjugate, which is complementary to the universal primer site present in the vector sequence. The blot is exposed to X-ray film for signal detection after incubation with the chemiluminescent substrate, AMPPD (see Fig. 1). There is no labeling of the primary probe. The secondary probe, which can be synthesized and conjugated in large quantities, is the same no matter what the target gene. This eliminates the problems encountered with the use of uncharacterized biotin or digoxigenin-labeled probes. Because the secondary probe is a direct oligonucleotide–alkaline phosphatase conjugate, none of the high backgrounds associated with the avidin or antidigoxigenin–alkaline phosphatase conjugates is seen. In practice, we consistently attain picogram levels of detection with very low backgrounds (see Fig. 2). This method should prove useful and reliable for genomic Southern and Northern applications requiring high sensitivity.

## A. Materials

### 1. Reagents

Casein (Bio-Rad, Richmond, California)
Urea (Bio-Rad)

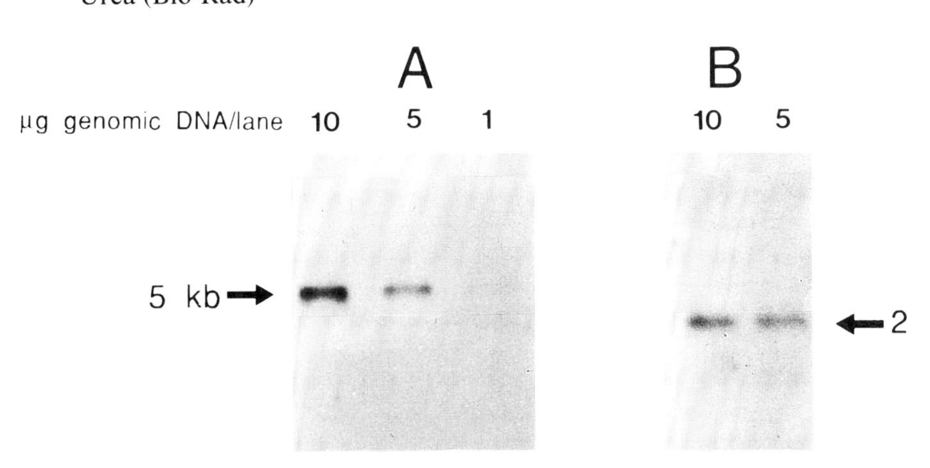


**Fig. 2** Detection of single-copy genes using the two-step hybridization method. (A) *Bgl*I-digested human genomic DNA was probed with single-stranded pTZ18u phagemid DNA (pTZFVIII) containing a 1.8-kb portion of a human Factor VIII cDNA clone (obtained from Genetics Institute), followed by a secondary hybridization with UP-AP. (B) *Eco*RI-digested human genomic DNA was probed with single-stranded pTZ18 phagemid DNA(pTZN-*myc*) containing a 1-kb fragment from a human N-*myc* cDNA (obtained from Oncor), followed by a secondary hybridization with UP-AP. Blots were incubated in AMPPD and exposed to X-ray film overnight.

SDS (Bio-Rad)
Triton X-100 (Bio-Rad)
25X Diethanolamine DEA buffer (Bio-Rad)
AMPPD substrate supplied as a 10 mg/ml stock solution (Bio-Rad or Tropix)
Universal primer-alkaline phosphatase conjugate (Bio-Rad)
Neutral nylon membrane Alpha-probe™GT Bio-Rad)
Water used to prepare buffers should be of the highest possible grade. We recommend using freshly deionized HPLC-grade water.

### 2. Instrumentation

Shaking water bath at 37°C; heat sealer; equipment for autoradiography.

## B. Hybridization Conditions

### 1. Reagent Preparation

#### *a. Stock Solutions*

10XPBS, pH 6.8
- add $Na_2HPO_4$ to a final concentration of 0.58 *M*
- add $NaH_2PO_4 \cdot H_2O$ to a final concentration of 0.17 *M*
- add NaCl to a final concentration of 0.68 *M*
- adjust pH to 6.8 with 85% $H_3PO_4$
- adjust volume with $ddH_2O$
- autoclave

50X Denhardt's
- 1% Ficoll
- 1% polyvinylpyrrolidine
- 1% bovine serum albumin (Fraction V)
- adjust volume with $ddH_2O$
- sterile filter and store at −20°C

20X SSC
- add NaCl to 3 *M*
- add sodium citrate to 0.3 *M*
- adjust pH to 7.0 with a few drops of concentrated acetic acid
- adjust volume with $ddH_2O$
- autoclave

20% SDS
- add 200g SDS to 800 ml autoclaved $88H_2O$, adjust volume to 1000 ml

1 M disodium phosphate, pH 7.2
- 1 *M* $Na_2HPO_4$ $7H_2O$
- adjust pH to 7.2 with 85% $H_3PO_4$

### *b. Preparation of Solutions for Hybridization, Stringency Washes, and Blocking*

Primary hybridization buffer
- add 20% SDS to a final concentration of 7%
- add 1 *M* disodium phosphate to a final concentration of 0.5 *M*

5% SDS, 0.04 *M* $NaPO_4$
- add 20% SDS to a final concentration of 5%
- add 1 *M* disodium phosphate to a final concentration of 0.04*M*

1% SDS, 0.04 *M* $NaPO_4$
- add 20% SDS to a final concentration of 1%
- add 1 *M* disodium phosphate to a final concentration of 0.4 *M*

2% casein solution
- add casein to a final concentration of 2% to rapidly stirring 1× PBS
- heat solution to 65°C while stirring (do not boil)
- stir at 65°C until casein is dissolved
- store on ice until ready to use
- make fresh solution for each experiment

1% Casein blocking solution
- dilute freshly made 2% casein to 1% with 1× PBS

Secondary hybridization buffer
- add urea to a final concentration of 6 *M*
- add 20× SSC to a final concentration of 2×
- add 50× Denhardt's to a final concentration of 5×
- add 2% casein to a final concentration of 1% casein
- adjust volume with $ddH_2O$

6 *M* urea, 1× SSC, 0.1% SDS
- add urea to a final concentration of 6 *M*
- add 20× SSC to a final concentration of 1×
- add 20% SDS to a final concentration of 0.1%

1× SSC, 0.1% triton
- add 20× SSC to a final concentration of 1×
- add Triton ×-100 to a final concentration of 0.1%

1× PBS

1× SSC

1× DEA (substrate) buffer
- dilute 25× DEA buffer to final concentration of 1×
- pH should be 9.8–10.0
- filter sterilize
- 1× DEA buffer contains 0.1 *M* DEA, 1m*M* $MgCl_2$, and 0.02% $NaN_3$

In practice, use sterile *nanopure* or HPLC-grade water when making up all buffers, and take reasonable precautions to prevent contamination of buffers with alkaline phosphatase. The membranes are always handled with gloved hands and forceps, on the outer edges only.

## 2. Procedures

### *a. Primary Hybridization*

1. Transfer digested genomic DNA to neutral nylon membrane according to standard Southern capillary transfer methods (Sambrook *et al.*, 1989). Use of positively charged membranes results in high backgrounds (see Fig. 3). We recommend UV-cross-linking of the DNA to the membrane (see Fig. 4).
2. Hybridize the genomic Southern blot overnight with 25–50 ng/ml hybridization buffer of single-stranded phagemid or M13 DNA containing the probe. Single-stranded phagemid or M13 DNA is prepared according to standard published procedures (Sambrook *et al.*, 1989). Any hybridization conditions may be chosen, but we recommend 7% SDS, 0.5 *M* $NaPO_4$ at 65°C overnight. It is important to use between 25 and 50 ng/ml of primary probe, because use of lower concentrations of primary probe results in reduced sensitivity (data not shown).
3. Perform stringency washes. If recommended hybridization conditions were used, wash twice with excess volumes of 5% SDS,

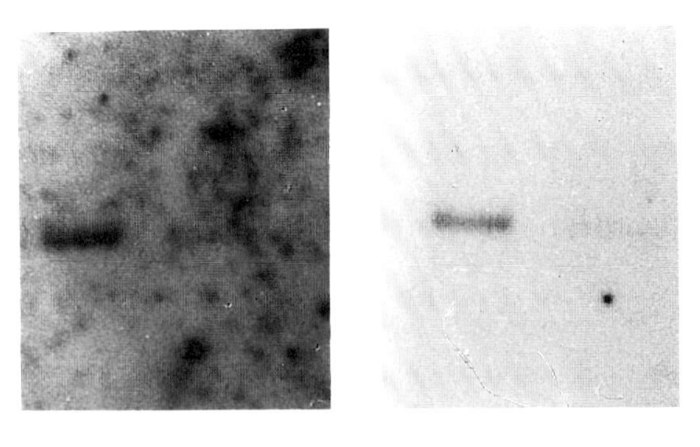


**Fig. 3** Comparison of signal to noise on charged versus neural nylon. *Bgl*I-digested human genomic DNA was transferred to either positively charged (Zeta-Probe®, Bio-Rad) or neutral nylon membrane (Alpha-Probe™GT, Bio-Rad). Primary hybridizations were done with pTZFVIII. Secondary hybridizations were done with the UP-AP. Blots were incubated in AMPPD and exposed to X-ray film overnight.

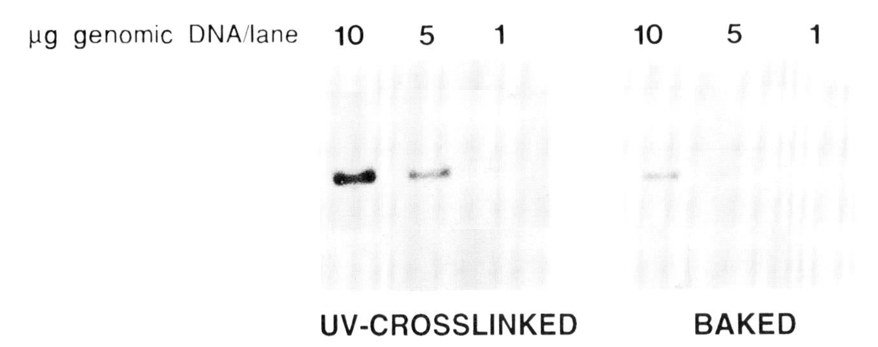


**Fig. 4** Effect of UV-cross-linking DNA to membrane on signal strength. *Bgl*I-digested human genomic DNA was transferred to neutral nylon membrane (Bio-Rad). The blots were either UV-cross-linked (150 mJ, GS Gene Linker®, Bio-Rad), or baked for 1 hr at 80°C. Primary hybridizations were done with pTZFVIII. Secondary hybridizations were done with UP-AP. Blots were incubated with AMPPD and exposed to X-ray film overnight.

0.04 *M* $NaPO_4$ for 30 to 60 min at 65°C, then twice with 1% SDS, 0.04 *M* $NaPO_4$ for 30 to 60 min at 65°C.

***b. Blocking***

1. Rinse membrane once with 1× PBS for 5 min at RT with gentle shaking. It is convenient to use individual large, disposable weigh boats for the blocking steps, but any clean container is acceptable. Multiple blots may be processed together, but sufficient buffer and blocker must be used to cover the blots and prevent them from sticking together.
2. Block the membrane for 30 to 60 min at RT with gentle shaking with 1% casein blocking solution.

***c. Secondary Hybridizations***

1. Prehybridize the membrane in a heat-sealed plastic bag for 10 to 15 min at 37°C with secondary hybridization buffer.
2. Remove prehybridization buffer and add fresh secondary hybridization buffer with 1.5 to 2.5 n*M* universal primer–alkaline phosphatase conjugate. Use between 50 and 75 µl hybridization buffer per $cm^2$ membrane.
3. Hybridize for 30 min at 37°C.
4. Wash twice with excess volumes of 6 *M* urea, 1× SSC, 0.1% SDS for 10 min at 37°C. Prewarm buffers.
5. Wash once with 1× SSC, 0.1% Triton for 5 min at 37°C.
6. Wash once with 1× SSC for 5 min at RT.

*d. Signal Detection*

1. Rinse membrane twice with 1× DEA buffer for 5 min at RT.
2. Transfer membrane to heat-sealable bag and add 100 μl per $cm^2$ of a 1:120 dilution of 10 mg/ml AMPPD in 1× DEA buffer.
3. Remove bubbles, seal, and incubate with gentle shaking or rocking for 30 min at RT.
4. Drain excess fluid from bag, remove bubbles, and reseal.
5. Expose to X-ray film overnight for signal detection. Additional exposures can be made. Signal should last for up to 3 days.

## C. Troubleshooting

The above procedures have been optimized for use with the universal primer–alkaline phosphatase conjugate available from Bio-Rad Laboratories. In addition, the secondary hybridization conditions were designed to give specific hybridization to target on human genomic Southern blots. If other conjugates are used, or other genomes are probed, secondary hybridization conditions may have to be optimized by increasing or decreasing the hybridization stringency (adjusting the NaCl concentration up or down). Exposure times may be decreased by increasing either the secondary probe concentration or the AMPPD concentration; however, the background levels will also increase.

# IV. CONCLUSIONS

Chemiluminescent detection of alkaline phosphatase-labeled DNA probes with AMPPD substrate routinely affords high sensitivities in shortened detection times. Furthermore, chemiluminescent analysis does not require sophisticated equipment, and results are simply imaged on X-ray or Polaroid instant film for a hard-copy record. The nonisotopic method is easily adaptable to various nucleic acid hybridization assays (e.g., Southern blots, Northern blots, colony hybridizations, and plaque lifts), as well as other ligand-binder assays (e.g., immunoassays and Western blots).

# REFERENCES

Bronstein, I., Edwards, B., and Voyta, J. C. (1989b). 1,2-Dioxetanes; novel chemiluminescent substrates: Applications to immunoassays. *J. Biolumin. Chemilumin.* **4,** 99–111.
Bronstein, I., Voyta, J. C., and Edwards, B. (1989a). A comparison of chemiluminescent and

colorimetric substrates in a hepatitis B virus DNA hybridization assay. *Anal. Biochem.* **180,** 95–98.

Bronstein, I., Voyta, J. C., Lazzari, K. G., Murphy, O., Edwards, B., and Kricka, L. K. (1990). Rapid and sensitive detection of DNA in Southern blots with chemiluminescence. *BioTechniques* **8,** 310–314.

Bronstein, I., and Voyta, J. C. (1989). Chemiluminescent detection of herpes simplex virus I DNA in blot and *in situ* hybridization assays. *Clin. Chem.* **35,** 1856–1857.

Church, G. M., and Gilbert, W. (1984). Genomic sequencing. *Proc. Natl. Acad. Sci. U.S.A.* **81,** 1991.

Dunn, and A. R., Hassell, J. A. (1977). A novel method to map transcripts: Evidence for homology between an adenorirus mRNA and discrete multiple regions of the viral genome. *Cell* **12,** 23–36.

Jablonski, E., Moomaw, E. W., Tullis, R. H., and Ruth , J. L. (1986). Preparation of oligodeoxynucleotide–alkaine phosphatase conjugates and their use as hybridization probes. *Nucleic Acids Res.* **14,** 115–6115–6128.

Kincaid, R. L., and Nightingale, M. S. (1988). A rapid nonradioactive procedure for plaque hybridization using biotinylated probes prepared by random primed labeling. *BioTechniques* **6,** 42–49.

Martin, R., Hoover, C., Grimme, S., Grogan, C., Holtke, J., and Kessler, C. (1990). A highly sensitive, nonradioactive DNA labeling and detection system. *BioTechniques* **9,** 762–768.

Sambrook, J., Fritsch, E. F., Maniatis, T. (1989). Molecular Cloning A Laboratory Manual. Coldspring Harbor Laboratory Press, Cold Spring Harbor, New York.

Southern, E. M. (1975). Detection of specific sequences among DNA fragments separated by gel electrophoresis. *J. Mol. Biol.* **98,** 503–517.

Tizard, R., Cate, R. L., Ramachandran, K. L., Wysk, M., Voyta, J. C., Murphy, O. J., and Bronstein, I. (1990). Imaging of DNA sequences with chemiluminescence. *Proc. Natl. Acad. Sci. U.S.A.* **87,** 4514–4518.

# 6 Detection of Alkaline Phosphatase by Colorimetry

Ayoub Rashtchian
Molecular Biology Research and Development
Life Technologies, Inc.
Gaithersburg, Maryland



## I. INTRODUCTION

Nucleic acid hybridization has been one of the most powerful techniques in molecular biology during the past 2 decades. The specificity and quantitative nature of annealing complementary strands of nucleic acids have been exploited in a variety of ways. The applications of hybridization range from determination of overall similarity between organisms (Brenner, 1989) to determination of a single-base mutation in a given gene. Although nucleic acid hybridizations can be performed in a variety of ways, basically three general techniques are used: (1) solution hybridization; (2) hybridization on membrane filters; and (3) *in situ* hybridization to cytological preparations.

Traditionally most hybridization techniques have used nucleic acid probes labeled with high-energy radioisotopes. Although generally short-lived and hazardous, radioactive labels can be detected in minute quantities and can be quantitated with relative ease by scintillation counting or autoradiographic methods. Although radio-labeled probes continue to be used by many researchers, the short half-life, problems with disposal, safety, and regulatory concerns have resulted in a widespread search for an ideal nonradioactive substitute. Consequently, a large number of nonradioactive labeling methods and appropriate signal-generation systems have been developed (Matthews and Kricka, 1988; see also Chapter 1). The properties of these systems vary greatly with regard to sensitivity of detection as well as the simplicity/complexity and requirements for specialized equipment. These properties, the hybridization format used, and ultimately the nature of the biological/biochemical question(s) being addressed by individual researchers affect the choice of the nonradioactive system.

## II. LABELING AND DETECTION STRATEGIES

The choice of a strategy for nonradioactive labeling of nucleic acid probes is heavily dependent on the type probe being used; however, the labeling strategies can generally be divided into two categories:

1. Indirect primary labels that require secondary recognition after hybridization, such as biotin and other haptens; and
2. Direct primary labels that are covalently attached to the nucleic acid probe and are directly detectable, such as enzymes and fluorophores.

A detailed discussion of labeling procedures can be found in Chapter 2 of this volume and basically can be divided into (1) *enzymatic* and (2) *chemical* methods. Regardless of the labeling strategy or procedure, alkaline phosphatase-based systems have been the most widely used for indirect detection of hapten-labeled probes (Langer *et al.*, 1981; Leary *et al.*, 1983) as well as direct labeling and detection of nucleic acid probes (Jablonski *et al.*, 1986; Renz and Kurz, 1984).

## A. Labeling of Nucleic Acid Probes with Biotin

The most common nonradioactive labeling method in recent years has used biotin as reporter group. In this system, probe is labeled with biotin and the presence of the labeled probe after hybridization is detected by streptavidin or avidin conjugated to alkaline phosphatase. The method used for labeling of nucleic acid probes is dependent on the type of probe used as well as the intended use of the probe. The most common methods of labeling DNA probes are by nick translation or random primer labeling (Feinberg and Vogelstein, 1983) using a suitable biotinylated nucleoside triphosphate (Gebeyehu *et al.*, 1987; Langer *et al.*, 1981). Both methods are efficient in incorporating biotin-labeled nucleoside triphosphates into DNA. Commercial kits are available for labeling DNA with biotin using these procedures (Life Technologies/BRL, Gaithersburg, Maryland). An alternative method for labeling uses photobiotin. This reagent contains a photoreactive aryl azide group attached to biotin through a charged linker arm. When irradiated with strong visible light in the presence of DNA, photobiotin covalently links biotin groups to the DNA (Forster *et al.*, 1985).

The choice of labeling method to be used for preparation of biotinylated probes depends on the specific use of the probe. While all three of the labeling methods mentioned above are suitable for detection of nucleic acids on membrane filters, only methods that yield probes of small size (50–500 bases) are suitable for use in *in situ* hybridization (Singer *et al.*, 1987). The most widely used method for preparation of labeled probe for *in situ* is nick translation (Singer *et al.*, 1987). Preparation of probes of optimal size for use *in situ* requires optimization of the concentration of DNase in the nick translation. A commercial kit specifically developed for preparation of probes suitable for *in situ* hybridization is available (Life Technologies/BRL, Gaithersburg, MD).

## B. Colorimetric Detection

There are four main types of detection methods for alkaline phosphatase: *colorimetric, chemiluminescent, bioluminescent,* and *fluorescent*. All of these detection methods have served as reliable and sensitive methods for nonradioactive detection of nucleic acid probes as well as solid-phase immunochemical assays (see Chapters 3,4, and 5). This chapter describes the colorimetric detection methods and thier application to variety of gene detection procedures.

Many chromogenic substrates have been developed for detection of alkaline phosphatase. Depending on the specific use and the format of the assay, different substrates can be used. While most enzyme immunoassays use substrates that result in colored soluble substrate, the filter-based assays and the *in situ* hybridization assays are often performed using substrates that result in a colored precipitate after dephosphorylation.

The substrate systems resulting in a colored soluble end product after dephosphorylation are *para*-nitrophenyl phosphate and the amplified enzyme cascade system based on nicotinamide-adenine dinuclestide phosphate (NADP) (ELISA Amplification System, Life Technologies/BRL; Gaithersburg Maryland). These substrates have been widely used in enzyme immunoassays (Self, 1985; Stanley *et al.*, 1985; Matthews and Kricka, 1988) or solution hybridization assays formatted to resemble immunoassays (Rashtchian *et al.*, 1990).

The most widely used colorimetric substrate system for detection of nucleic acids is 5-bromo-4-chloro-3-indolylphophate (BCIP) (McGadey, 1970). Reaction of this substrate with alkaline phosphatase produces a colored precipitate resulting from dephosphorylation of BCIP and subsequent oxidation with the dye nitroblue tetrazolium (NBT). The insolubility of the dephosphorylated product is ideal for the detection of nucleic acids on membranes and *in situ* hybridization procedures. For this reason, use of NBT/BCIP has become standard for nonradioactive detection of nuclei acids.

## C. Detection of Nucleic Acids on Membranes

Membrane hybridization is one of the most important techniques in molecular biology (Maniatis *et al.*, 1982). The initial report of membrane hybridization in its modern form was by Southern (1975), who transferred DNA from agarose gels to nitrocellulose membranes. Since then a wide variety of methods have been developed for transfer of nucleic acids to membrane filters. These include methods for transfer of RNA (Alwine *et al.*, 1977), and transfer of nucleic acids from bacterial colonies and bacteriophage plaques (Grunstein and Hogness, 1975) to membranes for subsequent analysis by hybridization. In addition, a variety of membranes suitable for hybridization have been developed, and have further advanced the technique of membrane hybridization. These include charged and uncharged nylon membranes as well as supported nitrocellulose. Blotting procedures and properties of membranes have been reviewed by Maniatis *et al.* (1982).

# III. HYBRIDIZATION OF BIOTINYLATED PROBES

Methods for hybridization of biotinylated DNA probes to membrane-bound DNA are essentially the same as those with radioisotope-labeled probes. Although the hybridization kinetics of biotinylated and radio-labeled probes are virtually identical (Langer *et al.*, 1981; Leary *et al.*, 1983), the melting temperature ($T_m$) of biotinylated probes is lower (Leary *et al.*, 1983). The hybridization principles and factors affecting hybridization have been extensively reviewed by Maniatis *et al.* (1982) and by Meinkoth and Wahl (1984). Leary *et al.* (1983) have adapted the hybridization technique of Wahl *et al.* (1979) for use with biotinylated DNA probes. A modification of these procedures is described below for nonradioacive detection using biotinylated probes.

## A. Reagents

*Prehybridization solution*
- 50% formamide
- 5 × SSC
- 5 × Denhardt's solution
- 25 m*M* sodium phosphate, pH 6.5
- 0.5 mg/ml freshly denatured herring sperm DNA

*Hybridization solution*
- 45% formamide
- 5 × SSC
- 1 × Denhardt's solution
- 20 m*M* sodium phosphate, pH 6.5
- 0.2 mg/ml freshly denatured sheared herring sperm DNA
- 5% dextran sulfate
- 0.1 to 0.5 μg/ml freshly denatured probe DNA

*20 × SSC*
- 3 M sodium chloride
- 0.3 M sodium citrate
- pH adjusted to 7.0

*50 × Denhardt's Solution*
- 1% (w/v) Ficoll
- 1% (w/v) polyvinylpyrrolidone
- 1% (w/v) bovine serum albumin, Fraction V

*Denatured sheared herring sperm DNA (10 mg/ml)*

Dissolve 250 mg herring sperm DNA in 25 ml water and allow to mix gently at 4°C overnight. Bring solution to room temperature and add to 50-ml syringe with plunger removed and an 18-gauge needle attached. Insert plunger all the way into syringe to force the DNA through the needle and shear the DNA. Repeat this procedure four times. Transfer the DNA to a glass tube and insert into boiling water bath for 10 min, cool on ice, and repeat. Store DNA in aliquots at −20°C.

## B. Hybridization Protocol for Nitrocellulose Membranes

1. After Southern blotting, bake the filter for 1 to 2 hr at 80°C under vacuum.
2. Soak the filter in 2× SSC until uniformly hydrated.
3. For prehybridization of the filter, denature the sheared herring sperm DNA by heating in a boiling water bath for 10 min followed by chilling on ice. Add the freshly denatured carrier DNA to the remaining prehybridization solution, mix and place in a hybridization bag along with the filter. Incubate the filter at 42°C in the sealed polypropylene bag with prehybridization solution for 2 to 4 hr. The volume of prehybridization solution used should be at least 20 to 100 $\mu$l per $cm^2$ of the filter.
4. For hybridization, heat-denature the probe and the herring sperm DNA for 10 min in a boiling water bath followed by chilling on ice. Add the denatured DNA to the hybridization solution just before use. Remove the prehybridization solution from the bag, and add the hybridization solution to the filter (20 to 100 $\mu$l per $cm^2$), remove air bubbles, and heat-seal bag. At probe concentrations of 100 ng/ml, the filter should be hybridized at 42°C overnight to achieve maximal sensitivity (Langer *et al.*, 1981)
5. The posthybridization washes are performed as follows.
   - Wash filters in 250 ml of 2× SSC/0.1% (w/v) SDS for 3 min at room temperature (20°C). Repeat once.
   - Wash filters in 250 ml of 0.2× SSC/0.1% (w/v) SDS for 3 min at room temperature. Repeat once.
   - Wash filters in 250 ml of 0.16× SSC/0.1% (w/v) SDS for 15 min at 50°C. Repeat once.
   - Briefly rinse filters in 2× SSC at room temperature. At this point the filter is ready for detection of the hybridized probe.

## C. Hybridization Protocol for Nylon Membranes

The protocol described for the detection of DNA on nitrocellulose filters may also be used on nylon filters, but parameters will vary with the type of membrane being used. We have found that with Biodyne A membranes, the addition of 5% dextran sulfate and 0.5% SDS in both the prehybridization and hybridization solutions decreases background and increases the sensitivity of the detection. Also, increased temperatures are required in the final posthybridization wash (e.g., 65°C). Other brands of nylon membranes also give satisfactory results. The sensitivity limits for detection of DNA on nylon membranes are equivalent to those obtained with nitrocellulose membranes, but higher backgrounds are sometimes observed.

## D. Reuse of Hybridization Mixture

The hybridization mixture containing the biotin-labeled probe may be reused. Store the mixture at +4°C for several days or at −20°C for longer periods. The probe should be denatured by placing the hybridization solution in a boiling water bath and cooling on ice just before use.

# IV. DETECTION OF BIOTINYLATED PROBES

The colorimetric detection of biotinylated DNA is achieved by reaction of a streptavidin–alkaline phosphatase conjugate with biotinylated DNA and subsequent detection of alkaline phosphatase with the colormetric substrate system, BCIP and NBT. The following procedure is used for detection of biotinylated probe hybridized to DNA on membrane filters such as Southern or dot blots. This procedure is based on use of the BluGene Nonradioactive Detection Systems (Life Technologies/BRL, Gaithersburg, Maryland).

## A. Reagents

The following reagents are provided with the BluGene Nonradioactive Detection System.

1. Streptavidin–alkaline phosphatase (SA-AP) conjugate (1 mg/ml) in 3 *M* NaCl, 1 m*M* $MgCl_2$, 0.1 m*M* Zn $Cl_2$, 30 m*M* triethanolamine (pH 7.6);

2. NBT (75 mg/ml) in 70% dimethylformamide; and
3. BCIP at a concentration of 50 mg/ml in dimethyl formamide.

The following buffers are required for colorimetric detection using the BluGene Systems and should be prepared by the user before start of the procedure.

| | |
|---|---|
| Buffer 1 | 0.1 *M* Tris-HCl (pH 7.5), 0.15 *M* NaCl. |
| Buffer 2 | 3% (w/v) bovine serum albumin (Fraction V) in Buffer 1 (3 g BSA/100 ml Buffer 1). |
| Buffer 3 | 0.1 *M* Tris-HCl (pH 9.5), 0.1 *M* NaCl, 50 m*M* $MgCl_2$. |
| 20 × SSC | 3.0 *M* NaCl, 0.3 *M* sodium citrate (pH 7.0). |
| 5% (w/v) SDS | SDS, 5 g/100 ml. |

## B. Procedure

**Caution:** Always wear gloves when handling filters and avoid excess pressure on filters. Points of contact will develop higher background color. Always handle the filters at the edges.

1. Wash and/or rehydrate filter for 1 min in Buffer 1.
2. Incubate filter for 1 hr at 65°C in Buffer 2.
   **Note:** If the temperature exceeds 65°C, the BSA in Buffer 2 might gel. This will not affect the sensitivity of the detection. This incubation may be performed in a sealed hybridization bag, or in a suitable tray.
3. If it is desired to store the filters at this stage, follow the steps 4 and 5. Otherwise proceed to step 6.
4. Gently blot filter between two sheets of 3 MM paper and dry the filter in vacuum oven at 80°C for 10 to 20 min.
5. Dried filters may be stored desiccated for months.
6. Thoroughly rehydrate filter in Buffer 2 for 10 min. Drain buffer.
7. In a polypropylene or siliconized glass tube, dilute (immediately before use) an appropriate volume of BRL streptavidin-alkaline phosphatase (SA-AP) conjugate to 1.0 μg/ml by adding 1 μl stock (1.0 mg/ml) solution per 1.0 ml of Buffer 1. Prepare approximately 7.0 ml solution per $100-cm^2$ filter. Incubate filter in diluted SA-AP conjugate for 10 min with gentle agitation, occasionally pipetting solution over filters.
8. Decant the conjugate solution and wash filter with Buffer 1 using at least 20- to 40-fold greater volume of Buffer 1 than employed in

step 7. Gently agitate filter for 15 min in Buffer 1. Decant solution and repeat this wash one more time.

9. Wash the filter once in Buffer 3 for 10 min.
10. In a polypropylene or glass tube, prepare approximately 7.5 ml of dye solution per 100-cm$^2$ filter. The dye solution is prepared by adding 33 $\mu$l NBT solution to 7.5 ml Buffer 3, gently mixing (by inverting the tube) and adding 25 $\mu$l BCIP solution, followed by gentle mixing. *The dye solution should be freshly prepared just before use.*
11. Incubate filter in the dye solution within a sealed hybridization bag or in a flat dish. Allow color development to proceed in the dark or in low light for 30 min to 3 hr. Maximal color development usually is obtained within 3 hr. Longer incubations may result in increased background.
12. Wash filters in 20 m*M* Tris (pH 7.5)/0.5 m*M* $Na_2EDTA$ (ethylenediaminetetraacetic acid) to terminate the color development reaction. Filters should be stored dry and should always be protected from strong light. To dry, bake at 80°C in vacuum oven for 1 to 2 min.
13. Nitrocellulose filters cannot be reused because the dyes, NBT and BCIP, appear to bind irreversibly to the nitrocellulose. The dyes can be removed from nylon filters. A procedure for reprobing has been described by Gebeyehu *et al.* (1987).
14. The bands on the dried nitrocellulose filters may be scanned by laser densitometry; both nylon and nitrocellulose filters may be photographed using a Kodak No. 5 yellow filter, or photocopied. Photocopying may be enhanced by using a yellow or blue transparency (those normally used in overhead projectors) between the filter and the photocopying machine.

## V. *IN SITU* HYBRIDIZATION

*In situ* hybridization has become an important tool in cellular and molecular biology. Conventional filter and solution hybridization methodologies rely on isolation of nucleic acids from a population of cells, thereby averaging the information from each individual cell with the total contribution of all cells. However, *in situ* hybridization provides a tool for studying the molecular information of individual cells within a tissue or cell population. This also allows the investigator to correlate the molecular information with the morphological markers in individual cells. *In situ* hybridization has been used extensively to reveal information regarding expression of particular genes (Lawrence and Singer, 1986), the presence of infectious agents (Hasse, *et al.* 1984), or the location of particular genes in chromosomes (Lawrence *et al.*, 1990).

Both radioactive and nonradioactive probes have been utilized for *in situ* hybridizations, (Singer *et al.*, 1987). Biotinylated probes are the most widely used nonradioactive system. In brief, a biotin-labeled probe is hybridized to target DNA or RNA in cells or tissues *in situ* on a microscope slide. A signaling group (alkaline phosphatase) covalently attached to streptavidin is then bound to the biotinylated probe. The hybridized probe is detected by incubating the samples with dye substrates for alkaline phosphatase, NBT and BCIP. Formation of a purple signal indicates the location of the hybridized probe.

## A. Design and Execution of *in Situ* Hybridization Experiments

The methods described below for use of biotinylated probes and streptavidin–alkaline phosphatase conjugate in *in situ* hybridization is based on a commercially available kit (*in situ* Hybridization and Detection System, Life Technologies/BRL, Gaithersburg, Maryland). A variety of factors affect the quality of results in *in situ* hybridization assays.

### 1. Optimizing Tissue Fixation

Selecting appropriate conditions for fixing tissue is essential. A number of fixatives have been used successfully, including cross-linking agents, such as paraformaldehyde and glutaraldehyde, and precipitating agents, such as ethanol/acetic acid. None of these agents has proved to be universally acceptable. The fixative that works best varies from tissue to tissue, and must be determined empirically. It is also important to determine the optimal length of the fixation step by investigating several time points for each new sample type. It is important to evaluate retention of cellular morphology and level of hybridization signal for each time point. Several recent reviews on *in situ* hybridization (Singer *et al.*, 1986, 1987; Smith, 1987) are available, and provide detail information for optimization of *in situ* hybridization assays.

### 2. Slide Preparation

Retention of cells on the slides requires pretreating the slides with an appropriate agent for increasing cell adhesion. The pretreatment process is known as *subbing* the slides. Samples often are lost from unsubbed slides, particularly during the hybridization step. A number of subbing agents have been used successfully, including gelatin, poly-L-lysine, and 3-aminopropyltriethoxysilane (AES) (Aldrich Chemical Company). Treatment of slides with AES, as outlined below, has generally been found satisfactory.

1. Wash slides in soapy water.
2. Rinse slides thoroughly in tap water.
3. Rinse slides in distilled water.
4. Dry slides in dust-free area.
   **CAUTION:** Before proceeding, be sure that slides are completely dry.
5. In a clean, dry staining dish, dip slides in 2% AES in acetone for 2 min.
6. Rinse slides in two changes of distilled water.
7. Dry slides at 37°C for 2 hr.
8. The treated slides can be stored at room temperature.
9. Handle slides only at the corners.

### 3. Deproteinization

The deproteinization step exposes the nucleic acid target molecules for hybridization. The amount of protein that must be removed depends on the tissue involved, the fixative used, and the target chosen (for example, DNA or RNA, nuclear or cytoplasmic). Excessive removal of protein may also release the target nucleic acids from the slide, especially with mRNA targets. Most *in situ* hybridization protocols use either proteinase K or 0.2 *M* HCl as deproteinizing agents. In most cases, proteinase K is satisfactory. For all deproteinizing agents, it is essential to run a series of test slides at varying digestion times and concentrations of deproteinizing agent. These slides should be evaluated for retention of morphology and hybridization signal. Since the time of digestion is a crucial factor, the digestion should be stopped promptly by rinsing the slides thoroughly in multiple changes of PBS.

### 4. Probe Preparation and Hybridization

A number of schemes have been used for the synthesis of biotinylated probes Forster *et al.*, 1985; Sheldon *et al.*, 1986; Singer *et al.*, 1987). For a biotinylated probe to be suitable for an *in situ* hybridization, it must meet the following criteria: (1) The number of biotin groups must give optimal signal without reducing the hybridization rate; (2) The probe must be small, 20 to 500 bases in length; (3) The linker arm between the probe and the biotin groups must be long enough to allow efficient binding of streptavidin. The BioNick Labeling System (Life Technologies/BRL, Gaithersburg, Maryland) has been optimized to meet these criteria.

Contaminants and unincorporated dNTPs in the probe preparation can cause high backgrounds. Purification of biotinylated probes by gel filtration on Bio-Gel P60 (from Bio-Rad Laboratories) in 10 m*M* Tris-HCl (pH 7.5), 1 m*M* EDTA is recommended.

For interpretation of results, proper hybridization temperature is critical. If the temperature is too low, the probe will hybridize nonspecifically, which will increase the background. On the other hand, too high hybridization temperatures will decrease the signal, and may cause loss of tissue morphology. Good results generally can be obtained with the recommended hybridization solutions (see below) at 42°C. For particular applications, the optimal hybridization temperature may vary from 37°C to 50°C.

## 5. Detection

Since the *in situ* methodology described here utilizes alkaline phosphatase as the signaling group, endogenous phosphatase can mask the hybridization signal. Certain tissues are high in endogenous phosphatase activity; it is, therefore, important to measure any endogenous activity with a control slide without the streptavidin–alkaline phosphatase conjugate. Endogenous phosphatase activity is reduced by baking the slide before hybridization (see Protocol for DNA Detection, Section V, C). If any phosphatase remains, it can usually be inhibited by adding 200 $\mu$g/ml levamisole to the alkaline-substrate solution. At that concentration, levamisole is a potent inhibitor of many endogenous alkaline phosphatases, but it has no effect on the streptavidin–alkaline phosphatase conjugate.

## 6. Counterstaining

Slides may be counterstained with nuclear fast red, methyl green, or acridine orange. This allows negative cells to be visualized and aids in identification of particular cell types.

## B. Reagents

The following reagents are provided with the *in situ* Hybridization and Detection System (Life Technologies/BRL, Gaithersburg, Maryland).

| | |
|---|---|
| 2× Hybridization buffer | 4× SSC, 0.2 *M* sodium phosphate (pH 6.5), 2× Denhardt's solution |
| 20% Dextran sulfate | 20% (w/v) dextran sulfate in formamide |
| Blocking solution | 50 mg/ml protein in 100 m*M* Tris-HCl (pH 7.8), 150 m*M* NaCl |
| Streptavidin–alkaline phosphatase conjugate | 40 μg/ml in conjugate dilution buffer |
| Conjugate dilution buffer | 100 m*M* Tris-HCl, 150 m*M* $MgCl_2$, 10 mg/ml bovine serum albumin |
| Nitroblue tetrazolium (NBT) | 75 mg/ml NBT in 70% (v/v) dimethylformamide |
| 4-bromo-5-chloro-3-indolylphosphate (BCIP) | 50 mg/ml in 100% dimethylformamide |
| Control slides | adenovirus type 2-infected HeLa cells, paraffin-embedded |
| Positive control probe | biotinylated adenovirus type 2 DNA in 2× hybridization buffer |

The following reagents or materials are not included and need to be obtained before start of experiments.

Treated microscope slides
Coplin jars or staining dishes
Xylene
2X SSC (300 m*M* NaCl, 30 m*M* sodium citrate, pH 7.0)
Phosphate-buffered saline (PBS) (137 m*M* NaCl, 2.7 m*M* KCl, 8.1 m*M* $Na_2HPO_4$, 1.5 m*M* $KH_2PO_4$, pH 7.4)
40 μg/ml proteinase K
4% (w/v) paraformaldehyde in PBS
50% (v/v) ethanol
70% (v/v) ethanol
90% (v/v) ethanol
100% ethanol
Biotinylated DNA or RNA probe

Heating block set at 100°C
Rubber cement
Glass coverslips
Hybridization oven, adjustable from 37°C to 65°C
Tris-buffered saline (TBS) (100 m*M* Tris base, 150 m*M* NaCl, pH 7.5)
Alkaline substrate buffer (100 m*M* Tris base, 150 m*M* $NaCl_2$, pH 9.5)
Mounting medium

## C. Protocol for DNA Detection

The protocol, has been optimized for the detection of DNA in the positive control slides provided with the *in situ* hybridization and detection kit. This general method may need further optimization in the pretreatment and hybridization steps.

*Pretreatment*

1. Label the side of the slide containing the tissue section with a diamond pen or other permanent marker. For the initial experiment, include three control slides.
2. Bake the slides, including the control slides, at 65°C for 1 hr to firmly bind the sections to the slide and to melt the paraffin of paraffin-embedded sections. If sections are not paraffin-embedded, go directly to step 4.
3. For the control slides and other paraffin-embedded sections, deparaffinize in two changes of xylene for 5 min each. Immerse slides in two changes of absolute ethanol for 1 min each. Air dry sections for 5 to 10 min.
4. If your tissue contains mRNA complementary to your probe, and hybridization to mRNA is not desired, the RNA may be removed by incubating slides in 40μg/ml RNase A in 2X SSC at 37°C for 1 hr. To prevent contamination of glassware with RNase, mark the container used for this step and use it solely for this purpose. Use sufficient RNase A solution to cover the slides. Rinse slides in PBS.
5. Incubate slides in prewarmed 40 μg/ml proteinase K in PBS at 37°C, making certain that the slides are completely covered. For the control slides, remove one slide after 5 min incubation, one after 10 min incubation, and one after 20 min incubation. Rinse each slide in PBS immediately after removing it from the proteinase K solution.

6. Immerse slides in fresh 4% paraformaldehyde in PBS at room temperature for 1 min. Rinse slides in PBS.
7. Dehydrate sections through a graded ethanol series (3 min each in 50%, 70%, 90%, and 100% ethanol). Air dry at room temperature for 5 to 10 min.

*Hybridization*

1. Probe DNA should be precipitated by ethanol and dissolved at twice the desired final concentration in 2X hybridization buffer. Final probe concentrations should be between 0.1 and 0.5 μg/ml. Mix 25 μl of this solution with 25 μl of 20% dextran sulfate solution for each slide to be probed.
2. Add 50 μl of probe solution to the appropriate section and cover with a coverslip. Be careful not to trap bubbles under the coverslip.
3. Denature the target and probe DNA by placing the slides flat on a heating block prewarmed to 100°C. Incubate for 5 min.
4. Slides may be hybridized from 3 hr to overnight. For overnight hybridization, seal the slide and coverslip with rubber cement by dripping rubber cement around the edge of the coverslip, being careful not to touch or jar the coverslip. Allow the rubber cement to dry for a few minutes before starting the incubation. Transfer the slides to a humid chamber and incubate at 42°C.
5. After hybridization, carefully remove rubber cement with forceps. Remove the coverslips and rinse the slides by immersing them in three changes of 0.2X SSC, using a minimum of 10 ml per slide. Be careful to avoid damaging the tissue section with the coverslip.
6. Wash the slides two times with 0.2X SSC, using a minimum of 10 ml per slide at room temperature for 15 min each.

*Detection*

1. Place the slides flat in the humid chamber. Cover each section with 100 to 300 μl of blocking solution. Incubate at room temperature for 15 min.
2. Prepare a working conjugate solution by mixing 10 μl of streptavidin–alkaline phosphatase conjugate with 90 μl of conjugate dilution buffer for each slide.
3. Remove the blocking solution from each slide by touching absorbent paper to the edge of the slide.
4. Cover each section with 100 μl of working conjugate solution and incubate it in the humid chamber at room temperature for 15 min. **NOTE:** Do not allow the sections to dry out after adding the conjugate.

5. Wash the slides twice in TBS for 15 min each at room temperature using a minimum of 10 ml per slide.
6. Wash slides once in alkaline-substrate buffer at room temperature for 5 min using a minimum of 10 ml per slide.
7. Prewarm 50 ml alkaline-substrate buffer to 37°C in a Coplin jar. Just before adding the slides, add 200 μl NBT and 166 μl BCIP. Mix well.
8. Incubate slides in the NBT/BCIP solution at 37°C until the desired level of signal is achieved (from 10 min to 2 hr). Check the color development periodically by removing a slide from the NBT/BCIP solution, covering the section and residual solution with a coverslip, and observing the section under the microscope. Be careful not to allow the sections to dry.
9. Stop the color development by rinsing the slides in several changes of deionized water. The sections may be counterstained at this point.
10. Dehydrate the sections through a graded ethanol series (50%, 70%, 90%, 100%) for 1 min in each concentration. Air dry at room temperature.
11. For permanent mounting, use a nonxylene medium such as Crystalmount available from Biomeda Corp. (Foster City, California).

## D. Detection of RNA Targets

Detection of RNA targets *in situ* requires optimizing the fixation, pretreatment, and hybridization steps specifically for RNA. Pretreatments that are optimal for exposure of DNA targets may be too harsh and may result in loss of RNA from the sample (Singer *et al.*, 1986). Shortening or eliminating the proteinase K step, using an alternate pretreatment, or using a different fixation method may be necessary for optimal exposure and retention of the RNA target (see Singer *et al.*, 1986, 1987).

Steps must be taken to protect the samples from RNases. The following general precautions should be observed when working with RNA.

Wear latex gloves at all times, including during preparatory steps, such as preparing solutions and washing glassware.

Practice good microbiological technique to prevent microbial contamination of reagents and utensils.

Treat solutions used before the hybridization step with 0.1% diethylpyrocarbonate (DEPC). Add the DEPC to the solution and leave at room temperature for 12 hr. Destroy the remaining DEPC by autoclaving for 15 min.

Bake glassware at 200°C for 8 hr before using (sterile disposable plasticware is generally RNase-free and can be used without treatment).

Proteinase K must be nuclease-free: predigest the proteinase K by incubating at 37°C for 40 min before using.

*Hybridization and Detection*

The hybridization protocol described for DNA target can be used for RNA targets with the following modifications.

1. Prepare the probe solution by ethanol precipitating and dissolving at 2.5 times the desired final concentration in 2X hybridization buffer. Final probe concentrations should range from 0.1 to 0.5 $\mu$g/ml. Mix 25 $\mu$l of this solution with 31 $\mu$l of 20% dextran sulfate solution for each slide to be probed. Denature the probe by boiling for 5 to 10 min. Chill on ice. Add 6$\mu$l of 200 m$M$ vanadyl ribonucleoside complex for each slide to be probed.
2. When detecting RNA targets, only the probe must be denatured, and slides are not placed on the 100°C heat block. All other steps in the hybridization and detection are identical to DNA detection.

## VI. CONCLUSIONS

Since nonradioactive methods for the detection of nucleic acids were described, a variety of labeling systems have been developed. Labeling of DNA with biotin still remains to be the major nonradioactive method in the research laboratory. Colorimetric detection of biotin-labeled probes is most often accomplished with use of alkaline phosphatase conjugates with NBT/BCIP. This technology is an effective nonradioactive method for most molecular biology applications. A chemiluminescent method for detection of biotinylated probes has been described, and commercial products based on this detection are available (PhotoGene, Life Technologies/BRL). However, the colorimetric method remains the only nonradioactive detection method suitable for *in situ* hybridization applications. Considerable progress has been made in sensitivity and simplicity of nonradioactive detection technology, and this trend is expected to continue. These improvements have made nonradioactive detection technology a good substitute for the isotopic detection methods used in research and clinical laboratories.

# ACKNOWLEDGMENTS

I would like to thank all my colleagues, especially Lenny Klevan, Dave Carlson, Mary Gaskill, and Gulilat Gebeyehu. I thank Jesse Mackey for constructive suggestions about the manuscript and Kathleen Blair for typing the manuscript.

# REFERENCES

Alwine, J. D., Kemp, D. J., and Stark, G. R. (1977). Methods for detection of specific RNAs in agarose gels by transfer to diazobenzyloxymethly paper and hybridization with DNA probes. *Proc. Natl. Acad. Sci. U.S.A.* **74,** 5350–5354.

Brenner, D. J. (1989). DNA hybridization for characterization, classification and identification of bacteria. *In* (B. Swaminathon and G. Prakash, (eds.).pp. 75–104.'' Nucleic Acid and Monoclonal Antibody Probes, Applications in Diagnostic Microbiology.'' Marcel Dekker, New York.

Feinberg, A. P., and Vogelstein, B. (1983). A technique for radiolabeling DNA restriction endonuclease fragments to high specific activity. *Anal. Biochem.* **132,** 6–13. (Addendum *Anal. Biochem.* **137,** 266–267.

Forster, A. C., McInnes, J. L., Skingle, D. C., and Symona, R. H. (1985). Non-radioactive hybridization probes prepared by the chemical labeling of DNA and RNA with a novel reagent, photobiotin. *Nucleic Acids Res.* **13,** 745–761.

Gebeyehu, G., Rao, P. Y., SooChan, P., Simms, D. A., and Klevan, L. (1987). Novel biotinylated nucleotide—analogs for labelling and calorimetric detection of DNA. *Nucleic Acids Res.* **15,** 4513–4534.

Grunstein, M., and Hogness, D. (1975). A method for the isolation of cloned DNAs that contain a specific gene. *Proc. Natl. Acad. Sci. U.S.* **72,** 3961–3966.

Hasse, A. T., Brakic, M., Stowring, L., and Blum, H., (1984). Detection of viral nucleic acids by *in situ* hybridization. *Methods Virol.* **7,** 189–226.

Jablonski, E., Moomaw, E. W., Tullis, R. H., and Ruth J. L. (1986). Preparation of oligodeoxynucleotide-alkaline phosphatase conjugates and their use as hybridization probes. *Nucleic Acids Res.* **14,** 6114–6128.

Langer, P., Waldrop, A., and Ward S. (1981). Enzymatic synthesis of biotin-labeled polynucleotides: Novel nucleic acid affinity probes. *Proc. Natl. Acad. Sci. U.S.A.* **18,** 6633–6637.

Lawrence, J. B., Singer, R. H., and McNeil, J. A. (1990). Interphase and metaphase resolution of different distances within the human dystrophin gene. *Science* **249,** 928–932.

Lawrence, J. B., and Singer, R. H. (1986). Intracellular localization of messenger RNA for cytoskeletal proteins. *Cell* **45,** 407–415.

Leary, J. J., Brigati, D. J., and Ward D. C. (1983). Rapid and sensitive colorimetric method for visualizing biotin-labeled DNA probes hybridized to DNA or RNA immobilized on nitrocellulose: Bio-blots. *Proc. Natl. Acad. Sci. U.S.A.* **80,** 4045–4049.

Maniatis, T., Fritsch, E. F., and Sambrook, J. (1982). "Molecular Cloning: A Laboratory Manual." Cold Spring Harbor Laboratory, Cold Spring Harbor, New York.

Matthews, J. A., and Kricka, L. J. (1988). Analytical strategies for the use of DNA probes. *Anal. Biochem.* **169,** 1–25.

McGadley, J. (1970). A tetrazolium method for non-specific alkaline phosphatase. *Histochemie* **23,** 180–184.

Meinkoth, J., and Wahl, G. (1984). Hybridization of nucleic acids immobilized on solid supports. *Biochemistry* **138,** 267–284.

Rashtchian, A., Schuster, D., Buchman, G., Berninger, M., and Temple, G. F. (1990). A non-radioactive sandwich hybridization assay using paramagnetic microbeads and unlabeled RNA probes. *Abstracts of 6th International Congress on Rapid Methods and Automation in Microbiology and Immunology*. Helsinki, Finland.

Renz M., and Kurz, C. (1984). A colorimetric method for DNA hybridization. *Nucleic Acids Res.* **12,** 3435–3444.

Self, C. H. (1985). Enzyme amplification, a general method applied to provide an immunoassisted assay for placental alkaline phosphatase. *J. Immunol. Methods*. **76,** 389–393.

Sheldon, E. L., Kellogg, D. E., Watson, R., Levenson, C. H., and Erlich, H. A. (1986). Use of nonisotopic M13 probes for genetic analysis: Application to HLA Class II loci. *Proc. Natl. Acad. Sci. U.S.A.* **83,** 9085–9089.

Singer, R. H., Lawrence, J. B., and Rashtchian, R. N. (1987). Toward a rapid and sensitive *in situ* hybridization methodology using isotopic and nonisotopic probes. *In*[11] *In Situ* Hybridization: Application to the Central Nervous System[11] (K. Valentino, J. Eberwine, and J. Barchas, eds.), pp. 71–96. Oxford University Press, New York.

Singer, R. H., Lawrence, J. B., and Villnave, C. (1986). Optimization of *in situ* hybridization using isotopic and non-isotopic detection methods. *BioTechniques* **4,** 230–250.

Smith, G. H., (1987). *In situ* detection of transcription in transfected cells using biotin-labeled molecular probes. *Methods Enzymol.,* **151,** 530–539.

Southern, E. M. (1975). Detection of specific sequences among DNA fragments separated by gel electrophoresis. *J. Mol. Biol.* **98,** 503–517.

Stanley, C. J., Johannson, A., and Self, C. H. (1985). Enzyme amplification can enhance both the speed and the sensitivity of immunoassays. *J. Immunol. Methods*. **83,** 89–95.

Wahl, G. M., Stern, M., and Stark, G. R. (1979). Efficient transfer of large DNA fragments from agarose gels to diazobenzloxymethyl paper and rapid hybridization by using dextran sulfate. *Proc. Natl. Acad. Sci. U.S.A.* **76,** 3683–3687.

# 7 Detection of Horseradish Peroxidase by Enhanced Chemiluminescence

Ian Durrant
Research and Development
Amersham International Place
Amersham, Buckinghamshire, England



## I. INTRODUCTION

This chapter describes the use of horseradish peroxidase (HRP)-catalyzed chemiluminescence as a detection procedure for molecular biology (Anon, 1990). To fully understand and capitalize on the advantages this detection method can offer, it is necessary also to consider the ways that HRP could be introduced into the experimental system.

The labeling of nucleic acid probes with haptens, such as digoxigenin and biotin, has been extensively reviewed (Pollard-Knight, 1990). After

*Nonisotopic DNA Probe Techniques*

hybridization these molecules are detected by use of immunochemical techniques. It is possible to envisage a system in which the final detection step involves the use of an HRP conjugate, which can in turn be detected using enhanced chemiluminescence (ECL).

However, an alternative procedure to reach this HRP/ECL endpoint involves the labeling of nucleic acid probes directly with HRP; this procedure will be described briefly here. The ECL detection reaction will be discussed in detail; the procedure is equally applicable to the detection of (1) directly labeled probes, (2) antihapten–HRP conjugates, and (3) antibody–HRP conjugates, as in Western blotting.

Nucleic acid probes can be divided into two broad classes: oligonucleotides (usually $<50$ bases in length) and long probes ($>50$ bases, but optimally $>300$ bases). Long probes can be prepared from single-stranded or double-stranded DNA, or RNA. The two classes of probes (oligonucleotide and long) require different labeling strategies and will be dealt with separately for the labeling and hybridization procedures.

## A. Labeling and Hybridization of Nucleic Acid Probes

### 1. Long Probes

The direct labeling of long probes, based on the method first described by Renz (Renz and Kurz, 1984), involves the labeling of single-stranded DNA (or RNA) with positive-charge modified HRP (Fig. 1). The derivatized HRP is prepared by cross-linking it to polyethyleneimine (PEI) via the cross-linking agent *para*-benzoquinone (PBQ) (Pollard-Knight *et al.*, 1990). The positive charges on the HRP complex interact electrostatically with the negative charges on the DNA phosphate backbone. This complex is then stabilized by the formation of covalent bonds through the action of glutaraldehyde (a short-range cross-linking agent). The cross-links are probably introduced between amino groups on PEI and the bases of the nucleic acid. The probes generated have approximately one complex for each 50 bases; as each complex consists of one, two, or three HRP molecules, the average distribution of HRP is approximately one every 30 bases. This level of attachment is similar to the optimal spacing of hapten nucleotides, and it does not significantly affect the melting temperature ($T_m$) or the hybridization stringency associated with a particular probe.

For probes labeled directly with HRP, one of the major factors influencing the hybridization and posthybridization conditions is the need to maintain enzyme activity. Traditionally, hybridizations have been performed at 42°C using 50% formamide as a denaturant, which effectively reduces the

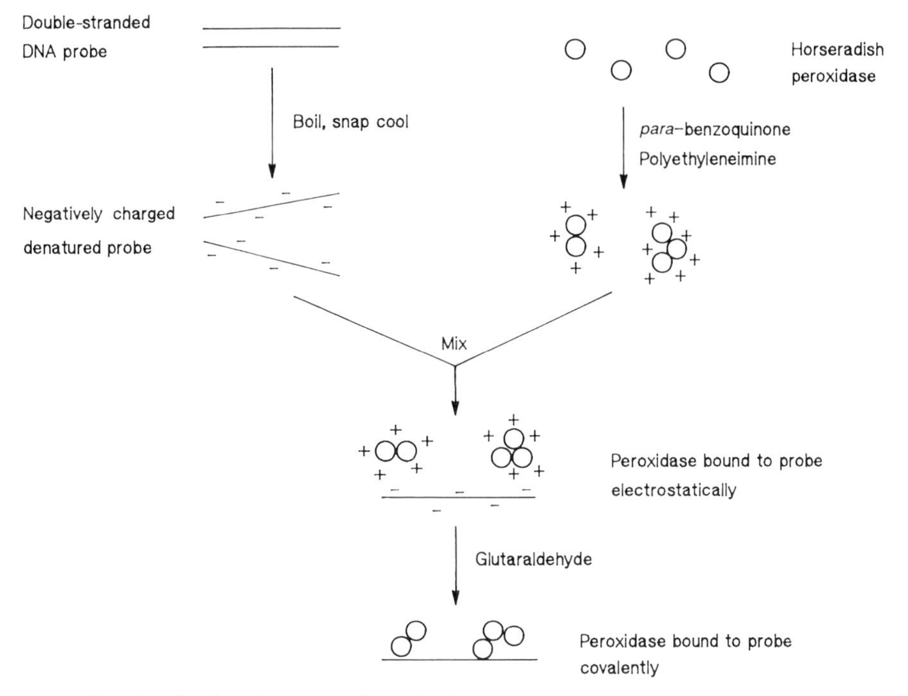


**Fig. 1** Outline of the procedures for the preparation of HRP-labeled probes.

$T_m$ of hybrids. However, 50% formamide was found to denature HRP during simulated hybridization incubations. A buffer has been developed based on 6 *M* urea as an alternative to formamide (Hutton, 1977). The $T_m$ of a typical long probe is reduced to around 67°C in this buffer. It is calculated that the optimal rate of hybridization occurs at a temperature 25°C lower than the $T_m$ (Britten and Davidson, 1985); for the urea-based buffer, this optimal temperature is 42°C. At this temperature the enzyme remains largely intact, even during an overnight incubation. The stability of the enzyme is aided by the inclusion of a protein-stabilizing agent in the buffer. Additional buffer contents include a rate enhancer (volume-excluding agent) and a proteinaceous membrane-blocking agent (especially important for blocking the binding of HRP to nylon membranes). The user defines the level of sodium chloride required to control the stringency of individual hybridizations.

The stringency of the system is best controlled, however, during the posthybridization wash steps. It must be appreciated that the temperature of these washes, as for the hybridizations, should not exceed 42°C. Consequently, urea is also included in the wash buffer, and the stringency is controlled by alteration of the sodium chloride concentration.

The probes are equally applicable to hybridizations on both nylon and nitrocellulose membranes. The sensitivity achievable is higher on nylon owing to the increased target-retention properties of this type of membrane.

### 2. Oligonucleotide Probes

In order to label an oligonucleotide with HRP, it is first necessary to synthesize an oligonucleotide derivatized with a chemically reactive group. The most convenient method is to introduce a thiol group at the 5′-end of the oligonucleotide (Connolly, 1985). The enzyme, in turn, is derivatized at a surface amine group using a bifunctional cross-linking agent. The free end of this agent reacts specifically with free thiol groups. The reaction is further simplified by presentation of the derivatized enzyme in a freeze-dried format. The enzyme is redissolved in the thiol-linked oligonucleotide such that there is a 5:1 molar ratio of HRP to oligonucleotide. This ensures a coupling efficiency of greater than 90% based on the oligonucleotide (Fowler *et al.*, 1990). The free enzyme does not interfere with subsequent hybridizations.

The maintenance of enzyme activity during hybridization is an important consideration. However, as oligonucleotide probes can be added at a greater molar concentration than can long probes, hybridization times can be reduced to as little as 1 hr. The hybridization is performed at 42°C, but the buffer is of a much more routine formulation than that required for long probes. The stringency is controlled during the posthybridization washes, by using any of the standard techniques such as altering the sodium chloride concentration and/or changing the temperature of the wash. The $T_m$ of an oligonucleotide labeled with HRP is not significantly different from that of the unlabeled sequence (Fowler *et al.*, 1990).

## B. Immunochemical Techniques

The labeling of antibodies with HRP has been reviewed in detail elsewhere (Hashida *et al.*, 1984). The detection of antibodies labeled with HRP, on Western blots by the use of the ECL reaction, represents probably the most sensitive and rapid technique currently available (Leong and Fox, 1988; Vachereau, 1989). Streptavidin–HRP conjugates may also be used. The level of sensitivity achievable reveals target antigens that other systems may not detect, and target antigens can be detected in far lower amounts of material than was previously possible. Western blots are usually performed on nitrocellulose membranes.

## C. Detection

The HRP-catalyzed oxidation of luminol to produce light is the basis for detection of the directly labeled probes (Durrant, 1990) (see Table I). However, the system is generic in that it can be applied without modification to the detection of antigens in immunoassays (Thorpe and Kricka, 1989) and to the detection of antigens on Western blots using HRP-linked antibodies or streptavidin as an end-point (Anon, 1990).

Oxidative degradation of various compounds leading to the emission of light has been reported for many years, and the luminol reaction was one of the first to be described (Roswell and White, 1978). It became clear that HRP could be used to drive this reaction by linking the oxidation to the action of HRP on peroxide (Schroeder *et al.*, 1978). However, the light output from this reaction was low and maintained for only a very short period. This made it unsuitable for general application. In 1983 it was reported (Whitehead *et al.*, 1983) that certain compounds could enhance the level and duration of light output, and that this improved system could be used to monitor immunoassays.

This work stimulated interest in the mechanism of HRP-catalyzed degradation of luminol, and searches were undertaken for improved enhancer molecules. A variety of such compounds have been identified, with the most successful being based on substituted phenol structures (Thorpe *et al.*, 1985). These are capable of stimulating the light output by >1000-fold and maintaining a significant level of light output for a number of hours.

The mechanism of the unenhanced reaction is generally accepted to follow the pathway illustrated in Fig. 2 (Thorpe and Kricka, 1987). Basically, the Fe(III) ion of the heme group at the center of HRP is oxidized by the degradation of peroxide to form HRP-I [Fe (V)]. The catalyst is re-formed in a two-step process, with an electron being gained at each

**Table I**
Suitability of ECL for Detection of Nucleic Acid Probes[a]

| Technique | Long probes | Oligonucleotides |
|---|---|---|
| Dot–slot blot | ++ | +++ |
| Northern blot | + | − |
| Southern blot | +++ | ++ |
| Sequencing gel | − | + |
| Colony screen | +++ | +++ |
| Plaque screen | +++ | +++ |
| RFLP analysis | +++ | +++ |
| PCR products | ++ | +++ |

[a] Probes directly labeled with horseradish peroxidase. ECL, enhanced chemiluminescence.

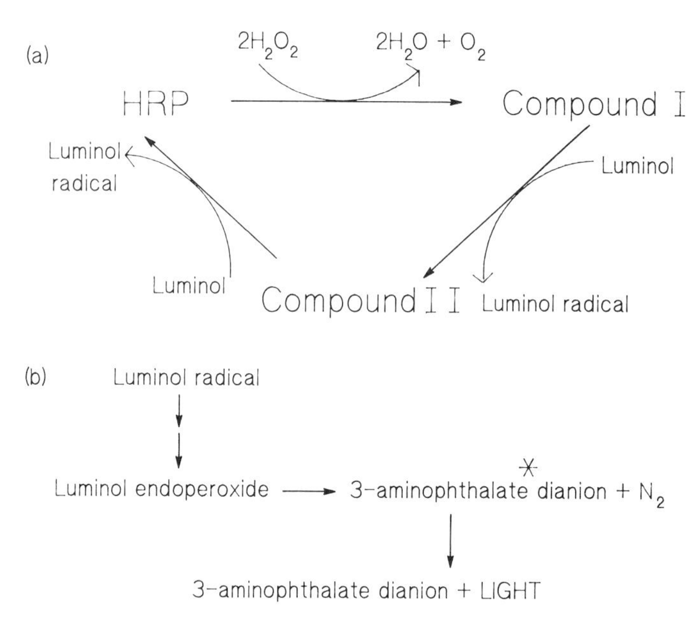


**Fig. 2** Proposed reaction scheme for the luminol-based enhanced chemiluminescent detection of horseradish peroxidase.

stage. These electrons are donated by luminol, and a luminol radical is formed at each stage. In this cycle, the rate-limiting step is the final one leading to the re-formation of the original enzyme. Consequently, the light output is low and quickly dies away because of the loss of active HRP and competing nonlight-producing reactions. In the enhanced reaction, the proposed scheme has the enhancer as a much more active electron donor, thus removing the rate-limiting step (Hodgson and Jones, 1989). The enhancer radical interacts with luminol to form luminol radicals so that the light-producing pathway from luminol is precisely the same for the enhanced and unenhanced reactions. This is confirmed by the fact that the quantum efficiency of the reaction is unchanged (the quantity of light emitted for each luminol degraded) and that the emission spectrum is identical in both cases (Thorpe and Kricka, 1986). The luminol radical undergoes a series of as yet undefined reactions that culminate in the formation of luminol endoperoxide. This compound degrades to release nitrogen and form a chemically excited form of 3-aminophthalate dianion. As this dianion falls back to the ground state, light is emitted.

In practice the peroxide is supplied as a peracid salt, and this salt and the enhancer/luminol mixture have to be kept separate until immediately before use, to avoid gradual chemical degradation of the substrates. Individually, the two substrate formulations can be stored at 4°C for up to 9 months.

In solution, the light output peaks within 1 min and then remains relatively steady for up to 20 min (Thorpe and Kricka, 1987), making it extremely convenient for immunoassay end-point determinations. On membranes, the light output increases rapidly over the first 2–3 min and finally peaks around 10 to 15 min after exposure of the substrates to HRP (Stone and Durrant, 1990). The light level decays with a half-life of approximately 1 hr. After 4 to 5 hr the light level decreases beyond detection. The cause of the decay in light output is uncertain but is most likely attributable to gradual loss of enzyme function due to a build-up of free radical damage. The quantity of light available during the first 2 hr gives a sensitivity of detection suitable for standard molecular biology application (Durrant *et al.*, 1990). For Western blots, typical exposure times may be only 1–10 min; high levels of nucleic acid target can be detected in 10 to 30 min, whereas low target levels may require exposure times of up to 60 min.

The loss of detectable signal after 4 to 5 hr simplifies the task of reprobing nucleic acid blots. There is no need to strip the previous probe from the membrane before rehybridizing with another probe (or even the same probe) (Durrant *et al.*, 1990). During ECL detection the first probe will not be visualized, as the enzyme label has been inactivated by the first detection process. For Western blots, the redetection with a second antibody can be done relatively quickly (the fist signal may still be detectable). If the position of this signal interferes with the detection of subsequent antigens, then it is possible to remove the first antibody system without damaging the target proteins (Kaufmann *et al.*, 1987).

The light output from the luminol reaction has a maximum wavelength of 428 nm. This makes it ideal for capture on standard X-ray film, which is designed to detect the blue light emitted by intensifying screens. Consequently, the system is characterized by producing a hard copy of the result, which can be quantified by densitometry.

# II. MATERIALS

## A. Reagents

The ECL gene-detection system (Amersham; RPN 3000/3001) contains the labeling reagent for the introduction of HRP onto long probes, hybridization buffer, and ECL detection reagents.

The ECL oligonucleotide labeling and detection system (Amersham; RPN 2113) contains the derivatized, freeze-dried HRP and ECL detection reagents. A thiol modifier reagent, for the derivatization of oligonucleotides, is also available from Amersham (RPN 2112); similar compounds are available elsewhere (i.e., Clontech and Cruachem).

The ECL detection reagents are available either separately (RPN 2105), in conjunction with Western blotting protocols (RPN 2106), or for immunasssay detection (RPN 190).

All the systems above are shipped at ambient temperature. The labeling and detection reagents, for whichever application, are stored at 4°C. Hybridization buffer for long probes is best stored aliquotted at −20°C. The user has to add the desired quantity of NaCl (typically 0.5 *M*) and membrane-blocking agent (for nylon blots).

The luminol used in preparation of the ECL detection reagents can be obtained commercially, but is specially purified for optimal results (Stott and Kricka, 1987).

The ECL systems have been optimized for use on Hybond membranes (Amersham), in particular Hybond-$N^+$ (nylon) and Hybond-ECL (nitrocellulose), but are compatible with some of the nylon and nitrocellulose membranes available from other sources.

The capture of the light emitted by the ECL reaction is most conveniently performed on X-ray film. The best results will be obtained with Hyperfilm-ECL (Amersham), but it is possible to use Hyperfilm-MP or Kodak-AR films. Blue-tinted films may not yield the maximal sensitivity of the system. Saran Wrap (Dow Chemical Company) is used to wrap membranes during exposure so that the detection reagents do not contact the film. This type of cling-wrap is especially suited for use in molecular biology owing to its ability to transmit light (including UV light). Other cling-wrap materials may be usable but require testing first. Saran Wrap is often available from local distributors.

## B. Instrumentation

As stated previously it is best to capture the light emitted by the ECL reaction on X-ray film. The recommended films can all be processed manually or, more conveniently, by using a commercial, X-ray film processing machine.

It is also possible to obtain images on a number of other systems. Polaroid film can be used in the format of a camera luminometer, particularly suitable for the detection of Western blots with a high level of signal. Camera luminometers are available from a number of sources. Alternatively, a countercurrent distribution (CCD) camera can be used; this is a cooled charge-coupled device that collects, records, and quantifies emitted photons. A number of such cameras, in a variety of formats, are available commercially. However, none is particularly dedicated to the detection of blots; they all require some adaptation by the user (Boniszewski *et al.*, 1990)

The thiol-linked oligonucleotides can be synthesized on any of the commercially available DNA synthesis machines (for example, those supplied by Applied Biosystems or Pharmacia). The thiol-modifier reagent can be used in conjunction with the standard small-scale synthesis program installed on such machines.

# III. PROCEDURES

The labeling and hybridization conditions, associated with either long probes or oligonucleotide probes, require different technologies; these methods will be dealt with individually. The procedures described in this section can be used for all the applications for which the probes are suitable; specific application-dependent variations will be discussed separately.

## A. Long Probes

*Labeling*

1. Dilute the sample of nucleic acid to be labeled to 10 ng/μl in sterile, distilled water. To obtain efficient labeling, the monovalent cation concentration should be less than 10 m*M*. Proceed to step 4 for single-stranded DNA or RNA samples.
2. Boil double-stranded DNA for 5 min. The boiling water bath should be vigorously boiling throughout; 95°C is not adequate to ensure complete denaturation; neither is the use of a heating block.
3. Cool on ice for 5 min.
4. Add an equal volume of labeling reagent. Mix gently by drawing up and down in the pipette tip.
5. Add a volume of 1.5% (v/v) glutaraldehyde equivalent to that of the labeling reagent used in the previous step.
6. Incubate for 10 min at 37°C.
7. Store the labeled probe on ice if to be used immediately. Store in 50% glycerol at −20°C for long-term storage (up to 12 months). Do not boil labeled probes before use.

*Hybridization*

1. Place blots in 0.25 ml hybridization buffer/$cm^2$ of membrane. Remember to include NaCl (0.5 *M* is suitable for most applications and gives the maximal rate of hybridization) and blocking agent

(for nylon membranes) in the hybridization buffer. Lower volumes can be used, particularly for hybridizations in bags or ovens or when a number of filters are hybridized together. The basic criterion is that each blot should move freely within the buffer.
2. Prehybridize for a minimum of 15 min at 42°C, in a shaking water bath.
3. Add labeled probe to the prehybridization buffer at a final concentration of 10 ng/ml (nylon) or 20 ng/ml (nitrocellulose). Avoid placing the probe directly onto the filter.
4. Incubate overnight at 42°C, in a shaking water bath.

*Stringency washes*

1. Place filters into 2 ml primary wash buffer/$cm^2$ of membrane. Incubate, in a shaking water bath, at 42°C for 20 min. Primary wash buffer controls the stringency. The basic recipe is 6 *M* urea, 0.5× standard saline citrate (SSC), 0.4% sodium dodecyl sulfate (SDS) and will be suitable for most applications; the SSC concentration can be altered to vary the stringency.
2. Replace the wash solution with fresh buffer and repeat the wash step.
3. Place filters in an excess volume of 2× SSC and incubate at room temperature, with agitation, for 5 min.
4. Replace the solution with fresh 2× SSC and repeat the wash step.
5. Blots can be stored in this buffer for up to 60 min before proceeding to the detection steps. For longer-term storage wrap the blots, still damp, in Saran Wrap and store at 4°C.

## B. Oligonucleotides

*Labeling*

1. Desalt 5 μg (100 μl) of thiol-linked oligonucleotide on a spun column (or similar) to remove the dithiothreitol present in the oligonucleotide storage buffer. Use the desalted oligonucleotide within 5 min to avoid dimerization of the thiol groups.
2. Add all 5 μg of the thiol-linked oligonucleotide to a tube containing derivatized, freeze-dried HRP. Redissolve the enzyme, and mix the reactants, by gently pipetting up and down in the pipette tip.
3. Incubate for 1 hr at room temperature or overnight at 4°C.
4. Dilute the labeled probe to an appropriate working concentration (for example, 10 ng/μl) in 0.1 *M* sodium phosphate, pH 7.0, and store on ice for immediate use. For long-term storage, add an equal volume of glycerol and place at −20°C (for up to 9 months).

*Hybridization*

1. Place blots in hybridization buffer (0.25 ml/$cm^2$ of membrane).
2. Prehybridize for a minimum of 15 min at 42°C, in a shaking water bath.
3. Add labeled probe to the prehybridization solution to a final concentration of 20 ng/ml. Avoid placing the probe directly onto the membrane.
4. Hybridize for 1 hr at 42°C, in a shaking water bath.

*Stringency washes*

1. Place blots in primary wash buffer (2 ml/$cm^2$ of membrane). Primary wash buffer controls the stringency and consists of a selected level of SSC and 0.1% SDS. Incubate at the required stringency for 15 min in a shaking water bath. The precise conditions for the stringency washes will be dictated by the $T_m$ of the probe and the presence or absence of sequences closely related to the specific target.
2. Replace the solution with fresh buffer and repeat the wash step.
3. Rinse in 2× SSC (2 ml/$cm^2$ of membrane) for 5 min at room temperature, with agitation.
4. Proceed to detection steps.

## C. Western Blots

*Antibody development*

1. Block the blotted membrane (typically nitrocellulose) in 5% dried milk in phosphate-buffered saline–Tween-20 solution (PBS-T) for 1 hr at room temperature or overnight at 4°C.
2. Rinse the membrane briefly, twice in PBS-T.
3. Wash the membrane in a fresh aliquot of PBS-T for 15 min at room temperature, with agitation.
4. Wash the membrane, twice, for 5 min at room temperature with fresh PBS-T solution.
5. Incubate in primary antibody, diluted as appropriate in PBS, for 1 hr at room temperature, with gentle agitation.
6. Wash the membrane as detailed in steps 2–4 above.
7. Incubate in secondary antibody (typically an HRP conjugate), diluted as appropiate in PBS, for 1 hr at room temperature, with gentle agitation.
8. Wash the membrane as detailed in steps 2–4 above.

## D. Detection

The detection procedure described in this section is equally applicable to the detection of HRP-labeled nucleic acid probes and HRP-labeled antibodies.

Read through this section carefully before starting. It is necessary to work quickly once the blots have been exposed to the detection reagents. The use of a darkroom is required beyond step 6. Do not let the blots dry out at any stage.

1. Mix equal volumes of the two detection reagents (one containing the peracid salt and the other containing the luminol/enhancer system). Use the freshly prepared substrate within 30 min of mixing. Discard unused substrate.
2. Drain the membrane after the final wash step. Lay the filters out on a clean surface, DNA/protein side up.
3. Cover with the freshly mixed ECL detection reagent (0.125 ml/$cm^2$ of membrane). Leave for precisely 1 min at room temperature.
4. Drain the blots and place DNA/protein side down onto a clean sheet of Saran Wrap. Wrap the blots, gently smoothing out air pockets. Ensure that the detection reagent cannot come into contact with the X-ray film.
5. Place the wrapped blots DNA/protein side up in a film cassette.
6. Switch off the light.
7. Place a piece of X-ray film onto the blots and close the cassette.
8. Remove the film after 1 min and replace with a second piece of film. Begin timer.
9. Develope first film and, on the basis of the signal and background, decide on exposure time required for second (or subsequent) films.

---

## E. Applications Guide

### 1. Long Probes

The absolute sensitivity of the system as described is approximately $1 \times 10^{-18}$ moles. This is sufficient to detect a single-copy gene in a loading of 2 $\mu$g of a human genomic DNA restriction enzyme digest (Durrant *et al.*, 1990). The best results are obtained for all applications, but particularly for genomic systems, with probes derived from DNA inserts that have been excised and purified away from the vector. It is also possible to label probes within agarose gel slices, provided that the gel consists of less than 0.7% agarose. Nylon membrane retains target better and is less prone to physical damage than nitrocellulose; nylon blots can be reprobed at least eight times (Fig. 3) without significant loss in sensitivity.

In applications in which significantly higher target levels are available, for instance, in the detection of colonies and plaques or DNA amplification products (Sorg *et al.*, 1990), it may be possible to reduce the probe concentration to as low as 2 ng/ml and to shorten the hybridization time to as little as 2 hr. Colony and plaque screening are particularly suited to use with this system. The results can be timed to give a clear distinction between

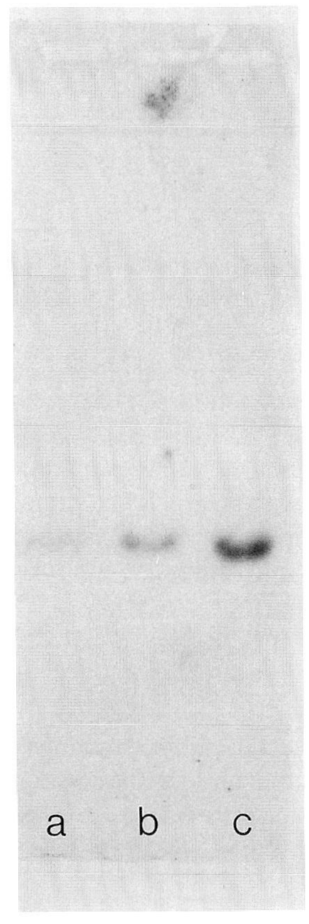


**Fig. 3** Enhanced chemiluminescent detection of a human, single-copy gene sequence. A 1.5-kb region of the human Nras protooncogene labeled with HRP was hybridized to ECORI-restricted human genomic DNA blotted on Hybond-N$^+$ (gel loadings of (a) 2 $\mu$g, (b) 5 $\mu$g and (c) 10 $\mu$g). This represents the eighth reprobing of the filter. Probe concentration, 10 ng/ml; exposure time, 30 min.

positive and negative signal. The presence of a limited level of signal on the negative colonies aids orientation of the filter with the original bacterial sample. It is important that the colonies are not too large and that the cellular debris is removed from the filter. After fixation of the colonies to the filter, the cell debris is removed by vigorous rinsing of the membrane in 2× SSC or by gentle rubbing of the wet filter. Failure to follow these suggestions can lead to difficulty in distinguishing positive and negative signal, especially if the exposure time is not kept to a minimum.

In general, the system performs well, as described, for all the major applications, with stringency controlled during the posthybridization washes. However, certain fingerprint probes are directed toward target sequences that are repeated throughout the genome. The stringency of the hybridization may need to be increased with these probes, by decreasing the NaCl concentration (to as low as 0.05 *M*), to avoid nonspecific binding of the probes.

## 2. Oligonucleotides

Stringency control is the feature that will require most optimization by the user. Typically, hybridizations will be performed at 42°C. For experimental systems in which closely related target sequences are not present, the required stringency may be achieved by using a primary wash of 3× SSC, 0.1% SDS at 42°C. However, in situations in which discrimination between perfectly matched and mismatched sequences is required, then the $T_m$ of the oligonucleotide should be calculated and washes performed about 3 to 5°C below this value (Fig. 4), (Fowler *et al.*, 1990); 0.1% SDS should be included in all stringency wash buffers. The NaCl concentration and the temperature of the wash are the two principal methods of controlling stringency. It must be remembered that the enzyme label will not survive long exposure to high temperature. Consequently, if using high temperature washes (up to a maximum of 60°C), the incubation time should be reduced to only 5 min.

The probe concentration can be reduced to 5 ng/ml for high target applications such as colony and plaque blotting but, as highlighted for long probes, it is vital that the cellular debris is removed, particularly for colony screening. The system has also been used successfully for fingerprint probes and to detect the products of DNA amplification reactions.

## 3. Western Blots

Because of the extremely high sensitivity and rapidity of result for the Western system (see Table II), it is often necessary to reevaluate the optimal antibody dilution. A significant saving in use of expensive primary antibody is often discovered. In addition, exposure times are sometimes

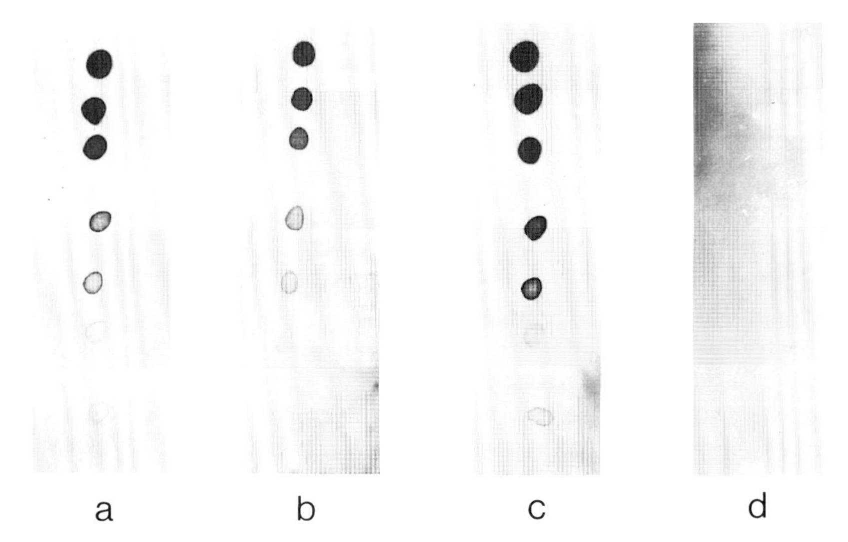


**Fig. 4** Enhanced chemiluminescent detection of HRP-linked oligonucleotides. A 16-base probe labeled with HRP was hybridized to target, derived from *in vitro* mutagenesis experiments, dot blotted on Hybond-N$^+$ (loadings of 40, 20, 10, 4, 2, 1, and 0.2 pg based on the level of the 16-base target). The target sequence was perfectly matched with the probe for blots (a) and (c) but contains one nucleotide different from the probe for blots (b) and (d). Stringency washes were performed in 3 × SSC, 0.1% SDS at 42°C for blots (a) and (b) and in the same buffer at 53°C for blots (c) and (d). Probe concentration, 20 ng/ml; exposure time, 30 min.

extremely short, and the initial 1-min exposure recommended may be too long. To overcome this, dilute the antibodies 10-fold or more, and reduce the exposure time to less than 1 min. If problems persist, then leave blots, after draining off the detection reagent, wrapped in Saran Wrap for 10 to 15 min before positioning the first film. This will allow some of the light to decay.

Filters can be redetected with, or without, removal of the first antibody; this is particularly useful for gaining as much information as possible from rare samples, for example, human protein samples. Blots can be stripped

**Table II**
Comparison of Western Blot Detection Systems

| | Light | Color | | Activity |
|---|---|---|---|---|
| Parameter | ECL | DAB | NBT/BCIP | 125-I |
| Sensitivity (dot blots) | <1 pg | 10 pg | 1 pg | 1–10 ng |
| Minimum detection time | 30 sec–10 min | 20 min–1 hr | 20 min–1 hr | Overnight |
| Antibody–conjugate concentration | 1:1000–1:100000 | 1:1000–1:5000 | 1:1000–1:5000 | 1:500–1:1000 |

by incubation in 60 m*M tris (hydroxymethyl)* aminomethane (Tris)-HCl, 2% SDS, 100 m*M* 2-mercaptoethanol for 30 min at 50°C (Davidson, 1991). This treatment does not affect the target material.

## F. Conclusion

The ECL detection system is universally applicable in molecular biology and immunology. It is a rapid system that has the sensitivity required to cover all the major applications of interest, producing a hard copy of the result. While it may not be a replacement for radioactivity in all applications, at least in sensitivity terms, the ECL system can offer a viable alternative for those users performing routine applications and for those without the facility to handle radiolabeled compounds.

## REFERENCES


Anon. (1990). "ECL Technical Manual." Amersham International, Amersham, UK.

Boniszewski, Z. A. M., Comley, J. S., Hughes, B., and Read, C. A. (1990). The use of charge-coupled devices in the quantitative evaluation of images, on photographic film or membranes, obtained following electrophoretic separation of DNA fragments. *Electrophoresis* **11,** 432–440.

Britten, R. J., and Davidson, E. H. (1985). Hybridization strategy. *In* "Nucleic Acid Hybridization: A Practical Approach" (B. D. Hames and S. J. Higgins, eds.), pp. 3–15. IRL Press, Oxford, England.

Connolly, B. A. (1985). Chemical synthesis of oligonucleotides containing a free sulphydryl group and subsequent attachment of thiol-specific probes. *Nucleic Acids Res.* **13,** 4485–4502.

Davidson, C. (1991). Structural changes in human articular cartilage proteoglycans. *Life Science,* **5,** 7–8.

Durrant, I. (1990). Light-based detection of biomolecules. *Nature (London)* **346,** 297–298.

Durrant, I., Benge, L. C. A., Sturrock, C., Devenish, A. T., Howe, R., Roe, S., Moore, M., Scozzafava, G., Proudfoot, L. M. F., and McFarthing, K. G. (1990). The application of enhanced chemiluminescence to membrane-based nucleic acid detection. *BioTechniques* **8,** 564–570.

Fowler, S. J., Harding, E. R., and Evans, M. R. (1990). Labeling of oligonucleotides with horseradish peroxidase and detection using enhanced chemiluminescence. *Technique* **2,** 261–267.

Hashida, S., Imagawa, M., Inoue, S., Ruan, K., and Ishikawa, E. (1984). More useful maleimide compounds for the conjugation of Fab′ to horseradish peroxidase through thiol groups in the huge. *J. Appl Biochem.* **6,** 56–63.

Hodgson, M., and Jones, P. (1989). Enhanced chemiluminescence in the peroxidase–luminol–$H_2O_2$ sysem: Anomolous reactivity of enhancer phenols with enzyme intermediates. *J. Biolumin. Chemilumin.* **3,** 21–25.

Hutton, J. R. (1977). Renaturation kinetics and thermal stability of DNA in aqueous solutions of formamide and urea. *Nucleic Acids Res.* **4,** 3537–3555.

Kaufmann, S. H., Ewing, C. M., and Shaper, J. H. (1987). The erasable Western blot. *Anal. Biochem.* **161,** 89–95.

Leong, M. M. L., and Fox, G. R. (1988). Enhancement of liminol-based immunodot and Western blotting assays by iodophenol. *Anal. Biochem.* **172,** 145–150.

Pollard-Knight, D. (1990). Current methods in nonradioactive nucleic acid labeling and detection. *Technique* **2,** 1–20.

Pollard-Knight, D., Read, C. A., Downes, M. J., Howard, L. A., Leadbetter, M. R., Pheby, S. A., McNaughton, E., Syms, A., and Brady, M. A. W. (1990). Nonradioactive nucleic acid detection by enhanced chemiluminescence using probes directly labeled with horseradish peroxidase. *Anal. Biochem.* **185,** 84–89.

Renz, M., and Kurz, C. (1984). A colorimetric method for DNA hybridization. *Nucleic Acids Res.* **12,** 3435–3444.

Roswell, D. F., and White, E. H. (1978). The chemiluminescence of luminol and related hydrazides. *Methods Enzymol.* **57,** 409–423.

Schroeder, H. R., Boguslaski, R. C., Carrico, R. J., and Buckler, R. T. (1978). Monitoring specific protein-binding reactions with chemiluminescence. *Methods Enzymol.* **57,** 424–445.

Sorg, R., Enczmann, J., Sorg, U., Kogler, G., Schneider, E. M., and Wernet, P. (1990). Specific nonradioactive detection of PCR-amplified HIV sequences with enhanced chemiluminescence labeling (ECL)—An alternative to conventional hybridization with radioactive isotopes. *Life Sci.* **2,** 3–4.

Stott, R. A. W., and Kricka, L. J. (1987). Purification of luminol for use in enhanced chemiluminescence immunoassay. *In* "Bioluminescence and Chemiluminescence: New Perspectives" (J. Scholmerich, R. Andreesen, A. Kapp, M. Ernst, and W. G. Woods, eds.), pp. 237–240. Wiley, New York.

Stone, T., and Durrant, I. (1990). Light-output capacity of the ECL-system. *Highlights* **1,** 3.

Thorpe, G. H. G., and Kricka, L. J. (1986). Enhanced chemiluminescent reactions catalyzed by horseradish peroxidase. *Methods Enzymol.* **133,** 331–353.

Thorpe, G. H. G., and Kricka, L. J. (1987). Enhanced chemiluminescent assays for horseradish peroxidase: Characteristics and applications. *In* "Bioluminescence and Chemiluminescence: New Perspectives" (J. Scholmerich, R. Andreesen, A. Kapp, M. Ernst, and W. G. Woods, eds.), pp. 199–208. Wiley, New York.

Thorpe, G. H., and Kricka, L. J. (1989). Incorporation of enhanced chemiluminescent reactions into fully automated enzyme immunoassays. *J. Biolumin. Chemilumin.* **3,** 97–100.

Thorpe, G. H. G., Kricka, L. J., Moseley, S. B., and Whitehead, T. P. (1985). Phenols as enhancers of chemiluminescent horseradish peroxidase–luminol–hydrogen peroxide reaction: Application in luminescence-monitored enzyme immunoassays. *Clin. Chem.* **31,** 1335–1341.

Vachereau, A. (1989). Luminescent immunodetection of Western blotted proteins from Coomassie-stained polyacrylamide gel. *Anal. Biochem.* **179,** 206–208.

Whitehead, T. P., Thorpe, G. H. G., Carter, T. J. N., Groucutt, C., and Kricka, L. J. (1983). Enhanced luminescence procedure for sensitive determination of peroxidase-labeled conjugates in immunoassay. *Nature (London)* **305,** 158–159.

# 8 Detection of Horseradish Peroxidase by Colorimetry

Peter C. Verlander
Department of Investigative Dematology
The Rockfeller University
New York, New York



## I. INTRODUCTION

### A. Historical Overview

Horseradish peroxidase (HRP) has been used extensively as a colorimetric marker in biological studies. A hemoprotein with a molecular weight of 40,000, HRP is an ideal detection reagent because of its stability, high turnover rate, and the availability of a wide assortment of colorimetric substrates. Although radioactive techniques are generally more sensitive, the sensitivity of peroxidase detection systems is adequate for many applications, and speed and safety considerations make peroxidase detection

*Nonisotopic DNA Probe Techniques*

perferable to radioactive techniques. HRP was originally used as a direct cytochemical marker for tracing protein uptake in animal tissues (Straus, 1964a,b), but development of immunological techniques and the ability to cross-link enzymes to antibodies (Avrameas, 1969) led to the more widespread use of peroxidase as an indicator for a variety of immunoassays, including western blots (Burnette, 1981) and enzyme-linked immunosorbent assays (ELISA) (Engvall and Perlman, 1972; Van Weemen and Schuurs, 1971).

The use of HRP as a colorimetric detection marker was later adapted to nucleic acid probe technology, with many of the assays directly derived from analogous immunoassays. The most commonly employed method takes advantage of the affinity of avidin and streptavidin for biotin (Langer *et al.,* 1981; Bayer and Wilchek, 1990). After hybridization of a biotinylated DNA probe to target, detection is achieved either by sequential washes with streptavidin (or avidin) and biotinylated peroxidase, or by a single wash with a preformed complex consisting of either streptavidin or avidin bound to biotinylated peroxidase (Fig. 1). Use of a streptavidin–peroxidase complex is generally preferable because of the speed of performing a single wash, and because it is usually more sensitive than sequential washes.

Other methods have included the direct cross-linking of peroxidase to oligonucleotide probes (Renz and Kurz, 1984), as well as immunologic detection of sulfonated probes (Morimoto *et al.,* 1987) or bromodeoxyuridine-labeled probes (Kitazawa *et al.,* 1989; Jirikowski et al., 1989). Immunologic detection can be accomplished either through use of biotinylated antibodies or through use of peroxidase directly coupled to antibodies. These detection techniques have been incorporated into a number of DNA probe assays, including *in situ* hybridizations, sandwich hybridization assays, and membrane hybridizations.

## I. *In Situ* Hybridizations

Horseradish peroxidase detection of nucleic acid probes has been most commonly and successfully applied to *in situ* hybridizations (Brigati *et al.,* 1983; Singer and Ward, 1982). Enzymatic detection is far more rapid than isotopic detection, with signal development in minutes rather than days, and amplification techniques have been described that rival the sensitivity of radioactive probes (Denijn *et al.,* 1990). Amplification usually involves the sequential application of antibodies in order to increase the ultimate number of binding sites for the streptavidin–peroxidase complex. Sensitivities sufficient to detect and localize mRNA in tissue sections have been claimed (Denijn *et al.,* 1990), while detection of 20 to 50 copies per cell are routine. Peroxidase has provided a convenient and sensitive marker for the detection of viral and bacterial infections in tissue samples, and has

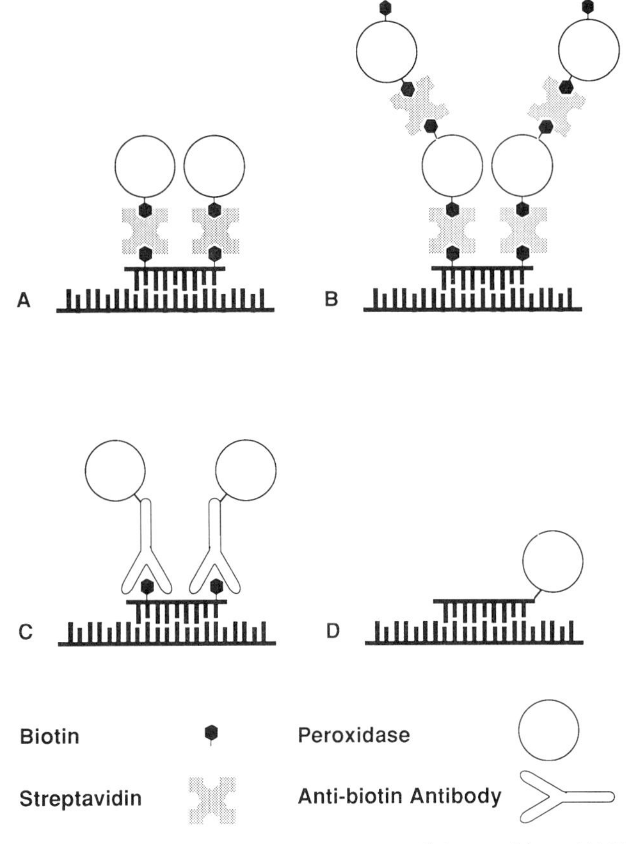


**Fig. 1** Detection of nucleic acid probes with horseradish peroxidase. (A) Hybridization with a biotinylated probe, followed by sequential washes with streptavidin and biotinylated peroxidase. (B) Hybridization with a biotinylated probe, followed by a single wash with a preformed complex of streptavidin and biotinylated peroxidase. (C) Hybridization with a biotinylated probe, followed by detection with antibiotin antibodies coupled to peroxidase. This technique also applies to detection of probes modified in other ways, such as by sulfonation. (D) Hybridization with a probe that is directly coupled to peroxidase.

been used for a variety of clinical diagnostic tests. Peroxidase detection has also been used for genetic testing applications, such as Y chromosome aneuploidy (Guttenbach and Schmid, 1990).

## 2. Sandwich Hybridizations

Sandwich assays are particularly well suited to peroxidase detection. These assays generally include a capture probe that is bound to a fixed matrix, such as nitrocellulose membranes (Dunn and Hassell, 1977; Ranki *et al.*, 1983), microtiter wells (Dahlen *et al.*, 1987; Keller *et al.*, 1989), or beads (Langdale and Malcolm, 1985; Polsky-Cynkin *et al.*, 1985). The target nucleic acid molecule hybridizes to the capture probe and is thereby bound to the matrix, while a second probe that is either directly or indirectly labeled with peroxidase hybridizes to an adjacent sequence on the target (Fig. 2). These techniques are particularly useful for the detection of polymerase chain reaction (PCR)-amplified nucleic acids, and have been used in assays for human immunodeficiency virus (HIV) (Kemp *et al.*, 1990; Keller *et al.*, 1989), β-thalassemia (Saiki *et al.*, 1988, 1989), and sickle-cell anemia (Saiki *et al.*, 1988) among others. Peroxidase detection is useful for these types of assays because of the high turnover rate of the enzyme, and because of the availability of a number of sensitive substrates for soluble assays.

## 3. Membrane Hybridizations

Peroxidase detection is perhaps less well suited to membrane hybridization techniques. The sensitivity of peroxidase detection does not match that of radioactive systems, although detection of subpicogram bands on Southern hybridizations have been claimed (Sheldon *et al.*, 1986, 1987). The lower stability of peroxidase as compared to alkaline phosphatase

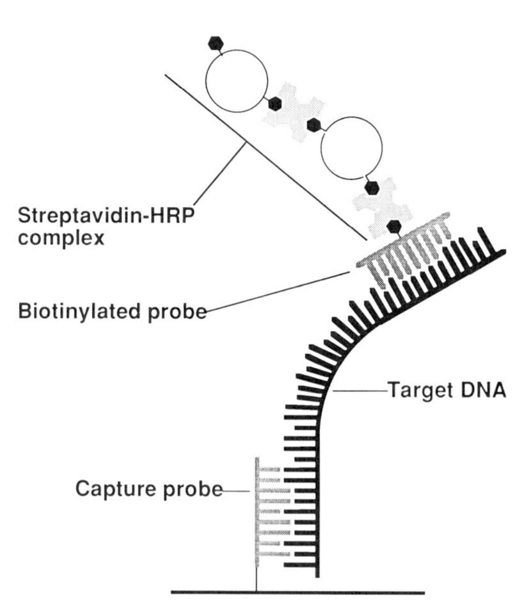


**Fig. 2** Sandwich assays. The target hybridizes to an immobilized probe and is captured to the solid matrix. A biotinylated probe hybridizes to adjacent sequences on the target and is detected by streptavidin–HRP.

limits the stringency of the wash conditions that can be used in reducing nonspecific background. Even with these limitations, peroxidase detection has been used in a variety of membrane-based assays (Saiki *et al.*, 1988, 1989). For example, sandwich assays have been developed that use a membrane-bound capture probe for characterizing polymorphisms at the histocompatibility locus in humans (Bugawan *et al.*, 1990).

## B. Substrates

A wide assortment of substrates is available for peroxidase assays, generating both soluble and insoluble products. Choice of a substrate is primarily based on the format of the assay and the sensitivity of the substrate, but other factors such as the stabilities of the substrate and the product, and the relative health risks of the substrates are also important considerations.

### 1. Insoluble Products

The substrates that generate insoluble products, appropriate for *in situ* hybridizations and membrane hybridizations, include 3-amino-9-ethylcarbazole (AEC), 3,3′-diaminobenzidine (DAB), and 4-chloro-1-naphthol. AEC and DAB are the most commonly used substrates because of their sensitivity. Techniques that claim very high sensitivities have been described for the use of 3,3′,5,5′-tetramethylbenzidine (TMB) in precipitable assays (Sheldon *et al.*, 1986; McKimm-Breschkin, 1990), but this is less common.

Initial studies using peroxidase as a colorimetric indicator employed benzidine as the substrate; unfortunately, the blue precipitate formed had a tendency to fade with time. Use of AEC as a substrate had the advantage of generating a stable, water-insoluble reaction product (Graham *et al.*, 1964). The red product generated by the oxidation of AEC is soluble in organic solvents, and this can be a problem in fixing slides from *in situ* hybridizations. Other drawbacks to the use of AEC include the difficulty of photographing the red product and the fact that AEC is toxic and is a suspected carcinogen.

DAB is a derivation of benzidine that overcomes the stability problems of the benzidine reaction product (Graham and Karnovsky, 1966). DAB is a very sensitive substrate with rapid development times, and the sensitivity of DAB can be improved further by using metal co-factors (Hsu and Soban, 1982). Some reports have found DAB to be a more sensitive substrate than AEC, but this may be owing to the better contrast provided by the darker precipitate as compared to the red product of AEC. Drawbacks to the use of DAB include (1) stock solutions must be prepared fresh; (2) the reaction can be difficult to control, creating background problems; and (3) DAB is very toxic and is a known carcinogen.

Chloronaphthol gives a blue-black oxidation product that, like AEC, is soluble in organic solvents (Nakane, 1968). Although less sensitive than DAB, chloronaphthol can be easier to use because of its slower development time, and is an acceptable substitute for DAB if high background is a problem (Harlow and Lane, 1988). However, the sensitivity of AEC and DAB make them the preferred substrates for assays requiring an insoluble product. Table I summarizes the characteristics of the substrates that generate insoluble products with peroxidase.

### 2. Soluble Products

3,3′,5,5′-Tetramethylbenzidine (TMB; Holland *et al.*, 1974; Hardy and Heimer, 1977), *o*-phenylenediamine (OPD; Voller *et al.*, 1979), and 2,2′-azinobis(3-ethylbenzthiazoline-6-sulfonic acid) (ABTS; Porstmann *et al.*, 1981) are all acceptable substrates for soluble peroxidase assays, such as microtiter plate assays. TMB is the most sensitive substrate for soluble assays; it is also has the advantage of being the safest substrate (Bos, 1981). It is oxidized more rapidly with less catalytic inactivation of the enzyme than other substrates (Sheldon *et al.*, 1986). TMB is oxidized by peroxidase to form a blue product that is changed to yellow on addition of

**Table I**
Substrates Generating Water-Insoluble Products

| Substrate | Product | Comments |
|---|---|---|
| 3-Amino-9-ethylcarbazole (AEC) | Red; soluble in organic solvents | Hazardous. Prepare 20 mg/ml stock solution in DMF and store at 4°C |
| 3,3′-Diaminobenzidine (DAB) | Brown | Hazardous. Prepare solution just before use |
| DAB/metal | Grey-black | Hazardous. Prepare solution just before use |
| 4-Chloro-1-naphthol | Blue-black; soluble in organic solvents | Hazardous. Prepare 30 mg/ml stock solution in absolute ethanol and store at −20°C |
| 3,3′,5,5′-Tetramethylbenzidine (TMB) | Blue; difficult to stabilize product | Nonhazardous. Prepare 5 mg/ml stock solution in DMF or DMSO and store at 4°C |

sulfuric acid to stop the reaction. Either OPD, which yields a brown reaction product, or ABTS, which yields a green reaction product, can be used if the high sensitivity of the TMB reaction leads to high background levels. Both ABTS and OPD have the disadvantage of being light sensitive and of posing a greater health hazard than does TMB. Table II summarizes the characteristics of the substrates that generate soluble products.

# II. MATERIALS

Most of the equipment required for these techniques is commonly available in modern laboratories. The *in situ* hybridizations described require a conventional light microscope, and the microtiter plate assays require a microtiter plate apectrophotometer for reading results. Peroxidase substrates are available from a variety suppliers including Aldrich, Boehringer-Mannheim, and Sigma. Dimethyl formamide (DMF; Fluka) should be deionized and stored in aliquots at −20°C. Stock solutions of AEC and TMB can be prepared in DMF and stored at 4°C. Hydrogen peroxide (30%, Sigma) can be stored at 4°C for up to 1 mo, or at −20°C for longer periods. Streptavidin–HRP detection complex (DETEK I-*hrp*) is available from ENZO Biochem, New York. The dilutions given in the following protocols are based on the recommendations of the manufacturer and may need to be adjusted to accommodate variations in experimental conditions.

# III. PROCEDURES

## A. *In Situ* Hybridizations

Streptavidin–HRP complex can be used for enzymatic detection of hybridized biotinylated DNA in fixed cells and tissues. Standard procedures for *in situ* hybridization are compatible with the use of biotinylated probes

**Table II**
Substrates Generating Water-Soluble Products

| Substrate | Product | $A_{max}$ | Comments |
|---|---|---|---|
| 2,2′-Azinobis (3-ethyl benzthiazoline-6-sulfonic acid) (ABTS) | Green | 410 nm<br>650 nm | Prepare solution just before use |
| *o*-Phenylenediamine (OPD) | Brown | 492 nm | Prepare solution just before use |
| 3,3′,5,5′-Tetramethylbenzidine (TMB) | Yellow | 450 nm | Prepare 5 mg/ml stock solution in DMF or DMSO and store at 4°C. |

and streptavidin–HRP complex (Brigati *et al.*, 1983; Burns *et al.*, 1988). Frozen or paraffin-embedded tissue sections or cells can be used, but paraffin-embedded sections must be deparaffinized through xylene and graded alcohols.

### 1. AEC Detection

*Reagents*

Streptavidin–HRP complex
3-Amino-9-ethylcarbazole, 20 mg/ml in DMF
1% $H_2O_2$ (prepared fresh from 30% stock)
0.1 *M* sodium acetate, pH 4.9

| | |
|---|---|
| Complex dilution buffer | 1× PBS, 1% BSA, 5 m*M* ethylenediaminetetraacetic acid (EDTA) (1× PBS is 130 m*M* sodium chloride, 10 m*M* sodium phosphate, pH 7.2) |
| Wash buffer | 1× PBS, 5 m*M* EDTA |
| Substrate mixture | 25 μl 1% $H_2O_2$ + 20 μl AEC + 1 ml 0.1*M* sodium acetate, pH 4.9 |

The AEC stock solution can be stored at 4°C; prepare and store this solution in an glass vial and protect from light. The substrate mixture should also be prepared in a glass tube and can be stored at 0 to 8°C in the dark for up to 1 week. The solution will darken with time, but will still perform as required.

*Procedure*

1. Dilute streptavidin–HRP complex 1:250 in complex dilution buffer.
2. Apply 100–200 μl of diluted complex to slide that has been hybridized and washed.
3. Incubate the slide at room temperature for 10 to 20 min. (For tissue sections, use 500 μl diluted complex, and incubate at 37°C for 15 min).
4. Rinse the slide with a steady stream of wash buffer from a squeeze bottle for 10 sec, taking care not to squeeze the solution directly onto the specimen. Tap excess mositure off the slide.
5. Add 300–400 μl of the substrate mixture to the specimen and incubate at room temperature for 10 to 15 min. (For tissue sections, incubate at 37°C for 15 to 30 min.
6. Rinse the slide as in step 4.
7. Soak the slide for 1 min in each of three Coplin jars filled with wash buffer at room temperature.

At this point, the slide is ready for viewing. Mount a coverslip on the slide, using water as the mounting medium. Do not use organic mounting agents, as the red precipitate is soluble in organic solvents. View at 40 to 400× magnification; light to brick-red deposits in the cells indicate positive reactivity (Fig. 3). Slides can be counterstained with 2% naphthol blue-black in water or with hematoxylin, if desired. Be careful to avoid heavy counterstaining, or positive results can be obscured.

***a. Potential Problems***

Do not allow the slides to dry out during the detection procedure, or nonspecific background signal may arise. Also, be sure to wash the slides well before and after application of the detection complex, as insufficient washing can also lead to elevated background. Problems with insufficient signal are rare and can usually be traced to the slide preparation and hybridization steps. Be sure to completely denature double-stranded DNA probes, or low signal may result. Insufficient protease treatment of the slide before hybridization can also lead to low signals by preventing access of the probe and detection reagents to the target DNA.

**2. DAB Detection**

*Reagents*

Streptavidin–HRP complex
Diaminobenzidine tetrahydrochloride
1% $H_2O_2$ (prepared fresh from 30% stock)

| | |
|---|---|
| Complex dilution buffer | 1× PBS, 1% BSA, 5 m*M* EDTA (1× PBS is 130 m*M* sodium chloride, 10 m*M* sodium phosphate, pH 7.2) |
| Wash buffer | 1× PBS, 5 m*M* EDTA |
| Dye solution | 500 $\mu$g/ml DAB in 1× PBS |
| Substrate mixture | 25 $\mu$l 1% $H_2O_2$ + 2.5 ml dye solution |

The DAB stock solution and the substrate mixture are unstable and should be prepared just before use.

*Procedure*

1. Dilute streptavidin–HRP complex 1:250 in complex dilution buffer.
2. Apply 100–200 $\mu$l diluted complex to slide that has been hybridized and washed.

3. Incubate the slide at room temperature for 10 to 20 min. (For tissue sections, use 500 $\mu$l diluted complex, and incubate at 37°C for 15 min).
4. Rinse the slide with a steady stream of wash buffer from a squeeze bottle for 10 sec, taking care not to squeeze the solution directly onto the specimen. Tap excess moisture off the slide.
5. Add 300–400 $\mu$l of the substrate mixture to the specimen and incubate at room temperature for 2 to 10 min.
6. Rinse the slide as in step 4.
7. Soak the slide for 1 min in each of three Coplin jars filled with wash buffer at room temperature.

At this point, the slide is ready for viewing. Mount a coverslip on the slide and view at 40 to 400× magnification; dark rust deposits in the cells indicate positive reactivity. As with AEC detection, slides can be counterstained with 2% naphthol blue-black in water if desired. Be careful to avoid heavy counterstaining, or positive results can be obscured.

## B. Microtiter Plate Assays

A variety of microtiter plate assays have been developed for use with colorimetric detection systems, most involving detection of PCR-amplified DNA. The assays generally differ in the method of capturing the hybridized target in microtiter wells; some of the approaches include covalent cross-linking of a capture probe to the wells, capture by specific DNA-binding proteins, and capture by antisulfonated DNA antibodies. The various protocols use standard hybridization conditions, which will not be described here.

### 1. TMB Detection

*Reagents*

Streptavidin–HRP complex
5 mg/ml TMB in 100% dimethylformamide
3% $H_2O_2$ (prepared fresh from 30% stock)

| | |
|---|---|
| Citrate–phosphate buffer | 0.2 *M* $Na_2HPO_4$, 0.1 *M* citric acid, pH 5.3 |
| Wash buffer | 1 × PBS, 0.1% Triton X-100 |
| Complex dilution buffer | 1× PBS, 0.1% Triton X-100, 1% bovine serum albumin (BSA) |

Substrate mixture — 100 μl 5 mg/ml TMB + 1 ml citrate–phosphate buffer + 33 μl 3% $H_2O_2$

Stop buffer — 2 *M* $H_2SO_4$

---

*Procedure*

1. Following hybridization, wash three times with wash buffer.
2. Dilute streptavidin–HRP complex 1:250 in complex dilution buffer.
3. Add 50 μl diluted complex to each well; seal wells and incubate at 37°C for 30 min.
4. Wash wells as in step 1.
5. Combine 100 μl 5 mg/ml TMB with 10 ml citrate–phosphate buffer and 33 μl 3% $H_2O_2$; add 100 μl of this substrate mixture to each well.
6. Incubate 30 min at room temperature. Blue reaction product should develop.
7. Stop the reaction by addition of 100 μl of stop buffer. Blue product should turn yellow.
8. Read $OD_{450}$ with a microtiter plate spectrophotometer.

---

***a. Potential Problems***

The major problem to contend with is high background levels. These problems can usually be traced to inadequate removal of excess probe or streptavidin–peroxidase complex. The level of background can vary depending on the microtiter plates used, the quality of the probe and the detection complex, and the conditions used to block nonspecific binding. While the conditions given should be sufficient to eliminate background in most cases, some experimentation may be necessary to optimize the conditions for a particular application. Other detergents, such as Tween-20, and blocking reagents, such as nonfat dry milk, have been used in place of Triton X-100 and BSA in other laboratories (Kemp *et al.*, 1990). If high background levels continue to be a problem, OPD can be used as a substrate in place of TMB.

### 2. OPD Detection

*Reagents*

Streptavidin–HRP complex
*o*-Phenylene diamine tablets (Sigma)
3% $H_2O_2$ (prepared fresh from 30% stock)

| | |
|---|---|
| Citrate–phosphate buffer | 0.2 *M* $Na_2HPO_4$, 0.1 *M* citric acid, pH 5.3 |
| Wash buffer | 1× PBS, 0.1% Triton X-100 |
| Complex dilution buffer | 1× PBS, 0.1% Triton X-100, 1% BSA |
| Stop buffer | 2 *M* $H_2SO_4$ |

*Procedure*

1. Following hybridization, wash three times with wash buffer.
2. Dilute streptavidin–HRP complex 1:250 in complex dilution buffer.
3. Add 50 μl diluted complex to each well; seal wells and incubate at 37°C for 30 min.
4. Wash wells as in step 1.
5. Dissolve 20.0 mg OPD in 10 ml citrate–phosphate buffer, and add 83 μl 3% $H_2O_2$. Add 100 μl of this to each well.
6. Incubate 30 min in the dark at room temperature. Brown reaction product should develop.
7. Stop the reaction by addition of 100 μl of 2 *M* $H_2SO_4$.
8. Read $OD_{492}$ with a microtiter plate spectrophotometer.

## C. Membrane Hybridizations

As mentioned earlier, colorimetric peroxidase detection techniques are generally not particularly useful for membrane applications such as Southern blots because of the lack of sensitivity of the technique, although the use of TMB as a substrate reportedly greatly increases the sensitivity (Sheldon *et al.*, 1986; McKimm-Breschkin, 1990). Even so, peroxidase detection has been used with blot hybridization of PCR-amplified material, as well as for sandwich assays with a membrane matrix. Also, the combination of different detection systems can be applied to mapping fragments, generating two colors on a single blot (Fig. 4).

### 1. AEC Detection

*Reagents*

Streptavidin–HRP complex
3-Amino-9-ethylcarbazole, 20 mg/ml in DMF
1% $H_2O_2$ (prepared fresh from 30% stock)
0.1 *M* Sodium acetate, pH 4.9

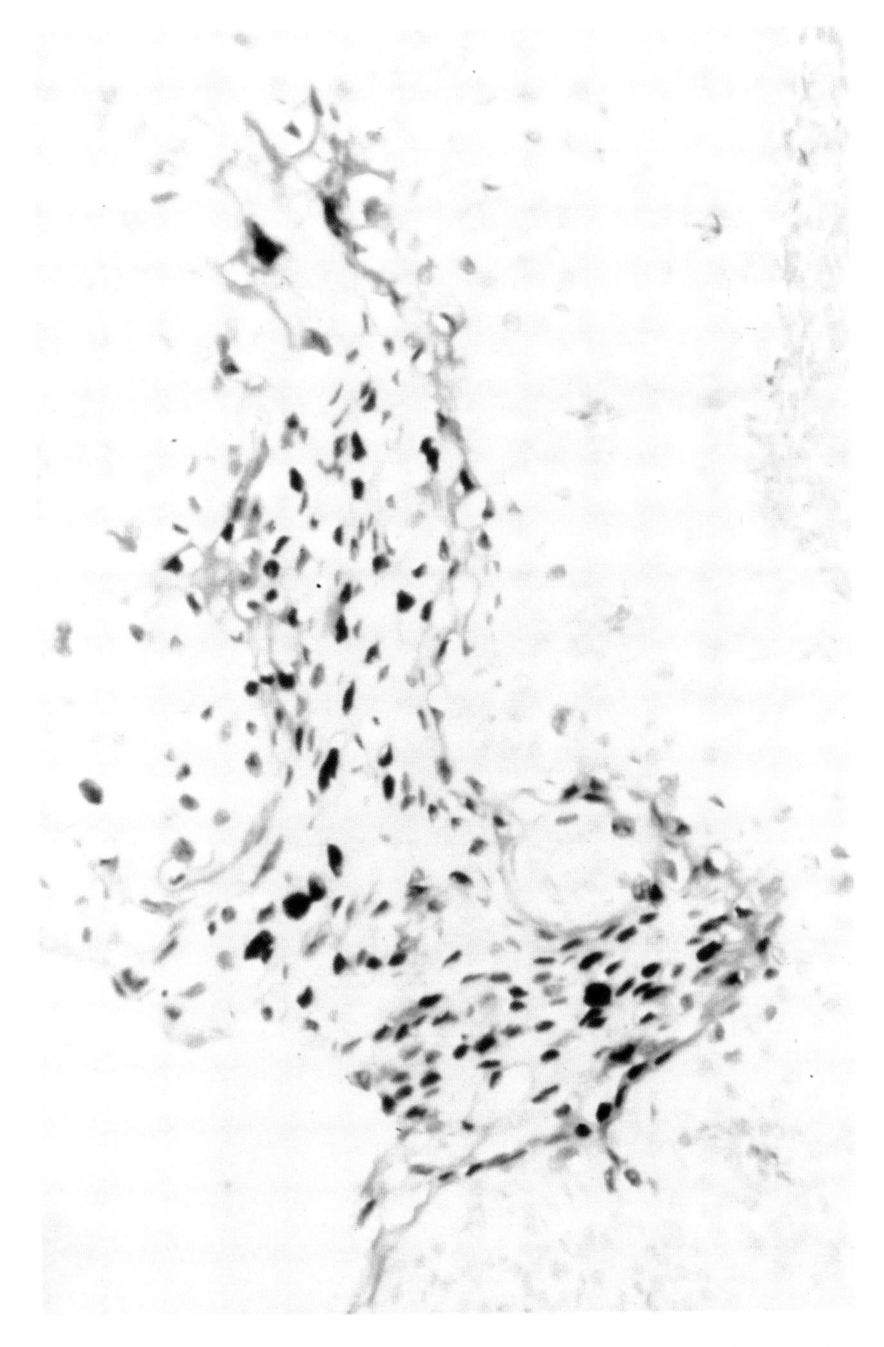

**Fig. 3** AEC detection of biotinylated probes in an *in situ* hybridization. A biopsy specimen hybridized with probes for human papillomavirus types 16 and 18. Deep reddish-brown color indicates a positive reaction. The tissue section was counterstained with hematoxylin and viewed at 64×. Courtesy of ENZO Biochem.

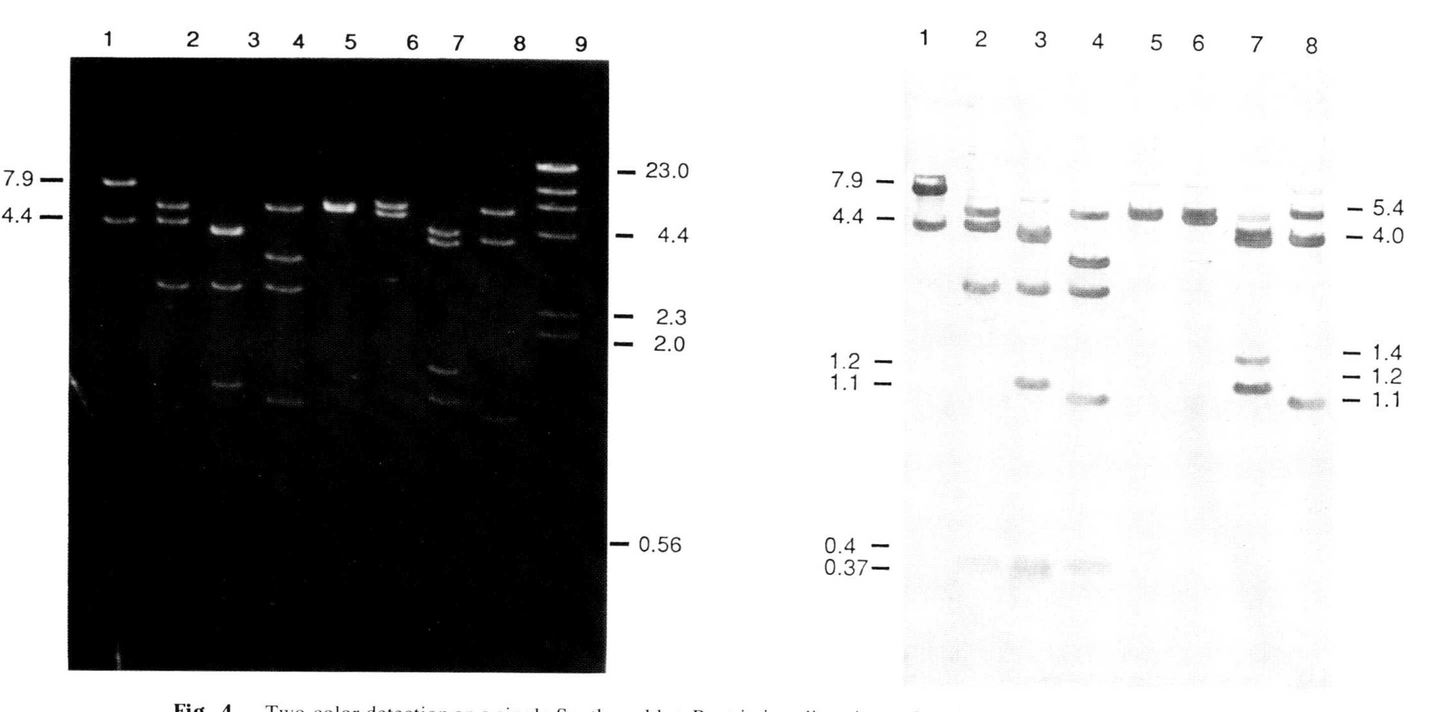


**Fig. 4** Two-color detection on a single Southern blot. Restriction digestions of a plasmid containing *Chlamydia* sequences were blotted and hybridized with two probes, a biotinylated pBR322 probe and a *Chlamydia* probe that had a poly dT tail added with terminal transferase. The pBR322 probe was detected first, using a streptavidin–alkaline phosphatase complex, and is indicated by the blue signal. The blot was then washed with biotinylated oligo-dA (BioBridge; ENZO Biochem) and detected with streptavidin–HRP, indicated by the red signal. Courtesy of Christine Braekel, ENZO Biochem.

| | |
|---|---|
| 20× SSC | 3.0 *M* sodium chloride, 0.3 *M* sodium citrate |
| 10× PBS | 1.3 *M* sodium chloride, 0.1 *M* sodium phosphate, pH 7.2 |
| Blocking buffer | 1× PBS, 2% BSA, 0.1% Triton X-100, 5 m*M* EDTA |
| Complex dilution buffer | 1× PBS, 1% BSA, 5 m*M* EDTA |
| High-salt washing buffer | 10 m*M* potassium phosphate, pH 6.5, 0.5 *M* sodium chloride, 0.05% Triton X-100, 0.1% BSA, 1 m*M* EDTA |
| Predetection buffer | 2× SSC, 0.1% BSA, 0.05% Triton X-100, 1 m*M* EDTA |
| Substrate mixture | 20 μl 20 mg/ml AEC stock solution + 1 ml 0.1 *M* sodium, acetate, pH 4.9 + 25 μl 1% $H_2O_2$ |

*Procedure*

1. After stringency washes, rinse the filter twice at room temperature in 0.2× to 2× SSC to remove SDS.
2. Transfer the filter to a heat-sealing bag and add ~0.1 ml blocking buffer per $cm^2$ membrane. Seal and incubate at room temperature for 15 min.
3. Dilute streptavidin–HRP complex 1:250 in complex dilution buffer.
4. Remove blocking buffer from bag, and replace with diluted detection complex. Use 0.012 ml/$cm^2$, and seal after removing bubbles. Incubate at room temperature for 30 to 60 min.
5. Remove the filter from the bag, transfer to a dish with high-salt washing buffer, and gently agitate for 5 min. Repeat three times.
6. Wash the filter twice with gentle agitation in predetection buffer for 5 min each.
7. Transfer the filter to a clean dish containing freshly prepared substrate mixture. Develop filter face up for up to 1 hr at room temperature. Color development is usually evident within 10 min.
8. Remove the filter, rinse with 1× PBS, and allow to dry.

Nylon membranes are preferable to nitrocellulose, as the color is retained on nylon, whereas the signal fades on nitrocellulose. However, the addition of water will restore the color to a nitrocellulose filter. As for microtiter plate assays, a variety of blocking agents can be used and should be tested with a given membrane if high background is a problem.

### 2. TMB Detection

The following protocol is adapted from Sheldon *et al.*, 1986, and is claimed to be sufficiently sensitive to detect restriction fragment length polymerphisms RFLPs in the human genome. Since the oxidation product of TMB is normally water soluble, the problem is to stabilize the signal; this is accomplished by the inclusion of dextran sulfate, which is thought to capture the blue product. Even so, the stability of the signal is not particularly good, and the developed filter can not be stored for long periods.

*Reagents*

Streptavidin–HRP complex
5mg/ml TMB in 100% dimethylformamide
3% $H_2O_2$ (prepared fresh from 30% stock)

| | |
|---|---|
| Complex dilution buffer | 5% Triton X-100, 2.7 m*M* KCl, 237 m*M* NaCl, 1.5 m*M* $KH_2PO_4$, 8 m*M* $Na_2PO_4$, pH 7.4. |
| Wash buffer | 1% dextran sulfate, 0.15*M* 1,1-diethylurea in complex dilution buffer |
| Substrate mixture | 0.1 mg/ml TMB, 5% ethanol, 10 m*M* sodium citrate, 10 m*M* sodium citrate, 10 m*M* EDTA, pH 5.0 |

*Procedure*

1. After probe removal, rinse in complex dilution buffer.
2. Dilute streptavidin–HRP complex 1:250 in complex dilution buffer.
3. Transfer the filter to a heat-sealing bag and add diluted detection complex. Use 0.012 ml/cm$^2$, and seal after removing bubbles. Incubate at room temperature for 40 min.
4. Wash five times for 5 min each in wash buffer to remove unbound detection complex.
5. Rinse in substrate mixture.
6. Incubate in the same buffer with the addition of $H_2O_2$ to 0.0014% for 30 to 60 min at room temperature. Read the results immediately.
7. Rinse four times in water, 30 to 60 min per rinse.

## IV. CONCLUSIONS

Colorimetric detection of HRP provides a rapid and convenient method for detection of DNA probes in a variety of assay formats. Although the sensitivity of other detection methods may sometimes be superior, peroxidase detection is sufficiently sensitive for many applications, and the speed of the reaction makes peroxidase detection the method of choice in many cases.

## ACKNOWLEDGMENTS

The author would like to thank Dr. Barbara Thalenfeld for supplying illustrations. Special thanks go to Dr. James Donegan for many helpful suggestions.

## REFERENCES

Avrameas, S. (1969). Coupling of enzymes to proteins with glutaraldehyde. Use of the conjugates for the detection of antigens and antibodies. *Immunochemistry* **6,** 43–52.

Bayer, E. A., and Wilchek, M. (1990). Biotin-binding proteins: Overview and prospects. *Methods Enzymol.* **184,** 49–67.

Bos, E. S., van der Doelen, A. A., van Rooy N., and Schuurs, A. H. (1981). 3,3′,5,5′ - Tetramethylbenzidine as an Ames test negative chromogen for horseradish peroxidase in enzyme-immunoassay. *J. Immunoassay* **2,** 187–204.

Brigata, D. J., Myerson, D., Leary, J. J., Spalholz, B., Travis, S. Z., Fong, C. K. Y., Hsiung, G. D., and Ward, D. C. (1983). Detection of viral genomes in cultured cells and paraffin-embedded tissue sections using biotin-labeled hybridization probes. *Virology,* **126,** 32–50.

Bugawen, T. L., Begovich, A. B., and Erlich, H. A. (1990). Rapid HLA-DPB typing using enzymatically amplified DNA and nonradioactive segvence-specific oligo nucleotide probes. *Immunogenet.* **32,** 231–241.

Burnette, W. N. (1991). Western blotting: Electrophoretic transfer of proteins from sodium dodecyl sulfate–polyacrylamide gels to unmodified nitrocellulose and radiographic detection with antibody and radioiodinated protein A. *Anal. Biochem.* **112,** 195–204.

Burns, J., Graham, A. K., and McGee, J. O'D. (1988). Nonisotopic detection of *in situ* nucleic acid in cervix: An updated protocol. *J. Clin. Pathol.* **41,** 897–899.

Dahlen, P., Syvanen, A. C., Hurskainen, P., Kwiatkowski, M., Sund, C., Ylikoski, J., Soderlund, H., and Lovgren, T. (1987). Sensitive detection of genes by sandwich hybridization and time-resolved fluorometry. *Mol. Cell. Probes* **1,** 159–168.

Denijn, M., De-Weger, R. A., Berends, M. J., Compier-Spies, P. I., Jansz, H., Van-Unnik, J. A., and Lips, C. J. (1990). Detection of calcitonin-encoding mRNA by radioactive and nonradioactive *in situ* hybridization: Improved colorimetric detection and cellular localization of mRNA in thyroid sections. *J. Histochem. Cytochem.* **38,** 351–358.

Dunn, A. R., and Hassell, J. A. (1977). A novel method to map transcipts: Evidence for homology between an adenovirus mRNA and discrete multiple regions of the viral genome. *Cell* **12,** 23–36.

Engvall, E., and Perlmann, P. (1972). Enzyme-linked immunosorbent assay (ELISA) quantitative assay of immunoglobulin G. *Immunochemisty* **8,** 871–879.

Graham, R. C., Jr., Lundholm, U., and Karnovsky, M. J. (1964). Cytochemical demonstration of peroxidase activity with 3-amino-9-ethylcarbazole. *J. Histochem. Cytochem.* **13,** 150–152.

Graham, R. C., Jr., and Karnovsky, M. J. (1966). The early stages of absorption of injected horseradish peroxidase in the proximal tubules of mouse kidney ultrastructural cytochemisty by a new technology. *J. Histochem. Cytochem.* **14,** 291–302.

Guttenbach, M., and Schmid, M. (1990). Determination of Y chromosome aneuploidy in human sperm nuclei by nonradioactive *in situ* hybridization. *Am. J. Hum. Genet.* **46,** 553–558.

Hardy, H., and Heimer, L. (1977). A safer and more sensitive substitute for diaminobenzidine in the light microscopic demonstration of retrograde and anterograde axonal transport of HRP. *Neurosci. Lett.* **5,** 235–240.

Harlow, E., and Lane, D. (1988). "Antibodies: A Laboratory Manual." Cold Spring Harbor Laboratories, New York.

Holland, V. R., Saunders, B. C., Rose, F., and Walpole, A. L. (1974). Safer substitute for benzidine in detection of blood. *Tetrahedron* **30,** 3299–3302.

Hsu, U.-M., and Soban, E. ((1982). Color modification of diaminobenzidine (DAB) precipitation by metallic ions and its application for double immunohistochemistry. *J. Histochem. Cytochem.* **30,** 1079–1082.

Jirikowski, G. F., Ramalho-Ortigao, J. F., Lindl, T., and Seliger, H. (1989). Immunocytochemistry of 5-bromo-2′-deoxyuridine labeled oligonucleotide probes. A novel technique for *in situ* hybridization. *Histochemisty* **91,** 51–53.

Keller, G. H., Huang, D. P., and Manak, M. M. (1989). A sensitive nonisotopic hybridization assay for HIV-1 DNA. *Anal. Biochem.* **177,** 27–32.

Kemp, D. J., Smith, D. B., Foote, S. J., Samaras, N., and Peterson, M. G. (1989). Colorimetric detection of specific DNA segments amplified by polymerase chain reactions. *Proc. Natl. Acad. Sci. U.S.A.* **86,** 2423–2427.

Kemp, D. J., Churchill, M. J., Smith, D. B., Biggs, B. A., Foote, S. J., Peterson, M. G., Samaras, N., Deacon, N. J., and Doherty, R. (1990). Simplified colorimetric analysis of polymerase chain reactions: Detection of HIV sequences in AIDS patients. *Gene* **94,** 223–228.

Kitazawa, S., Takenaka, A., Abe, N., Maeda, S., Horio, M., and Sugiyama, T. (1989). *In situ* DNA–RNA hybridization using *in vivo* bromodeoxyuridine-labeled DNA probe. *Histochemistry* **92,** 195–9.

Langdale, J. A., and Malcolm, A. D. B. (1985). A rapid method of gene detection using DNA bound to Sephacryl. *Gene* **36,** 201–210.

Langer-Safer, P. R., Levine, M., and Ward, D. C. (1982). Immunological method for mapping genes on Drosophila polytene chromosomes. *Proc. Natl. Acad. Sci. U.S.A.* **79,** 4381–4385.

Langer, P. R., Waldrop, A. A., and Ward, D. C. (1981). Enzymatic synthesis of biotin-labeled polynucleotides: Novel nucleic acid affinity probes. *Proc. Natl. Acad. Sci. U.S.A.* **78,** 6633–6637.

Lee, J. J., Warburton, D., and Robertson, E. J. (1990). Cytogenetic methods for the mouse: Preparation of chromosomes, karyotyping, and *in situ* hybridization. *Anal. Biochem.* **189,** 1–17.

McKimm-Breschkin, J. L. (1990). The use of tetramethylbenzidine for solid-phase immunoassays. *J. Immunol. Methods.* **135,** 277–280.

Morimoto, H., Monden, T., Shimano, T., Higashiyama, M., Tomita, N., Murotani, M., Matsuura, N., Okuda, H., and Mori, T. (1987). Use of sulfonated probes for *in situ* detection of amylase mRNA in formalin-fixed paraffin sections of human pancreas and submaxillary gland. *Lab. Invest.* **57,** 737–741.

Nakane, P. K. (1968). Simultaneous localization of multiple tissue antigens using the peroxidase-labeled antibody method: A study of pituitary glands of the rat. *J. Histochem. Cytochem.* **16,** 557–560.

Polsky-Cynkin, R., Parsons, G. H., Allerdt, L., Landes, G., Davis, G., and Raschtchian, A. (1985). Use of DNA immobilized on plastic and agarose supports to detect DNA by sandwich hybridization. *Clin. Chem.* **31,** 1438–1443.

Porstmann, B., Porstmann, T., and Nugel, E. (1981). Comparison of chromogens for the determination of horseradish peroxidase as a marker for enzyme immunoassay. *J. Clin. Chem. Clin. Biochem.* **19,** 435–439.

Ranki, M., Palva, A., Virtanen, M., Laaksonen, M., and Soderlund, H. (1983). Sandwich hybridization as a convenient method for the detection of nucleic acids in crude samples. *Gene* **21,** 77–85.

Renz, M., and Kurz, C. (1984). A colorimetric method for DNA hybridization. *Nucleic Acids Res.* **12,** 3435–3444.

Saiki, R. K., Chang, C. A., Levenson, C. H., Warren, T. C., Boehm, C. D., Kazazian, H. H., Jr., and Erlich, H. A. (1988). Diagnosis of sickle-cell anemia and beta-thalassemia with enzymatically amplified DNA and nonradioactive allele-specific oligonucleotide probes. *N. Engl. J. Med.* **319,** 537–541.

Saiki, R. K., Walsh, P. S., Levenson, C. H., and Erlich, H. A. (1989). Genetic analysis of amplified DNA with immobilized sequence-specific oligonucleotide probes. *Proc. Natl. Acad. Sci. U.S.A.* **86,** 6230–6234.

Sheldon, E. L., Kellogg, D. E., Watson, R., Levenson, C. H., and Erlich, H. A. (1986). Use of nonisotopic M13 probes for genetic analysis: Application to HLA class II loci. *Proc. Natl. Acad. Sci. U.S.A.* **83,** 9085–9089.

Sheldon, E., Kellogg, D. E., Levenson, C., Bloch, W., Aldwin, L., Birch, D., Goodson, R., Sheridan, P., Horn, G., and Watson, R. (1987). Nonisotopic M13 probes for detecting the beta-globin gene: Application to diagnosis of sickle-cell anemia. *Clin. Chem.* **33,** 1368–1371.

Singer, R. H., and Ward, D. C. (1982). Actin gene expression visualized in chicken muscle tissue culture by using *in situ* hybridization with a biotinated nucleotide analog. *Proc. Natl. Acad. Sci. U.S.A.* **79,** 7331–7335.

Straus, W. (1964a). Factors affecting the cytochemical reaction of peroxidase with benzidine and the stability of the blue reaction product. *J. Histochem. Cytochem.* **12,** 462–469.

Straus, W. (1964b). Factors affecting the state of injected horseradish peroxidase in animal tissues and procedures for the study of phagosomes and phago-lysosomes. *J. Histochem. Cytochem.* **12,** 470–480.

Van Weemen, B. K., and Schuurs, A. H. W. M. (1971). Immunoassay using antigen–enzyme conjugates. *FEBS Lett.* **15,** 232–236.

Voller, A., Bidwell, D., and Bartlett, A. (1976). Microplate enzyme immunoassays for the immunodiagnosis of virus infections. *In* "Manual of Clinical Immunology" (Rose, N. R., and Friedman, H.) pp. 506–512. American Society for Microbiology.

# 9 Detection of Glucose 6-Phosphate Dehydrogenase by Bioluminescence

Jean-Claude Nicolas, Patrick Balaguer,
Béatrice Térouanne, Marie Agnès Villebrun,
and Anne-Marie Boussioux
INSERM Unité 58, Montpellier, France



## I. INTRODUCTION

Nucleic acid hybridization has become a fundamental technique in biochemistry and molecular biology for detecting specific DNA sequences. Sensitive detection of the hybrid has been obtained using probes labeled with radioisotopes such as $^{32}P$, which unfortunately has a short half-life and is not well suited to routine application. Other labels have been developed for this purpose, such as enzymatic incorporation of biotinylated (Langer *et al.*, 1981; Lo *et al.*, 1988) or digoxigenin analogs (Lion

*Nonisotopic DNA Probe Techniques*

and Haas, 1990) or chemical coupling of haptens (Lebacq *et al.*, 1988; Hopman *et al.*, 1987), biotin (Reisfeld *et al.*, 1987; Forster *et al.*, 1985) or enzymes (Renz and Kurtz, 1984; Saiki *et al.*, 1988; Urdea *et al.*, 1989).

We used glucose 6-phosphate dehydrogenase (G6PDH) as a bioluminescent label in nucleic acid hybridization reactions, because this enzyme can be detected at the attomole level and has already been used successfully for bioluminescent immunoassays (Térouanne *et al.*, 1986). G6PDH is coupled with streptavidin or oligonucleotides, and detection is carried out using a bioluminescent chain reaction catalyzed by flavine mononucleotide (FMN) oxidoreductase and luciferase from *Vibrio harveyi*.

Dot blot analysis on nitrocellulose filters can be performed using biotin as a label, and streptavidin-G6PDH to reveal labeled samples. Light emission is detected using photographic films or a photon-counting camera, which can provide a quantitative analysis.

Some steps of this hybridization procedure can be eliminated by direct enzyme labeling of oligonucleotides; using luminescence as a detection system, several probes can be detected on the same filter. We have used this strategy for the rapid identification of papillomaviruses after polymerase chain reaction (PCR). Several papillomavirus-specific DNA sequences can be amplified using general oligonucleotide primers (Resnick *et al.*, 1990; Van den Brule *et al.*, 1990), and identified by hybridization with specific oligonucleotide probes. These DNA probes can be labeled with different enzymes; using luminescence reagents, it is possible to detect on the same dot these oligonucleotides labeled with peroxidase, alkaline phosphatase, or G6PDH. The papillomavirus type is rapidly identified by sequential revelation of each enzyme activity.

This technique can also be used for identification of different mutations or polymorphism in DNA sequences. However, if the number of variants is too high, as is the case for human leukocyte antigen (HLA) typing, identification can be readily accomplished by *reverse hybridization* (Saiki *et al.*, 1989) between sequence-specific oligonucleotide probes immobilized on the filter and biotinylated amplified DNA samples. This process can be used to screen a sample for all known allelic variants at an amplified locus. Oligonucleotide probes are elongated by chemical synthesis rather than enzymatically, and elongation with 20 nucleotides is sufficient to immobilize probes on Hybond-C-Extra (Amersham). DNA amplification is carried out with biotin-dUTP (uridine 5′-triphosphate (Lo *et al.*, 1988) instead of biotinylated primers, and biotinylated PCR-amplified sequences are detected using streptavidin G6PDH and a bioluminescent reaction.

For clinical analysis the nitrocellulose filter format is inconvenient. We thus developed a rapid bioluminescent assay, which can be performed in tubes using an oligonucleotide immobilized on Sepharose, and another labeled with G6PDH. This procedure has been optimized to obtain high specificity, and it can be adapted for screening of numerous samples to identify pathogenic agents.

Homogeneous systems using fluorescence energy have been described to quantify PCR sequences by hybridization in solution (Morrison *et al.*, 1989). In order to improve sensitivity, we have developed a sensitive bioluminescent assay to monitor hybridization occurring in solution. DNA sequences are synthesized with two different labels (biotin and fluorescein) carried on the 5′ end of each strand. These molecules are captured on a bioluminescent adsorbent comprising antifluorescein antibodies, FMN oxidoreductase, and luciferase coimmobilized on Sepharose. Streptavidin-G6PDH is involved in an enzyme-channeling reaction and used to quantify biotinylated DNA sequences retained on the immunoadsorbent. This method can be used to quantify DNA sequences obtained after the amplification reaction. In certain conditions, the exchange of DNA strands between a double-labeled probe and a homologous target is specific enough to detect the presence of a single mutation in a 150-base pair sequence. It can also be applied to screen a great number of samples and used to identify sequences that are slightly different from reference probes. We used this technique for HLA typing, studying apolipoprotein polymorphism, and identifying different homologous viruses.

# II. MATERIALS

## A. Reagents

Dithiothreitol (DTT), decanal, *S*-acetyl mercapto-succinic anhydride (SAMSA), fluorescein isothiocyanate (FITC), gelatin, polyvinyl pyrolidone, and bovine serum albumin are obtained from Sigma Chemical Co., (St. Louis, Missouri). Ficoll, Sephadex G 50 Medium, CNBr-activated Sepharose 4B and Sephadex G25 PD10 columns are purchased from Pharmacia LKB Biotechnology, Inc. (Piscataway, New Jersey). Glucose 6-phosphate (G6P), glucose 6-phosphate dehydrogenase (G6PDH) from *Leuconostoc mesenteroides* (EC 1-1-1-49) (500 IU/mg at 25°C), nicotinamide adenine dinucleotide (NAD), flavine mononucleotide (FMN) and biotinyl ε-amino caproic acid *N*-hydroxy succinimide ester (biotin × NHS) are obtained from Boehringer. Para-nitro blue tetrazolium (pNBT), phenazine methoxy sulfate (PMS) and bromo-chloro-indolyl phosphate (BCIP) are Aldrich Chemical Co., (Millwauke, Wisconsin) products. Succinimidyl 4-(*N*-maleimido-methyl) cyclohexane-1-carboxylate (SMCC) are from Pierce Chemical Co. (Rockford, Illinois).

FMN oxidoreductase and luciferase are purified from *Vibrio harveyi* according to the method described by Jablonski and DeLuca (Jablonski and DeLuca, 1977). Enzymes are kept at −20°C in 20% glycerol. Luciferase and FMN oxidoreductase of *Photobacterium fischeri* can also be used to detect NAD(P)H and are available from Boehringer.

Nitrocellulose filters (BA 83) are from Schleicher and Schuell, and Hybond C extra from Amersham (Little Chalfort, UK). Polaroid photographic films (20,000 ASA) are used for detection of bioluminescent reactions.

### 1. Buffers

- 1 × SSC buffer: 150 m*M* NaCl, 15 m*M* $Na_3$ citrate, pH 6.8
- 1 × FPGe solution: 0.02% gelatin, 0.02% polyvinylpyrolidone, 0.02% Ficoll.
- 1× hybridization buffer: 2× SSC, 5× FPGe, 25 m*M* $KH_2PO_4$, pH 7, 2.5 m*M* EDTA, salmon sperm DNA (0.2 mg/ml).
- Washing buffer: Tris-HCl 100 m*M*, pH 7.5, NaCl 150 m*M*, Tween-80 0.2%
- Blocking buffer: washing buffer and 3% gelatin.
- Incubation buffer: Tris-HCl 100 m*M*, pH 7.5, $MgCl_2$ 10 m*M*.
- Bioluminescent reagent for filter hybridization: incubation buffer plus $10^{-3}$ *M* NAD, 3 × $10^{-3}$ *M* G6P, $10^{-5}$ *M* FMN, 6 × $10^{-5}$ *M* decanal, 50 μg/ml luciferase, 5 mIU/ml FMN oxidoreductase and 1g/liter BSA.
- Bioluminescent reagent for channeling reactions on Sepharose: $10^{-3}$ *M* NAD, 3 × $10^{-3}$ *M* G6P, 3 ×$10^{-3}$ *M* sodium pyruvate, $10^{-5}$ *M* FMN, 6 × $10^{-5}$ *M* decanal and 30 mIU/ml LDH in 0.1 *M* phosphate buffer, 0.15 *M* NaCl, and 1g/liter BSA.

### 2. Labeled Probes

#### *a. Labeling of Plasmids*

Plasmids are labeled with biotin dUTP by nick translation (BRL Kit). Biotin dUTP is used at a final concentration of 20 μ*M* in the absence of dTTP.

1. Add to an Eppendorf tube:
   - 5 μl dATP, dCTP, dGTP 0.2 m*M* in Tris-HCl 500 m*M*, pH 7.5, $MgCl_2$
   - 50 m*M*, 2-mercaptoethanol 100 m*M*, and BSA 100 μg/ml
   - 1 μg DNA in 2 μl $H_2O$
   - 2.5 μl biotin-11-dUTP 0.4 m*M*, or digoxigenin dUTP 0.06 m*M* plus 0.14 m*M* dTTP

- 5 $\mu$l DNA polymerase 0.4 IU/ml, plus DNA nuclease I 40 pg/ml
- 36.5 $\mu$l $H_2O$

2. Centrifuge the tube and incubate for 2 hr at 15°C.
3. Stop the reaction by adding 5 $\mu$l EDTA 0.1 m*M* and 1.25 $\mu$l SDS 1%.

In these conditions the average length of fragments is 500 pb, and 10 to 20% of thymidine is replaced by biotin-11-dUTP.

#### *b. Labeling of Oligonucleotides*

Oligonucleotides are enzymatically labeled with biotin-dUTP or dUTP digoxigenin using the deoxynucleotide terminal transferase.

*Enzymatic Labeling*

1. Add to an Eppendorf tube:
   - 1 $\mu$l oligonucleotide 0.2 m*M*
   - 2 $\mu$l potassium cacodylate 1 *M*, Tris base 250 m*M*, adjusted to pH 7.6 $CoCl_2$ 10 m*M* and DTE 2 m*M*
   - *5 $\mu$l biotin-11-dUTP 0.4 mM*
   - 4 $\mu$l terminal transferase 25 IU/$\mu$l
   - 8 $\mu$l water
2. Incubate for 2 hr at 37°C.
3. Stop by heating 10 min at 95°C.

Two to four labeled molecules are incorporated per mole of oligonucleotide. Labeled oligonucleotides are recovered by gel filtration on a Sephadex G 25 column equilibrated with 0.1% SDS.

Oligonucleotides with an amino group at the 5′ end are coupled to to biotin *N*-hydroxysuccinimide ester or fluorescein isothiocyanate.

*Chemical Labeling*

1. Add 500 $\mu$l of a freshly prepared solution of *N*-biotinyl-6-aminocaproic *N*-hydroxysuccinimide ester (10 mg/ml in DMF) or FITC (15 mg/ml in acetone) to 100 nanomoles oligonucleotides in 500 $\mu$l carbonate/bicarbonate 0.2 *M*, pH 9.5.
2. Incubate overnight at room temperature.
3. Remove excess reagent from the labeled oligonucleotides on a Sephadex G 25 PD 10 column, in 20 m*M* Tris-HCl buffer, pH 8.

A final purification can be done by high-pressure liquid chromatography (HPLC) on a C8 reverse phase column.

To immobilize oligonucleotides on nitrocellulose filters, 20-mer tail of polyadenosine deoxynucleotides is added to the oligonucleotide during chemical synthesis.

### 3. Enzyme Conjugates

*Measurement of G6PDH activity*

1. In a spectrophotometer cuvette, add 910 μl Tris-HCl pH 8, 0.05 *M*, $MgCl_2$ 3 $10^{-3}$*M*, 60 μl NAD, 0.1 *M*, 30 μl G6P, 0.3 *M*, and 10 μl enzyme solution.
2. Measure the absorbance increase at 340 nm for 1 min.

*Incorporation of maleimide groups into G6PDH*

1. Incubate 1.5 mg G6PDH in 500 μl of 0.1 *M* sodium phosphate, pH 7.5, containing $10^{-4}$*M* NADP, and $3 \times 10^{-3}$ *M* G6P with 25 μl of SMCC ($10^{-2}$ *M* in DMF) for 45 min at room temperature.
2. Purify the modified enzyme on Sephadex G25 (PD 10) using 20 m*M* sodium phosphate, pH 6.4, containing $5 \times 10^{-3}$ *M* EDTA, and collect fractions having the highest activity.
3. Use immediately after preparation.

*Preparation of thiolated streptavidin*

1. Dissolve 1 mg streptavidin in 300 μl 0.1 *M* sodium phosphate buffer, pH 7.5, and incubate with 6 μl of SAMSA ($10^{-1}$ *M* in acetonitrile) for 1 hr.
2. Add 30 μl hydroxylamine (1 *M*, pH 7.5) and 5 μl of DTT (0.1 *M*) and incubate for 20 min.
3. Purify thiolated streptavidin by gel filtration on Sephadex G 25 (PD 10) in 20 m*M* sodium phosphate, pH 6.4, containing $5 \times 10^{-3}$ *M* EDTA.
4. Collect the most concentrated fractions and add immediately after preparation to the modified G6PDH.

*Preparation of streptavidin–G6PDH conjugate*

1. Mix thiolated streptavidin and maleimide G6PDH fractions.
2. Concentrate the mixture using an Amicon 30 to 100 to 200 μl.
3. Incubate for 18 hours in the dark at room temperature.
4 Purify the conjugate by FPLC on Superose 12 in 0.1 *M* sodium phosphate, pH 7.4, containing $2 \times 10^{-3}$ *M* 2-mercapto-ethanol and $2 \times 10^{-3}$ *M* EDTA.
5. Store the conjugate fractions in 1% BSA and 20% glycerol at −20°C.

---

G6PDH modified with SMCC incorporates five maleimide groups, and streptavidin modified with SAMSA incorporates at least three thiol groups.

SMCC produces the weakest inhibition of G6PDH, but inhibits the binding of biotin to streptavidin. SAMSA is the best reagent for streptavidin since it results in less than 10% inhibition of biotin binding.

The reaction of thiolated streptavidin with maleimido-G6PDH produces conjugates in good yield: 60% of the G6PDH activity is bound to streptavidin. Separation by fast purification by liquid chromatography (FPLC) on Superose 12 shows a high-molecular-weight fraction well separated from the free enzyme (Fig. 1). Fraction I gives the best results for detection on a filter, and fraction II gives lowest background when used for hybridization in solution.

*Preparation of enzyme oligonucleotide conjugates*

Oligonucleotides with a thiol group at the 5′ end are synthesized directly using a C6 thiol modifier from Clontech. After deprotection, oligonucleotides are incubated in $10^{-2}$ *M* DTT for 30 min, and excess reducing agent is eliminated by chromatography on a G25 column equilibrated with 20 m*M* sodium phosphate, pH 6.4, containing $5 \times 10^{-3}$ *M* EDTA.

1. Mix thiolated oligonucleotide and maleimide-modified G6PDH fractions.
2. Evaporate to 100 $\mu$l under vacuum at room temperature.
3. Incubate overnight.

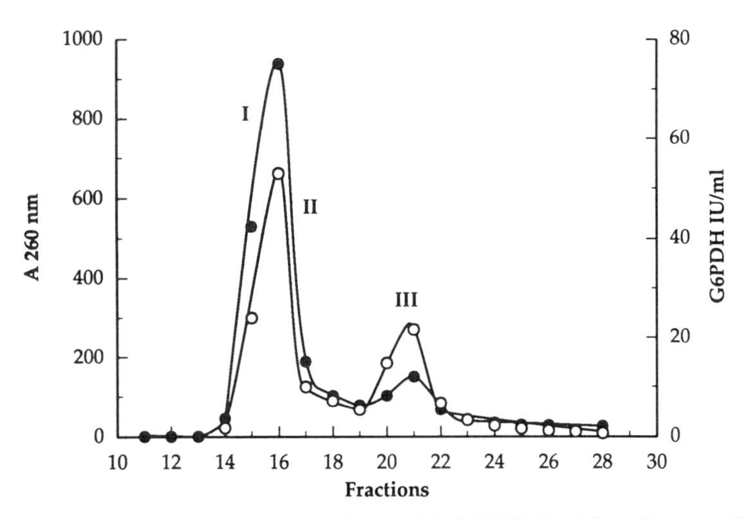


**Fig. 1** Profiles of the reaction products of maleimide-G6PDH eluted from Superose 12 with thiolated streptavidin. Open circles (○) indicate G6PDH activity and closed circles (●) absorbance at 280 nm.

4. Purify by FPLC on an anion exchange column (DEAE Spherodex).
5 Store fractions at −20°C in 1% BSA and 20% glycerol.

The reaction of thiolated oligonucleotides and maleimide G6PDH produces conjugates with a yield of 30 to 60%. The conjugate elutes from DEAE Spherodex column at 0.35 *M* NaCl and is well separated from free enzyme or oligonucleotides (Fig. 2).

## 4. DNA Amplification

Symmetric DNA amplification is performed (Saiki *et al.*, 1988) in a 100-μl reaction mix containing 2 μg DNA, 0.2 m*M* dATP, dCTP, dGTP, dTTP, 0.5 μ*M* of each primer in 1× reaction buffer (10 m*M* Tris-HCl pH 9, 50 m*M* KCl, 1.5 m*M* $MgCl_2$, and 0.1 g/liter gelatin, Triton 0.1%). Samples are heated for 1 min at 95°C to denature the DNA. Two units of Taq polymerase (Promega) is added to each sample, which is incubated at 55°C for 1 min (annealing) and 2 min at 72°C for primer elongation. Further rounds, including 1-min denaturation, 1-min annealing and 2-min synthesis, are continued for 30 cycles. Asymmetric PCR is performed with a 10-n*M* concentration of the limiting primer.

Labeling of DNA sequences during the amplification process is per-

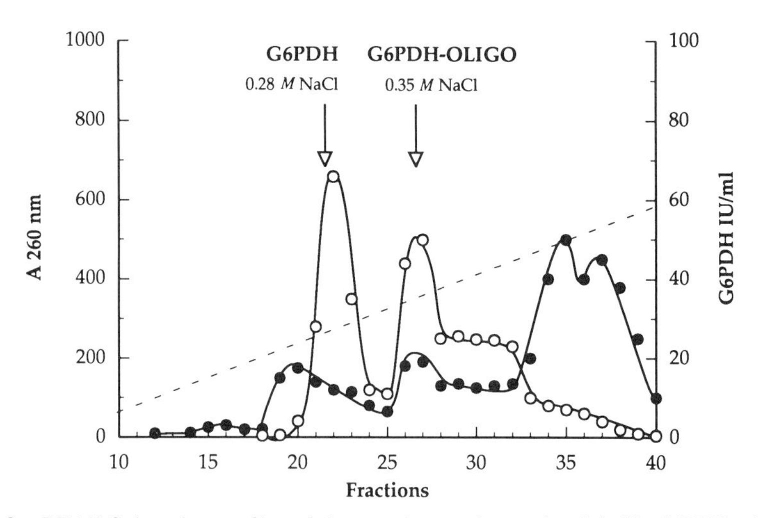


**Fig. 2** DEAE Spherodex profiles of the reaction products of maleimide-G6PDH with a thiolated oligonucleotide. Open circles (○) indicate G6PDH activity and closed circles (●) absorbance at 280 nm.

formed at a final concentration of 100 $\mu M$ for biotin-dUTP and 10$\mu M$ for dTTP. The incubation time for primer extension is extended to 3 min.

Double labeling by biotin and FITC of a specific DNA sequence is done using two labeled primers. Primers are purified on a C8 reverse phase by HPLC. The amplified material (10 × 10 $\mu$l) is electrophoresed in a 12% polyacrylamide gel and the double-labeled sequence is extracted as described by Maniatis *et al.*, (1982). The slices are crushed in 1 ml elution buffer (0.5 *M* ammonium acetate, 10 m*M* magnesium acetate, 1 m*M* EDTA, pH 8, 0.1% SDS), and the acrylamide gel is discarded after centrifugation.

### 5. Bioluminescent Adsorbent

*Oligonucleotide Adsorbent*

Bioluminescent adsorbents are prepared by coupling luciferase, FMN oxido-reductase and 5′ amino oligonucleotide to Sepharose 4B.

1. Wash CNBr-activated Sepharose (1g) with 200 ml $10^{-3}$ *M* HCI.
2. Suspend Sepharose in 4 ml 0.1 *M* potassium pyrophosphate buffer, pH 8.2, containing 3 mg luciferase, 1 IU FMN oxidoreductase, and 0.1 mg 5′ amino oligonucleotide.
3. Shake the suspension gently for 2 hr at room temperature and then overnight at 4°C.
4. Wash the Sepharose with 0.1 *M* phosphate buffer, pH 7.2, and incubate with the same buffer containing $10^{-3}$ *M* DTT under gentle shaking at room temperature. This buffer is changed three times, and the immunoadsorbent is stored at 4°C in 0.1 *M* phosphate buffer, pH 7.2, $10^{-3}$ *M* DTT, and 0.1% $NaN_3$.

*Antifluorescein Adsorbent*

Mouse anti-FITC monoclonal antibodies are precipitated from ascitic fluid using 50% ammonium sulfate, and the immunoglobulin solution is chromatographed on DEAE trisacryl M.

The luciferase oxidoreductase enzyme system is coimmobilized on Sepharose 4B with an anti-FITC $\gamma$-globulin antibody according to the following protocol:

1. Wash CNBr-activated Sepharose 4B (1 g) with 200 ml $10^{-3}$ *M* HCl.
2. Suspend Sepharose in 5 ml 0.1 *M* potassium pyrophosphate buffer, pH 8.2, containing 5 mg luciferase, 1 IU oxidoreductase and 2 mg antibody. Shake the suspension gently for 2 hr at room temperature and then overnight at 4°C.

3. Wash the Sepharose with 200 ml 0.1 *M* phosphate buffer, pH 7.2, and incubate with the same buffer containing $10^{-3}$ *M* DTT under gentle shaking at room temperature. This buffer is changed three times and the immunoadsorbent is stored at 4°C in 0.1 *M* phosphate buffer, pH 7.2, $10^{-3}$ *M* DTT and 0.1% $NaN_3$.

## B. Instrumentation

Oligonucleotides are made on an Applied Biosystem synthesizer model 391 using the phosphoramidite procedure. Hybridizations are performed in plastic bags and incubated under gentle shaking on a rotating shaker. PCR reactions are done with a Techne or MJ Research thermocycler. The latter instrument is also used for hybridizations in solution that require precise decreases of incubation temperatures.

Light emission is visualized using a video camera system (Argus 100) that allows direct quantification of photon emission. Accurate photon counts for each spot can be performed; they show a good linear relationship between counts and amounts of target DNA. The photon-counting system provides higher sensitivity than photographic films and is limited only by background emission due to nonspecific adsorption of the label on the filter.

Light emitted from bioluminescent adsorbent contained in polypropylene tubes is measured with an automated luminometer (LKB 1251).

# III. PROCEDURES

## A. Dot Blot Hybridization

### 1. Biotin-Labeled DNA Probe

Dot-blot hybridization is performed with biotinylated probes, which are detected using a streptavidin–G6PDH conjugate. Bound conjugate is detected by either a colorimetric or a bioluminescent assay (photographic detection). Bioluminescence is rapid and provides multiple exposure for single dot blots, which is valuable for identification of allele variations or mutations.

1. Denature DNA samples in $H_2O$ at 98°C for 10 min.
2. Cool on ice for 5 min.

3. Add an equal volume of cold 20 × SCC.
4. Apply 1 μl of sample onto a nitrocellulose filter.
5. Dry at room temperature.
6. Bake at 90°C for 1 hr in an oven.
7. Place the filter in a plastic bag, add 1 ml of 1 × hybridization buffer per 5 $cm^2$ of filter, and incubate at 65°C overnight.
8. Change hybridization buffer.
9. Add denatured labeled probe (final concentration 100 ng/ml).
10. Incubate at 65°C overnight.
11. Wash the filter with 0.2 × SSC 0.2% Tween 80, twice at room temperature and twice at 65°C (10 min per wash).
12. Incubate in blocking buffer for 1 hr at 65°C.
13. Incubate with blocking buffer containing 1 μg/ml streptavidin–G6PDH at 37°C for 15 min.
14. Rinse with washing buffer three times under gentle shaking (10 min per wash).
15. Rinse with incubation buffer and blot the membranes to remove excess buffer.

*Colorimetric Detection of G6PDH*

1. Put the filter into a bag with $3 \times 10^{-3}$ *M* NAD, $3 \times 10^{-3}$ *M G6P*, 0.33 mg/ml pNBT, 0.02 mg/ml PMS in 50 m*M* Tris-HCl, pH 7.5, 150 m*M* NaCl, 10 m*M* $MgCl_2$.
2. Incubate 3–10 hr and then wash the filter with water and dry.

Serial dilutions of λ DNA nick-translated with biotin-11-dUTP spotted onto nitrocellulose and detected by streptavidin–alkaline phosphatase or by streptavidin-G6PDH show similar detection limits for the two labels, but color development is faster with alkaline phosphatase than G6PDH (1 hr instead of 3 hr).

*Bioluminescent Detection*

1. Place the filter on a polyethylene sheet and add the bioluminescent reagent (10 μl/$cm^2$). Cover the membrane with a second polyethylene sheet. Spread the reagent and remove air bubbles and excess of reagent.
2. Seal the edge and place in contact with photographic film or under a video camera.

Using highly sensitive polaroid film, DNA-bound G6PDH is rapidly detected. After 30 min, the sensitivity is similar to that of alkaline phosphatase detected by colorimetry, and since the nitrocellulose filter is not stained, it can be reused. Filters are reused after bioluminescent detection after removing the bound probe by heating at 100°C for 15 min in 0.1 × SSC, 0.1% SDS. The filter is then successfully rehybridized to a second probe. When the filter is to be reused, Hybond-C extra is chosen as the membrane, since it is stronger than nitrocellulose. Nylon filters result in excessively high background with G6PDH.

Photographic detection is useful for studying the relative intensity of various spots. Since the film is highly sensitive, the exposure range is narrow, and contrast is very high. These properties are used to detect base mutations in human genomic DNA from several donors. We spotted four different HLA-DQβ sequences obtained after PCR amplification of human DNA and hybridized with a specific 17-mer oligodeoxynucleotide. The 4 DNA samples show 0, 1, 2, and 3 base mismatches with the oligodeoxynucleotide. Photographic detection of a few minutes reveals only DNA with 0 base mismatch (Fig. 3).

## 2. Enzyme-Labeled Oligonucleotides

1. Denature DNA samples in $H_2O$ at 98°C for 10 min.
2. Cool on ice for 5 min.
3. Add an equal volume of cold 20× SCC.
4. Apply 1 μl of sample onto a nitrocellulose filter.
5. Dry at room temperature.
6. Bake at 90°C for 1 hr in an oven.
7. Place the filter in a plastic bag, add 1 ml 1× hybridization buffer per 5 $cm^2$ of filter and incubate at 37°C for 1 hr.

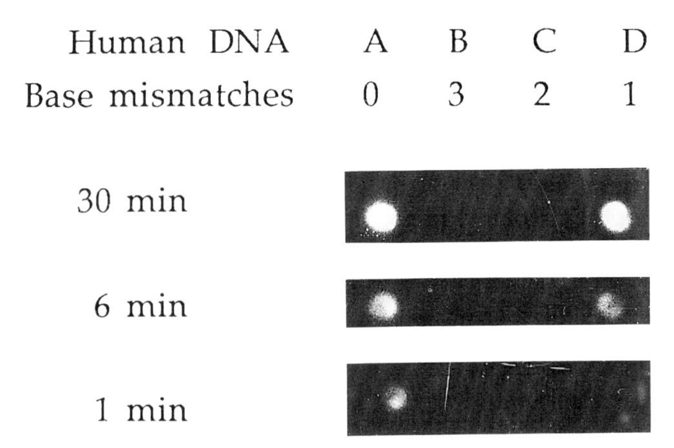


**Fig. 3** PCR-amplified sequences of human DNA were hybridized with an oligonucleotide tailed with biotin dUTP. Polaroid film (20,000 ASA) were exposed for 1, 6, and 30 min.

8. Change hybridization buffer.
9. Add specific labeled oligonucleotides (final concentration: 10 ng/ml).
10. Incubate at 37°C 1 hr.
11. Wash the filter with 0.2 × SSC 0.2% Tween-80, twice at room temperature and twice at 37°C (10 min per wash).
12. Rinse with G6PDH incubation buffer and blot the membranes to remove excess buffer.
13. Place the filter on a polyethylene sheet and add the bioluminescent reagent (10 $\mu l/cm^2$). Cover the membrane with a second polyethylene sheet. Spread the reagent and remove air bubbles and excess reagent.
14. Seal the edge and place in contact with photographic film or under a video camera and expose 1 to 10 min.
15. Rinse with peroxidase or alkaline phosphatase incubation buffer, blot the membranes to remove excess buffer, and repeat steps 13 and 14 with the corresponding luminescent reagent. Peroxidase is detected with the ECL reagent from Amersham and alkaline phosphatase with adamantyl dioxetane phosphate (AMPPD) (Photogene from Gibco BRL). Since AMPPD remains adsorbed on nitrocellulose, alkaline phosphatase detection must be performed last.

Using general primers it is possible to amplify several homologous human papillomaviruses (HPVs). Cells obtained from cervix scrapes are lyzed in boiling water, and 10 $\mu l$ of the supernatant is amplified. After PCR, several labeled probes can be used for identification. The multidetection system avoids having to perform several hybridizations; moreover, the competition between the probes improves specificity of the hybridization reaction. Using oligonucleotide probes labeled with alkaline phosphatase, G6PDH, or peroxidase and a luminescent assay reagent, it is possible to identify papillomavirus 18, 16 and 11, or 6. From the same blot, different viruses are identified, and quantitative results are obtained as shown in Table I.

## B. Reverse Dot Blot Hybridization

Amplification is performed in the presence of 100 $\mu M$ dUTP biotin and 20 $\mu M$ dTTP. Amplified sequences are heat denatured and diluted 100-fold in 1× hybridization buffer.

**Table I**
Luminescence Values after Successive Detection of Enzyme-Labeled Oligonucleotides[a]

| Samples | G6PDH HPV 16 (23-mer) | Peroxidase HPV 11 (20-mer) | Phosphatase HPV18 (20-mer) |
|---|---|---|---|
| 1 | 152,000 | 0 | 0 |
| 2 | 0 | 248,000 | 0 |
| 3 | 75,000 | 35,000 | 0 |
| 4 | 0 | 0 | 36,000 |
| Mix HPV (10 pg) | 180,000 | 355,000 | 62,000 |

[a] Amplified sequences (1 μl) are spotted on nitrocellulose filter and hybridized with the oligonucleotide mix. Background values for each detection system are subtracted. Counts from HPV references are not identical since hybridization yields and sensitivity of detection systems are different.

1. Apply 1 μl of each oligonucleotide (10 ng/μl in 10× SSC) on a nitrocellulose filter.
2. Dry at room temperature.
3. Bake at 90°C for 1 hr in an oven.
4. Place the filter in a plastic bag, add 1 ml 1× hybridization buffer per 5 $cm^2$ filter and incubate at 65°C overnight.
5. Add denatured labeled sample (final concentration: 100 ng/ml).
6. Incubate at 55°C overnight.
7. Wash the filter with 0.2× SSC 0.2% Tween-80, twice at room temperature and twice at 55°C (10 min per wash).
8. Incubate in blocking buffer for 1 hr at 55°C.
9. Incubate with blocking buffer containing 1μg/ml streptavidin-G6PDH at 37°C for 15 min.
10. Rinse with washing buffer three times under gentle shaking (10 min per wash).
11. Rinse with incubation buffer and blot the membranes to remove excess buffer.
12. Place the filter on a polyethylene sheet and add the bioluminescent reagent (10 $\mu l/cm^2$). Cover the membrane with a second polyethylene sheet. Spread the reagent and remove air bubbles and excess reagent.
13. Seal the edge and place in contact with photographic film or under a video camera.

Reverse dot blots are particularly valuable for assays in which the number of potential sequences to be analyzed exceeds the number of samples, as for HLA typing. Chemical tailing (20 bases) of oligonucleotides seems to be sufficient to immobilize probes on the filter. Luminescent detection allows better quantification and faster detection than colorimetric detection, and the filters can be reused for other assays.

## C. Sandwich Hybridization

Amplified DNA sequences are commonly detected using oligonucleotide probes by conventional filter hybridization, but for rapid analysis a simplified format is suitable. We have developed a rapid sandwich hybridization technique using a channeling enzymatic reaction on a bioluminescent support and a small amount of labeled oligonucleotide. When hybridization is performed in solution, reannealling of the double-strand DNA target leads to a strand-displacement reaction, reducing the hybridization yield with oligonucleotide probes. To solve this problem, an asymmetric PCR protocol can be used to produce single-strand DNA. We described a rapid and sensitive assay of viral DNA that does not require denaturation and separation steps.

Hybridization is performed at 37°C using two different oligonucleotides, one labeled with G6PDH and the other immobilized on bioluminescent Sepharose (luciferase and oxido-reductase immobilized on Sepharose). The enzyme bound to the hybrid is detected by a channeling reaction that leads to light emission. This process does not require any separation step since the unbound enzyme does not lead to light emission. In the supernatant the unbound label produces NADH, which is oxidized by lactate dehydrogenase-pyruvate. In the solid phase, NADH, formed by the G6PDH-oligonucleotide bound to the target, is oxidized by FMN oxidoreductase and FMN. The immobilized luciferase oxidizes $FMNH_2$ in the presence of an aldehyde and oxygen and emits light (Fig. 4). Thus without any separation step, the photons count is related to the amount of bound label.

---

*Hybridization and Detection Procedure*

1. Asymmetric amplification (30 cycles) is performed on 0.1 to 1 $\mu$g DNA using an oligonucleotide primer ratio of 50 in order to obtain a single-strand DNA complementary to the oligonucleotides used for capture and detection.
2. Incubate 10 $\mu$l of the amplification medium with 20 femtomoles

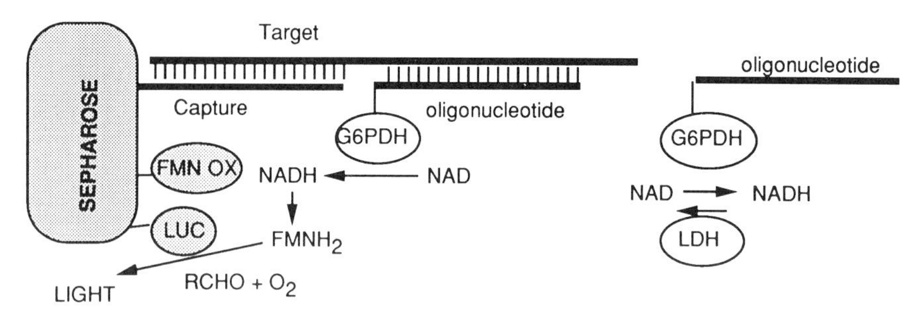


**Fig. 4** Principle of the bioluminescent DNA assay. The hybrid was captured by oligonucleotide immobilized on Sepharose. Luminescent signal is proportional to NADH produced by the hybridized probe.

G6PDH oligonucleotide and 2 mg bioluminescent Sepharose oligonucleotide for 2 hr at 37°C under gently shaking.

3. Add 1 ml bioluminescent reagent containing $10^{-3}$ *M* NAD, $3 \times 10^{-3}$ *M* G6P, $3 \times 10^{-3}$ *M* sodium pyruvate, $10^{-5}$ *M* FMN, $6 \times 10^{-5}$ *M* decanal and 30 mIU/ml lactate dehydrogenase (LDH) in 0.1 *M* phosphate buffer, 0.15 *M* NaCl, and 1g/liter BSA.
4. After a few minutes measure the luminescence (10 sec intervals).

Hybridization of double-strand DNA in solution is limited by the competition reaction between the complementary target strand and the probe. Thus, asymmetric amplification is recommended for routine analysis since it produces the best hybridization yield and a greater signal variation with different amounts of target (Fig. 5). Even at 37°C the method is very specific; the labeled oligonucleotide does not bind to nonspecific amplified DNA or sequences showing 3 base mismatches with the G6PDH probe. This is probably related to the low concentration of probe and target and to the bulky substitution produced by enzyme labeling or immobilization, which can affect the hybridization reaction rate or stability of the hybrid.

Oligonucleotide lengths must be chosen according to their stability at the concentration used (14 to 20 bases depending on the G–C content).

The small size and the low probe concentration allows us to use a bioluminescent channeling reaction that was initially described for immunoassays (Térouanne *et al.*, 1986); thus, no separation step is required. The detection limit of the bioluminescent assay is 20 attomoles of single-strand DNA ($10^7$ molecules). It is possible to detect less than 100 viral bodies from 2 $\mu$g total DNA coupled to symmetric PCR, and using asymmetric amplification, the detection limit is 1000 copies. This method can be automated for rapid and specific identification of PCR-amplified products.

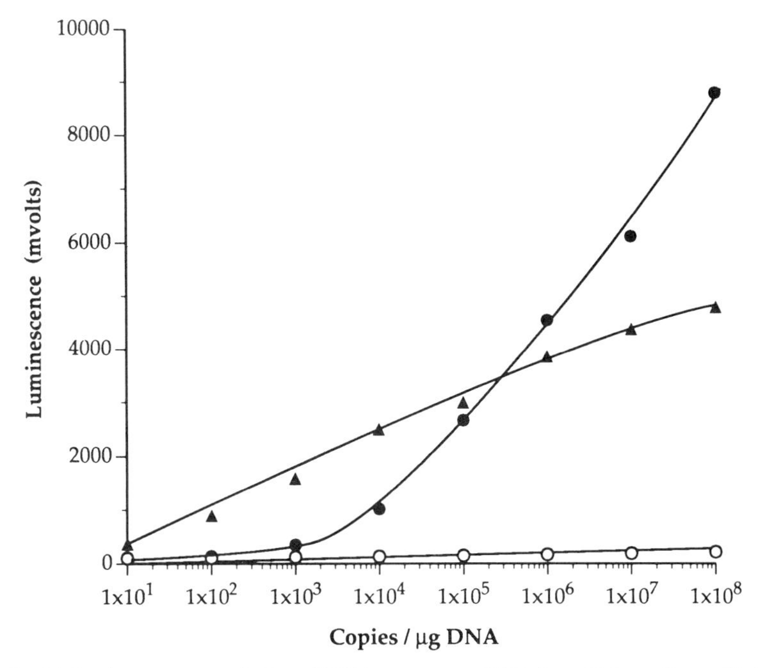


**Fig. 5** Bioluminescent quantitation of DNA sequences. DNA from HeLa cell was diluted in human DNA, and 2 $\mu$g of the mix was amplified. (▲) symmetric amplification and denaturation (●) asymmetric amplification, (○) nonspecific sequence.

## D. DNA Strand-Exchange Assay

Double-labeled sequences synthesized by PCR with biotinylated and FITC oligonucleotides primers are used as DNA probes. These probes are denatured and hybridized in solution in the presence of DNA sequences to be studied (homologous, different in size or sequence). The remaining reformed double-labeled hybrid is captured with a bioluminescent immunosorbent and detected with a streptavidin–G6PDH conjugate (Fig. 6). A similar method, using fluorescence and energy transfer, has been described for quantitative analysis of PCR sequences (Morrison et al., 1989), and we have modified the hybridization procedure in order to identify single base mutations or deletions. In equilibrium conditions, homology in sequence and in size between probes and targets is necessary to achieve complete disappearance of the signal. This new bioluminescent assay can be used for a rapid and nonradioactive analysis of sequences obtained by PCR.

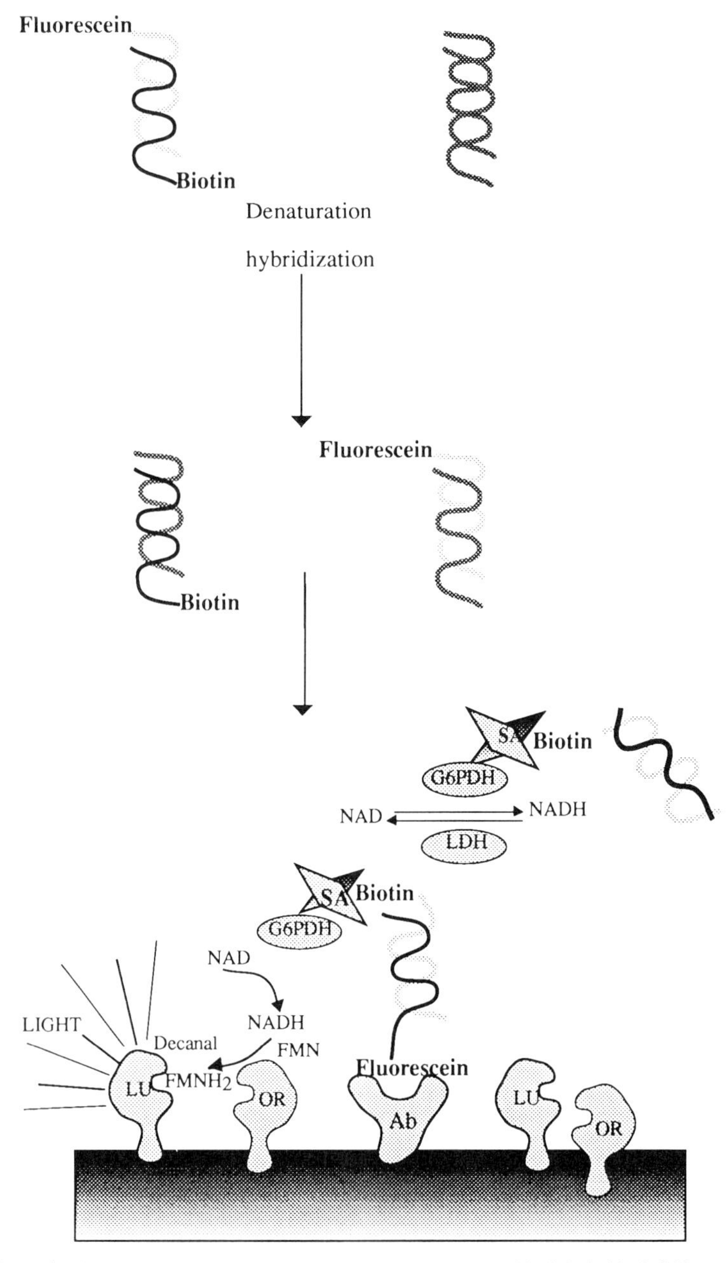


**Fig. 6** Principle of the DNA strand-exchange assay. The double-labeled hybrid is measured using a bioluminescent immunosorbent.

### 1. Quantitative Analysis of PCR Sequences

#### *a. Quantitation of Double-Labeled Probes*

Concentrations of double-labeled probes are determined by a sandwich assay. A standard curve is obtained using a known concentration of a synthetic double strand (a 20-mer biotinylated oligonucleotide hybridized with its 20-mer complementary oligonucleotide labeled with FITC). All dilutions and incubations are carried out in buffer A: 0.1 *M* phosphate buffer, pH 7.2, with 0.5 *M* NaCl, 0.1% gelatin, and denatured sonicated salmon sperm DNA (10 $\mu$g/ml).

1. Incubate 50 $\mu$l biotin-fluorescein probe dilutions ($10^{-9}$ to $10^{-12}$ *M*) with 50 $\mu$l bioluminescent anti-FITC adsorbent (1 g wet adsorbent suspended in 20 ml buffer A) and 50 $\mu$l streptavidin-G6PDH (2 mIU/ml) at room temperature for 2 hr under constant shaking.
2. Add the luminescent reagent: $10^{-3}$ *M* NAD, $3 \times 10^{-3}$ *M* G6P, $3 \times 10^{-3}$ *M* sodium pyruvate, $10^{-5}$ *M* FMN, $6 \times 10^{-5}$ *M* decanal, and 30 mIU/ml LDH in 0.1 *M* phosphate buffer, 0.15 *M* NaCl, and 1 g/liter BSA.
3. Measure photon counts for 10 sec.

Figure 7 shows a typical standard curve obtained with a double-labeled hybrid.

#### *b. Quantitation of Unlabeled Amplified Sequences*

1. Add 1 to 10 $\mu$l unlabeled sequences obtained by amplification with 10 femtomoles of the homologous sequence labeled with biotin and FITC in 100 $\mu$l buffer A. Complete with amplification buffer, since small amounts of detergent (Triton ×-100) are contained in this buffer and can lead to some inhibition of the bioluminescent reaction.
2. Incubate for 10 min at 95°C, cool rapidly to 60°C, and incubate for 2 hr.
3. Centrifuge the samples and add 50 $\mu$l of the medium to luminometer tubes with 50 $\mu$l streptavidin-G6PDH and 50 $\mu$l bioluminescent adsorbent and incubate as described above.

A standard curve must be edited using known amounts of target DNA (Fig. 7). Luminescence values obtained for biological samples are plotted on this curve to determine DNA sequence concentrations. If sequences

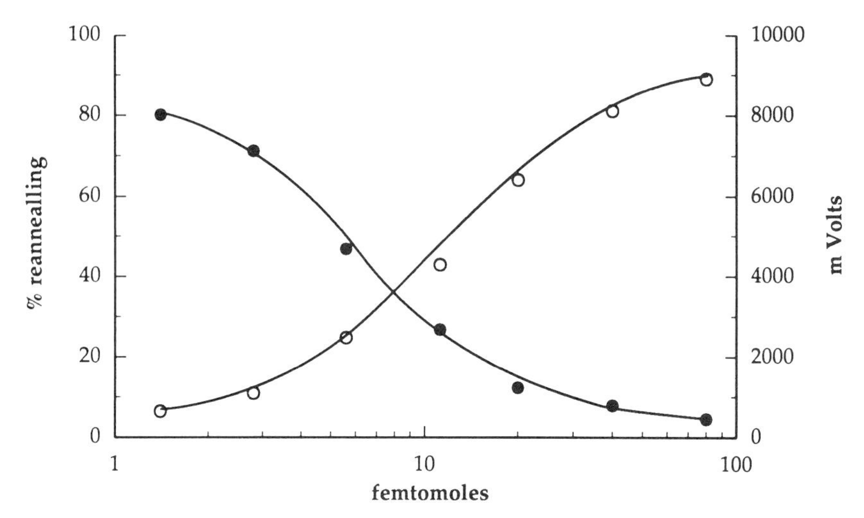


**Fig. 7** Standard curves using a bioluminescent immunosorbent. (○) Luminescence values for increasing amounts of double-labeled hybrid. (●) Reannealling percentage as a function of nonlabeled target amounts added to 10 femtomoles of double-labeled probe.

are identical and if the probe concentration [P] is known, the target concentration [T] can be determined from the equation:

$$[T] = [P] \times (1 - Rp)/Rp)$$

where Rp is the reannealling percentage of the probe.

In competitive assays the measurement range of the concentration is more restricted than in the sandwich assay. The advantage of the competitive assay is its more precise estimation of target amounts, since reannealling of the DNA target does not lead to underestimation, as does the sandwich assay.

## 2. Qualitative Analysis of PCR Sequences

If rehybridation is performed in conditions that hinder the formation of heteroduplex, hybridation occurs only between fully complementary sequences.

1. Add 1 μl unlabeled sequences obtained by amplification with 10 femtomoles of sequence labeled by biotin and FITC in 100 μl buffer A.
2. Incubate for 10 min at 95°C and then slowly decrease the incubation temperature (1° per 3 min) to 60°C.

3. Centrifuge the samples and add 50 $\mu$l of the incubation medium to luminometer tubes with 50 $\mu$l streptavidin–G6PDH and 50 $\mu$l bioluminescent adsorbent, and proceed as described above.

These hybridization conditions reduce the formation of heteroduplex. If luminescence values are identical to those obtained using a constant incubation temperature of 60°C, this indicates that labeled and nonlabeled sequences are identical. On the contrary, if the sequence differs from the unlabeled sequence by a single base pair, the signal is greater, as shown in Table II. The analysis of a polymorphic part of apolipoprotein A2 gene showing a single mismatch reveals very significant differences between the procedures used during hybridization assays.

# IV. CONCLUSIONS

Like alkaline phosphatase or peroxidase, glucose 6-phosphate dehydrogenase can be used to detect DNA probe on filters. We have developed various dot blot procedures to identify known DNA sequences. Glucose 6-phosphate dehydrogenase shows the same sensitivity as alkaline phosphatase, and luminescent reagent is commercially available. Only small amounts of reagents are required to soak the nitrocellulose filter; thus, this detection method is not expensive. It can be used at the same time as other detection systems, and several probes hybridized on the same sample can be detected successively.

**Table II**
Double-Labeled Probe Percentages Obtained after Hybridization under Various Conditions

| Incubation conditions[a] | Gradient of temperature | | Fixed temperature | |
|---|---|---|---|---|
| DNA | Probe 1 (%) | Probe 2 (%) | Probe 1 (%) | Probe 2 (%) |
| 1 | 6 | 41 | 5 | 4 |
| 2 | 7 | 42 | 6 | 6 |
| 3 | 6 | 39 | 4 | 5 |
| 4 | 40 | 8 | 8 | 6 |
| 5 | 38 | 5 | 3 | 4 |
| 6 | 14 | 16 | 7 | 8 |

[a] Amplified sequences (1 $\mu$l) are incubated with 0.1 $\mu$l of double-labeled probe. Probe 1 and 2 are 105-bp sequences that differ by a single bp in position 72. PCR sequences of DNA 1, 2, 3 are identical to probe 1, DNA 4 and 5 are identical to probe 2, and DNA 6 is heterozygous and shares both sequences.

Direct labeling of oligonucleotides by enzymes provides a rapid method for screening PCR-amplified products, but for routine analysis in clinical laboratories, the filter hybridization procedure is difficult to automate, and hybridization in solution is the preferred format. G6PDH can be used either for sandwich or DNA strand-exchange assays and to identify and quantify PCR products. Whereas the methods are straightforward and do not require any separation steps, the specificity is excellent. Sandwich assays can be used when asymmetric PCR is performed for rapid quantitation of viruses in biological samples. The DNA strand-exchange assay is a new method that allows good sequence quantification and is more precise than sandwich assays, but the measurement range is more limited. This method is very specific and, if the reaction is performed in conditions that favor the formation of homoduplex, a single base mismatch can be identified even in a long 200-bp sequence. This method can be used to compare PCR sequences to reference sequences and can be a method for rapid screening of allele variations or single base mutations.

## REFERENCES

Forster, A. C., McInnes, J. L., Skingle, D. C., and Symons, R. H. (1985). Nonradioactive hybridization probes prepared by the chemical labeling of DNA and RNA with a novel reagent, photobiotin. *Nucleic Acids Res.* **13,** 745–761.

Hopman, A. H. N., Wiegant, J., and Van Dujin, P. (1987). Mercurated nucleic acids probes, a new principle for nonradioactive *in situ* hybridization *Exp. Cell. Res.* **169,** 357–368.

Jablonski, E., and DeLuca, M., (1977). Purification and properties of the NADH- and NADPH-specific FMN oxidoreductase from *Beneckea harveyi. Biochemistry,* **16,** 2932.

Langer, P. R., Waldrop, A. A., and Ward, D. C. (1981). Enzymatic synthesis of biotin-labeled polynucleotides: Novel nucleic acid affintiy probes. *Proc. Natl. Acad. Sci. U.S.A.* **78,** 6633–6637.

Lebacq, P., Squalli, D., Duchenne, M., Pouletty, P., and Joannes, M. (1988). A new sensitive nonisotopic method using sulfonated probes to detect picogram quantities of specific DNA sequences on blot hybridizations. *J. Biochem. Biophys. Methods.* **15,** 255–266.

Lion, T., and Haas, O. A. (1990). Nonradioactive labeling of probe with digoxigenin by polymerase chain reaction. *Anal. Biochem.* **188,** 335–337.

Lo, Y-M. D., Mehal, W. Z., and Fleming, K. A. (1988). Rapid production of vector-free biotinylated probes using the polymerase chain reaction. *Nucleic Acids Res.* **16,** 8719.

Maniatis, T., Fritsch, E. F., and Sambrook, J. (1982).''Molecular Cloning. A Laboratory Manual.'' Cold Spring Harbor Laboratory, Cold Spring Harbor, New York.

Morrison, L. E., Halder, T. C., and Stols, L. M. (1989). Solution-phase detection of polynucleotides using interacting fluorescent labels and competitive hybridization. *Anal. Biochem.* **183,** 231–244.

Reisfeld, A., Rothenberg, J. M., Bayer, E. A., and Wilchek, M. (1987). Nonradioactive hybridization probes prepared by the reaction of biotin hydrazide with DNA. *Biochem. Biophys. Res. Commun.* **142,** 519–526.

Renz, M., and Kurtz, C. (1984). A colorimetric method for DNA hybridization. *Nucleic Acids Res.* **12,** 3435–3444.

Resnick, R. M., Cornelissen, M. T. E., Wright, D. K., Eichinger, G. H., Jan ter Schegget, H. S. F., and Manos, M. (1990). Detection and typing of human papillomavirus in archival cervical cancer specimens by DNA amplification with consensus primers. *J. Natl. Cancer. Inst.* **82,** 1477–1484.

Saiki, R. K., Gelfand, D. H., Stofffel, S., Scharf, S. J., Higuchi, R., Horn, G. T., Mullis, K. B., and Erlich, H. A. (1988). Primer-directed enzymatic amplification of DNA with a thermostable DNA polymerase. *Science* **239,** 487–494.

Saiki, R. K., Chang, C-A., Levenson, C. H., Warren, T. C., Boehm, C. D., Kazazian, H. H., and Erlich, H. A. (1988). Diagnosis of sickle-cell anemia and b-thalassemia with enzymatically amplified DNA and nonradioactive allele-specific oligonucleotides probes. *N. Engl. J. Med.* **319,** 537–541.

Saiki, R. K., Walsh, P. S., Levenson, C. H., and Erlich, H. A. (1989). Genetic analysis of amplified DNA with immobilized sequence-specific oligonucleotides probes. *Proc. Natl. Acad. Sci. U.S.A.* **86,** 6230–6234.

Térouanne, B., Carrie, M. L., Nicolas, J. C., and Crastes de Paulet, A. (1986). Bioluminescent immunoabsorbent for rapid immunoassays. *Anal. Biochem.* **154,** 118–125.

Urdea, M. S., Kolberg, J., Clyne, J., Running, J. A., Besemer, D., Warner, B., and Sanchez-Pescador, R. (1989). Application of a rapid nonradioisotopic nucleic acid analysis system to the detection of sexually transmitted disease-causing organisms and their associated antimicrobial resistances. *Clin. Chem* **35,** 1571–1575.

Van den Brule, A. J. C., Snijders, P. J. F., Gordijn, R. L. J., Bleker, O. P., Meijer, C. J. L. M., and Walboomers, J. M. M. (1990). General primer-mediated polymerase chain reaction permits the detection of sequenced and still unsequenced human papillomavirus genotypes in cervical scrapes and carcinomas. *Int. J. Cancer* **45,** 644–649.

# 10 Detection of Lanthanide Chelates by Time-Resolved Fluorescence

Timo Lövgren, Department of Biochemistry, University of Turku, Turku, Finland
Pertti Hurskainen and Patrik Dahlén, Wallac Biochemical Laboratory, Turku, Finland



*Nonisotopic DNA Probe Techniques*

# I. INTRODUCTION

The use of fluorescent labels has previously been limited by the high background fluorescence always present in the measurements. This sensitivity limitation has been overcome by the use of time-resolved fluorometry and labels with a long fluorescent decay time. In time-resolved fluorometry, the lifetime of the emission that occurs after a pulsed light excitation is followed at a certain wavelength. The actual measurement starts after the background fluorescence has decayed, which provides a substantial increase in sensitivity. The principle of the measurement is outlined in Fig. 1. The applications of time-resolved fluorescence of lanthanide labels in biotechnology has been extensively reviewed elsewhere (Soini and Lövgren, 1987).

The high specific activity of europium chelates, when measured by time-resolved fluorometry, provides a sensitivity that gives the label technology great potential in both immunoassays (Lövgren and Pettersson, 1990) and nucleic acid hybridizations. This chapter focuses on the recent status of time-resolved fluorometry and europium fluorescence when applied in various nucleic acid hybridization assays in our laboratory.

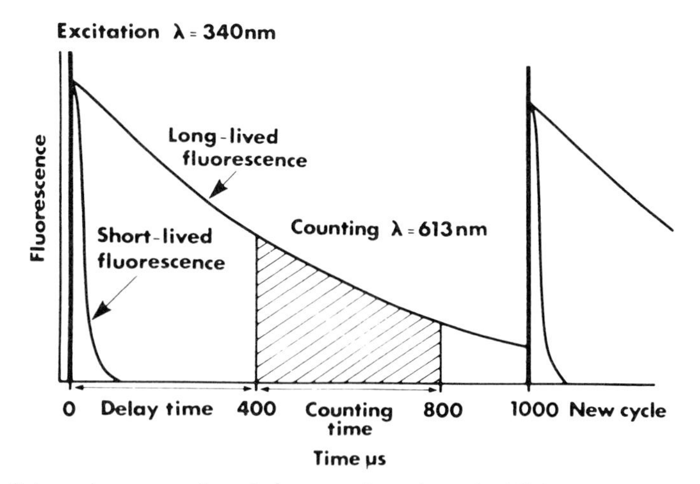


**Fig. 1** Schematic presentation of the operation of a pulsed-light source time-resolved fluorometer. For the measurement of europium using the DELFIA Enhancement Solution, the cycle time is 1 msec; a pulsed excitation occurs at the beginning of each cycle. The delay time after the pulsed excitation is 400 $\mu$sec; the actual counting time within the cycle has the same duration. The total measurement time is 1 sec.

## II. INDIRECT LABELING

The most widely used nonradioactive DNA markers today are indirect labels, i.e., the DNA probe can be modified with a marker that in turn can be detected by a secondary detection system. The biotin–streptavidin affinity pair is most commonly used, but hapten–antibody systems are also available for the indirect nonradioactive labeling of DNA. The advantage of the biotin–streptavidin system lies with the high affinity of biotin for streptavidin (dissociation constant $k_d = 10^{-15}$ M). The fact that streptavidin can bind four biotin molecules has made it possible to amplify the signal from the biotin molecule quite easily by making sandwich detection systems (Leary *et al.*, 1983). A wide variety of methods for the biotinylation of DNA have been developed. The biotin molecule can be incorporated into DNA probes either enzymatically (Langer *et al.*, 1981; Takahashi *et al.*, 1989) or chemically (Forster *et al.*, 1985; Viscidi *et al.*, 1986; Sheldon *et al.*, 1986; Reisfeld *et al.*, 1987). When detecting a biotin-labeled probe, the detection procedure usually involves the use of enzyme-conjugated streptavidin, and thus the final step is an enzymatic reaction resulting in a colorimetric, chemiluminescent, or fluorescent end product.

We have used europium-labeled (Eu-labeled) streptavidin and time-resolved fluorometry for the detection of biotinylated DNA probes (Dahlén, 1987; Dahlén *et al.*, 1987). In addition, an antibody labeled with Eu has been used for the detection of hapten-labeled DNA probes (Syvänen *et al.*, 1986). The use of time-resolved fluorometry in the detection of these non-radioactive probes provides a sensitive and simple detection methhod. The principle of using Eu-labeled streptavidin in the detection of biotinylated DNA probes is shown in Fig. 2. Since the Eu fluorescence is read in solution after an enhancement step (Hemmilä *et al.*, 1984), optimal results are obtained when a plastic surface, ideally a microtitration plate, is used as a solid support in the hybridization assay.

Europium-labeled streptavidin was applied for the detection of biotinylated probes in both dot-blot and sandwich hybridization assays (Dahlén, 1987; Dahlén *et al.*, 1987). Viral DNA (adenovirus type 2) was detected by attaching the target DNA onto microtitration wells before the hybridization reaction. The detection limit was 10 pg of target DNA added to the well, and quantitative results were obtained (Dahlén, 1987). In this method the target DNA must be extensively purified before immobilization onto the solid support, and the coating of the target is an overnight procedure. The limitations inherent in the direct hybridization format on plastic surfaces are omitted in sandwich hybridization. The $\beta$-lactamase gene was detected by this assay (Dahlén *et al.*, 1987). The detection limit was $4 \times 10^5$ molecules or 1.9 pg of target DNA. The performance of the test was simple and rapid, and quantitative results were obtained.

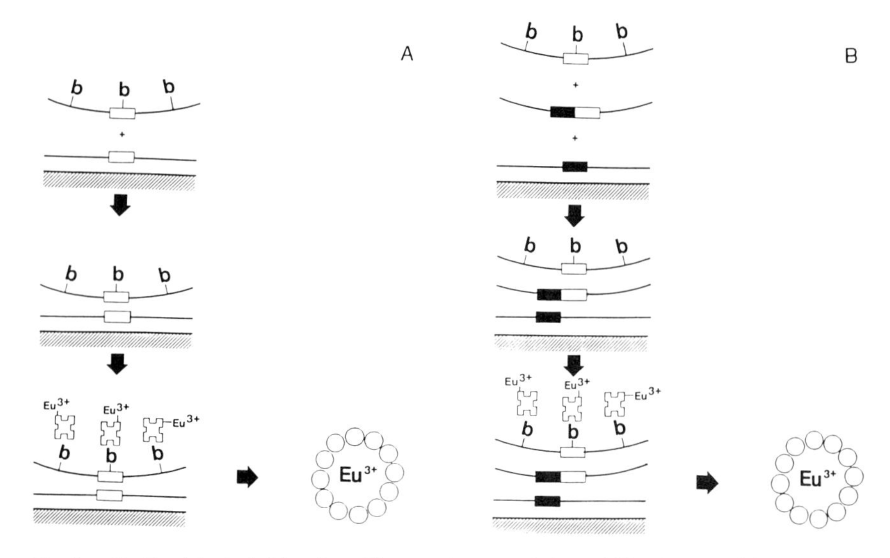


**Fig. 2** (A) Dot blot hybridization. The sample DNA is immobilized on a solid support. A biotinylated DNA probe is allowed to hybridize to its target sequences. The bound biotin-probe is detected by Eu-labeled streptavidin. The Eu-fluorescence is developed by adding Enhancement solution. (B) Sandwich hybridization. The sample is added simultaneously with the biotinylated DNA probe to the hybridization reaction. The captive probe, which has been attached to the solid support before the assay, and the biotinylated probe will hybridize to the target DNA. The bound biotin-probe is detected by using Eu-labeled streptavidin as above.

# A. Materials

## 1. Reagents

All reagents used should be of highest possible purity. Use nuclease-free compounds whenever available. The DNA probes needed can be prepared following the strategy outlined in previous publications (Ranki *et al.*, 1983; Dahlén *et al.*, 1987). Eu-labeled streptavidin was from Wallac Oy, (1244-360) and lysozyme was from Boehringer-Mannheim (Cat No. 107,255).

### *a. Solutions*

Coating buffer — 20 m*M* *tris*(hydroxymethyl)aminomethane (Tris)-HCl buffer, pH 7.5, containing 100 m*M* $MgCl_2$ and 150 m*M* NaCl.

| | |
|---|---|
| Hybridization buffer | 20 m*M* Tris-HCl, pH 7.5, containing 0.9 *M* NaCl, 90 m*M* Na-citrate, 50% formamide, 0.1% bovine serum albumin (BSA), 0.1% Ficoll, 0.1% polyvinyl pyrrolidone (PVP), 0.5% sodium dodecyl sulphate (SDS), 100 $\mu$g/ml denatured and sheared herring sperm DNA, and 3% polyethylene glycol (PEG). |
| Hybridization wash solution | 10 m*M* Tris-HCl, pH 7.5, containing 0.1% SDS, 50 $\mu M$ ethylenediaminetetraacetic acid (EDTA), and 50 m*M* NaCl. |
| Other solutions | 3 *M* NaOH, 0.1 *M* NaOH, 3 *M* HAc, 0.1 *M* HOAc, 10% SDS, 0.5 *M* Tris-HCl, pH 7.5, and 10 m*M* Tris-HCl, pH 7.5, containing 1 m*M* EDTA, DELFIA[1] Assay Buffer (Wallac Oy, Cat. No. 1244-106), DELFIA Wash Concentrate (Wallac Oy, Cat. No. 1244-500), DELFIA Enhancement Solution (Wallac Oy, Cat. No. 1244-105). |

***b. Biotinylation Reagents***

Several commercial kits are available for labeling of DNA probes with biotin, e.g., Bio-Nick, BRL Cat No. 8247SA. Please note that single-stranded DNA probes are biotinylated using either random priming or chemical labeling procedures.

**2. Instrumentation**

***a. Time-Resolved Fluorometer***

A TR-fluorometer is available from Wallac Oy, the DELFIA Research Fluorometer (Cat. No. 1234-001). The operational principle of this TR-fluorometer is described in Fig. 1.

***b. Other Laboratory Equipment***

Microtitration plates and equipment for handling of the microtitration plates, specially designed for the TR-fluorometry technology, are available from Wallac Oy. Microtitration Plate (Cat. No. 1244-550), DELFIA Plateshake (Cat No. 1296-001), DELFIA Platewash (Cat. No. 1296-024), DELFIA Plate Dispense (Cat. No. 1296-043).

[1] DELFIA is a registered trade mark of Wallac Oy, Finland.

## B. Procedures

### 1. Sandwich Hybridization on Microtitration Plates

The principle of sandwich hybridization is outlined in Fig. 2. Two DNA probes (a capture and a detector probe) complementary to two adjacent regions of the target DNA are needed. The capture probe is immobilized onto a solid support before the assay. In the hybridization reaction the target DNA, when present in the sample, forms a bridge between the two probes, and the detector probe is captured onto the solid support. The presence of the detector probe can be determined and thus also the presence of target DNA. When a biotinylated detector probe is used, it can be detected using Eu-labeled streptavidin.

#### *a. Coating of the Solid Support*

1. Linearize the capture probe (single-stranded) before attaching it to the solid support. Boil the capture probe in 0.1 *M* NaOH at 100°C for 10 min. Neutralize the probe solution with an equal volume 0.1 *M* HOAc and dilute the capture probe in the coating buffer to a concentration of 250 ng/ml.
2. Pipette 200 $\mu$l of the coating solution containing the capture probe into the wells of microtitration plates. Use Wallac microtitration plates, since these are of high binding quality, which ensures efficient binding of the DNA. If desired several probes can be used as capture probes simultaneously in one well. The amount of capture probe can be varied to determine the optimal coating concentration for any particular probe.
3. Incubate the strips overnight at room temperature (RT). Aspirate the wells, air dry the strips under a hood for 3 hr. Dry DNA-coated microtitration strips covered with adhesion tape are stable when stored at 4°C for at least 3 mo.

#### *b. Sample Pretreatment*

The samples added to the hybridization reaction can be relatively crude and yet they will not cause background problems. The sample pretreatment protocol given here is for bacterial cells (up to $5 \times 10^6$ cells can be used).

1. Pellet the cells by brief centrifugation ($1000 \times g$, 4 min). Suspend the cells in 10 $\mu$l 10 m*M* Tris-HCl, pH 7.5, containing 1 m*M* EDTA.
2. Lyse the cells by adding 2 $\mu$l lysozyme (10 mg/ml); incubate for

30 min at 37°C. Add 2 μl 10% SDS and 2 μl 3 *M* NaOH; boil the cell samples for 10 min at 100°C. Quickly cool on ice. Add 2 μl 3 *M* HOAc and 4 μl 0.5 *M* Tris-HCl, pH 7.5.

***c. The Hybridization Reaction***

1. Prehybridize the microtitration strips coated with the capture probe using 200 μl hybridization buffer at 42°C for 1 hr, then aspirate the strips.
2. Boil the biotinylated DNA probe at 100 °C for 10 min to ensure proper denaturation. Cool the probe on ice. Add the probe to fresh hybridization buffer at a probe concentration of 400 ng/ml.
3. Pipette 100 μl of this probe-containing hybridization buffer to each well. Add 10–20 μl of the lysed cell samples to individual wells. Include wells with similarly treated proper positive and negative DNA controls. The hybridization reaction is allowed to proceed overnight at 42°C.
4. Aspirate the wells and wash them by incubating with 200 μl of the hybridization wash solution at 65°C for 5 min. Repeat the washing step three times. Finally, rinse the wells at RT with the DELFIA Assay Buffer.

***d. Incubation with Eu-Labeled Streptavidin***

1. Dilute the Eu-labeled streptavidin in DELFIA Assay Buffer to a concentration of 1 μg/ml. Pipette 100 μl of the dilution to the wells and incubate with continuous shaking at RT for 1 hr.
2. Wash the strips with DELFIA wash solution (made from the DELFIA wash concentrate) six times, using the plate washer.

***e. Fluorescence Measurement***

1. Dispense 200 μl Enhancement Solution to the wells.
2. Shake at RT for 5 min, and measure the resulting fluorescence using the DELFIA Research Fluorometer. See Table I for typical results.

---

## 2. Useful Hints

***a. Sample Pretreatment***

It is important to ensure the proper neutralization of the sample solution. This can be checked by spotting 0.5 μl of sample solution on a pH paper. Where the samples contain precipitated material, this may cause increased background signals. A precipitate can be avoided by decreasing the

**Table I**
Detection of the β-Lactamase Gene[a]

| Controls[b] Copy number | Fluorescence (cps) | |
|---|---|---|
| 0 | 2300 ± 130 | |
| $5 \times 10^5$ | 3000 ± 150 | |
| $5 \times 10^6$ | 7100 ± 280 | |
| $5 \times 10^7$ | 31000 ± 1100 | |
| | Fluorescence | |
| Strains | (cps) | (+/−) |
| IHE 11045 | 2657 | − |
| IHE 11065 | 2563 | − |
| IHE 11039 | 2527 | − |
| IHE 11104 | 2477 | − |
| IHE 11071 | 2559 | − |
| KS 71 | 2677 | − |
| IHE 11002 | 2515 | − |
| IHE 12131 | 2561 | − |
| IHE 11020 | 6357 | + |
| IHE 11239 | 33637 | + |
| IHE 11051 | 7795 | + |
| IHE 11046 | 24048 | + |
| IHE 11026 | 45016 | + |
| IHE 11072 | 58200 | + |

[a] From colonies (n = 3) of *Escherichia coli* strains by sandwich hybridization and $Eu^{3+}$-labeled streptavidin (Dahlén *et al.*, 1987). The mean of the background + 3 SD (2700 cps) was used as the cut-off value.
[b] Pure plasmid pBR 322 DNA samples

amount of cells used in the test, by prolonging the lysozyme incubation time, or by centrifuging the samples after pretreatment and carefully taking only the supernatant for the hybridization test.

If other than bacterial cells are to be analyzed, the protocol given can be modified by including Proteinase K and 0.5% SDS rather than lysozyme. It is also possible to use alkaline denaturation alone. A 300 m*M* NaOH solution containing 1.9% SDS is used when boiling the cells for 15 min at 100°C. The samples are neutralized using an equal volume of 300 m*M* HOAc.

### *b. Coating of the Capture DNA*

It has been reported that UV light improves the binding of DNA to plastic surfaces (Dahlén, 1987). UV light does not improve binding to the microtitration plates used here, but with other plastic qualities, the use of UV light is often necessary.

Several captive probes may be used simultaneously. These may represent probes from other adjacent regions or may be probes that capture both polarities of the target DNA. Optimization of the amount of any particular capture probe is necessary.

The quality of the capture probe is important. The probe should be linearized, but not too sheared. If a suitable restriction site is available in the vector, it is recommended that the probes be linearized by specific cleavage using restriction enzymes. Another aspect is the presence of possible homologous regions with the detector probe. The vector and probing sequences must under no circumstances be homologous within the pair of capture and detector probes. Possible contaminating sequences may also give rise to a hybridization reaction directly between the detector probe and the solid phase capture probe.

#### c. *Hybridization Reaction*

The stringency of the hybridization reaction can be controlled by the amount of formamide added, the temperature used, etc. The kinetics by which the hybridization proceeds can be improved by including more PEG, or by increasing the amount of probe.

The stringency of the washes can be adjusted by altering washing temperatures or the salt concentration. Temperatures above 75°C should be avoided, since the strips can be physically damaged.

#### d. *The Detection Step*

The optimal concentration of Eu-labeled streptavidin is 0.2–2 $\mu$g/ml. It is recommended that the Enhancement Solution be dispensed using a dedicated dispenser to avoid any contamination with Eu from pipettes.

## III. CHEMICAL EUROPIUM LABELING OF DNA PROBES

Labeling of DNA probes with stable Eu chelates allows the development of simple and sensitive DNA hybridization assays employing time-resolved fluorometry. The sensitivity of Eu-labeled DNA probes depends on the labeling degree and hybridization efficiency of the labeled probe. Thus the Eu-labeling chemistry should allow the incorporation of a large number of Eu chelates to DNA without sacrificing its hybridization properties.

To overcome the chemical inertness of DNA, a two-phase approach is used to label DNA chemically with Eu chelates (Hurskainen *et al.*, 1991). First, DNA is transaminated to introduce primary aliphatic amino groups into DNA (Fig. 3). Cytosine bases undergo a transamination reaction in the presence of sodium bisulfite and an amine (Shapiro and Weisgras, 1970). By using a diamine such as ethylenediamine, free aliphatic amino groups are incorporated into single-stranded DNA (Viscidi *et al.*, 1986). Subsequently the amino-modified DNA is reacted with an isothiocyanate derivative of the Eu chelate (Hurskainen *et al.*, 1991). The Eu-labeling process including purifications is a multistep procedure but, on the other hand, large amounts of DNA can be labeled.

The transamination reaction is a very efficient method of modifying DNA, and allows the Eu labeling of practically all deoxycytidine residues. However, Eu labeling affects DNA by lowering the thermal stability and hybridization efficiency (Hurskainen *et al.*, 1991). The incorporation of 4 to 8 Eu chelates per 100 bases is the optimum degree of labeling that makes sensitive and specific hybridization assays possible. Eu-labeled DNA probes are directly detectable, making the detection step simple and time saving compared to the detection of biotinylated and hapten-labeled probes. In addition, interpretation of numerical results obtained with Eu-labeled probes is simple.

Eu-labeled DNA probes have been used in the detection of adenovirus DNA in clinical specimens (Dahlén *et al.*, 1988). After pretreatment, specimens were fixed to a nitrocellulose filter and hybridized against Eu-labeled adenovirus DNA. Simple and sensitive sandwich (Ranki *et al.*, 1983) and solution hybridization assays (Syvänen *et al.*, 1986) with Eu-DNA probes have also been made in microtitration wells for the detection of cytomegalovirus and human papillomavirus (Huhtamäki *et al.*, in prepa-

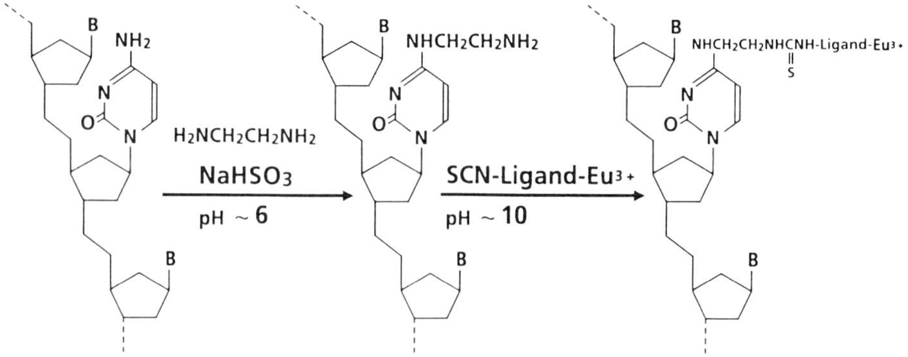

**Fig. 3** Chemical labeling of DNA with a europium chelate. In the first reaction the cytosine bases in DNA are transaminated in the presence of sodium bisulfite and ethylenediamine. In the second reaction the modified DNA is reacted with an isothiocyanate derivative of the europium chelate.

ration). The detection limit in these assays has been $5 \times 10^5$ molecules of target DNA.

## A. Materials

### 1. Reagents

The following reagents were used: sodium metabisulfite (Aldrich, Cat. No. 25555-6), ethylenediamine (Aldrich, Cat. No. E2626-6), hydroquinone (Aldrich, Cat. No. H1790-2), Eu chelate of 4-[2-(4-isothiocyanatophenyl)-ethyl]-2,6-*bis*[*N*,*N*-*bis*(carboxymethyl)aminoethyl]pyridine (Wallac Oy), photobiotin (long arm) (Vector, Cat. No. SP-1020), streptavidin (Porton Products, United Kingdom).

#### *a. Solutions*

Sample pretreatment solution: 2% SDS in 0.3 *M* NaOH

Neutralization solution: 0.3 *M* HOAc in 0.1 *M* Tris-HCl pH 8.5.

DNA coating buffer: 20 m*M* Tris-HCl, pH 7.5, containing 100 m*M* $MgCl_2$ and 150 m*M* NaCl.

2x Sandwich hybridization solution: 140 m*M* HEPES, pH 8.0, containing 0.1% SDS, 1.7 *M* NaCl, 7x Denhardt's solution, 200 μg/ml herring sperm DNA, 200 μg/ml sodium polyacrylate (average MW 90,000), 0.2 m*M* EDTA, and 3% PEG. Incubate the solution at 40 to 50°C for 30 min and filter through a sterile membrane (0.45 μm). Store the solution at RT.

Coating buffer for biotinylated protein: 50 m*M* $K_2HPO_4$, pH 9.0, containing 0.05% Na-azide and 0.9% NaCl.

2 x Solution hybridization mixture: 140 m*M* HEPES, pH 8.0, containing 0.1% SDS, 1.7 *M* NaCl, 7 x Denhardt's solution, 150 μg/ml herring sperm DNA, 0.2 m*M* EDTA, and 3% PEG. Filter the solution through a sterile membrane (0.45 μm) and store at RT.

Hybridization wash solution: 10 m*M* Tris-HCl, pH 8.5, containing 0.1% SDS, 0.05 m*M* EDTA, and 50 m*M* NaCl.

Other solutions: Assay Buffer (Wallac Oy, Cat. No. 1244-106), DELFIA Wash Concentrate (Wallac Oy, Cat. No. 1244-500), Enhancement Solution (Wallac Oy, Cat. No. 1244-105).

### 2. Component Preparation

#### *a. Coating of DNA on the Solid Support*

See the procedure in Section II,B,1.

***b. Biotinylation of Protein (IgM)***

1. Add a 50-fold molar excess of biotin-amidocaproate *N*-hydroxysuccinimide ester in 200 μl dry *N,N*-dimethylformamide to 10 mg of IgM in 1 ml 50 m*M* sodium carbonate pH 9.0 and incubate for 3 hr.
2. Remove the unreacted biotin by running the reaction mixture twice on PD-10 columns (Pharmacia) using 20 m*M* pH 7.5, containing 0.05% Na-azide and 0.9% NaCl as eluent. Store the biotinylated protein at 4°C.

***c. Coating of Biotinylated Protein***

1. Use microtitration plates of high-binding quality (Wallac Oy, Cat. No. 1244-550). Dilute the biotinylated IgM in the coating buffer (200 ng/ml) and pipette 200 μl to each microtitration well. Cover the plate with adhesive tape and incubate at 42°C overnight.
2. Wash the plate with DELFIA Wash Solution three times.
3. Prepare a 1 μg/ml solution of streptavidin in DELFIA Assay Buffer and add 200 μl/well. Incubate at RT for 3 hr. Wash the microtitration plate four times with DELFIA Wash Solution. Store the coated plate at 4°C covered with adhesive tape.

***d. Photobiotinylation of DNA***

Follow the manufacturer's (Vector) instructions.

***e. Eu Labeling of DNA***

1. Prepare the transamination solution just before use. Dissolve 475 mg sodium metabisulfite in 1 ml water in a glass tube. Place the tube on ice and add 1 ml ethylenediamine. Titrate the solution to pH 6–6.2 by cautiously adding 2.1–2.3 ml concentrated HCl. Add 50 μl hydroquinone (500 m*M* in ethanol), and adjust the final volume to 5 ml with water. Keep the transamination solution at 40 to 50°C to prevent precipitation.
2. Transaminate naturally single-stranded DNA (e.g., M13 DNA) as follows. Add nine volumes of transamination solution (450 μl) to 50 μg DNA (1 μg/μl) in an Eppendorf tube. Vortex carefully and incubate in a water bath set to 42°C for 1 hr 45 min.
3. After the reaction, place the tube on ice and add 28 μl 6 *M* NaOH to stop the transamination reaction. Dialyze the reaction mixture overnight at RT against four exchanges of water. If a precipitate is formed in the dialysis tube, add 250 μl 6 *M* NaOH/liter of dialysis solution to dissolve the precipitate.
4. After dialysis, concentrate the solution to about 300 μl in a vacuum centrifuge and precipitate the DNA with ethanol. Dissolve

the DNA in 100 $\mu$l of 0.1 *M* $Na_2CO_3$, pH 9.8, containing 0.1 m*M* EDTA.

5. Start the Eu-labeling process of double-stranded DNA (e.g., plasmid DNA) by denaturing (98°C, 10 min). Then add nine volumes of transamination solution (450 $\mu$l) to the cooled DNA solution (50 $\mu$g, 1 $\mu$g/$\mu$l). Vortex the reaction mixture well and incubate at 98°C for 5 min. Plasmid DNA requires denaturing conditions during transamination because it tends to renature at 42°C. Stop the reaction and purify the transaminated DNA as above.
6. Add the transaminated DNA in 100 $\mu$l 0.1 *M* $Na_2CO_3$, pH 9.8, containing 0.1 m*M* EDTA to a small test tube containing 1 mg of an isothiocyanate derivative of the Eu chelate. After dissolving the chelate, incubate at RT overnight.
7. Purify the Eu-labeled DNA probe on a Sephacryl S-400 column (47 × 0.7 cm) equilibrated and eluted with 10 m*M* Tris-HCl, pH 8.5, containing 0.1 *M* NaCl. The labeled DNA is eluted in the void volume and is well separted from the unreacted chelate. Follow the chromatography spectrophotometrically and measure, in addition, the Eu content in each fraction (1 ml) by diluting aliquots (e.g., 1:1000) in 200 $\mu$l Enhancement Solution. Pool the fractions containing Eu-labeled DNA, measure $A_{260}$ and determine the Eu concentration against an $EuCl_3$ standard solution. The contribution of the Eu chelate to the measured $A_{260}$ is

$$A_{260}\ \text{correction} = 0.0166 \times [\text{Eu}]$$

where Eu concentration is in $\mu$moles/liter. Correct the obtained $A_{260}$ value and use it to calculate the DNA concentration. An $OD_{260} = 1$ of naturally single-stranded DNA is equivalent to a concentration of 37 $\mu$g/ml, and $OD_{260} = 1$ of plasmid DNA is equivalent to 40 $\mu$g/ml after Eu labeling. Subsequently, calculate the concentration of nucleotides by dividing the DNA concentration with the average molecular weight of nucleotide residue in DNA (330 g/mole). The labeling degree is obtained using the following formula:

$$\text{Labeling degree (\%)} = \frac{\text{Eu concentration } (\mu M)}{\text{nucleotide concentration } (\mu M)} \times 100$$

The procedure typically gives labeling degrees between 6 and 8%. The yield of DNA varies between 40 and 70%.

The stability of the Eu-labeled DNA probe is improved with EDTA. Adjust the EDTA concentration to be 12-fold compared to the Eu concentration. Eu-labeled DNA probes are stable for at least 18 months at 4°C or −20°C.

### 3. Instrumentation

See Section II,A,2.

## B. Procedures

### 1. Sandwich Hybridization

The principles of sandwich hybridization have been described in Section II.

#### *a. Sample Pretreatment*

Crude samples can be analyzed with Eu-labeled DNA probes in sandwich hybridization assays. The following pretreatment protocol is suitable, e.g., for cervical scrapes.

1. The cells from the spatula or cytobrush are suspended in phosphate-buffered saline (PBS) and briefly centrifuged (5 min, 1000 × g). Suspend the pelleted cells in 25 μl PBS, add 25 μl 0.3 *M* NaOH containing 2% SDS, and boil for 12 min.
2. After cooling on ice and spinning down, neutralize the samples with 25 μl 0.3 *M* acetic acid in 0.1 *M* Tris-HCl, pH 8.5. Check that pH is 7–8.5 by applying 0.5 μl of the neutralized sample on an indicator paper.

#### *b. Hybridization Reaction*

The sandwich hybridization can be done either at 42°C in the presence of formamide or at 65°C without formamide. The problem in the hybridization at 42°C is the fact that only small sample volumes (10–20 μl) can be used in the total hybridization volume of 200 μl. On the other hand, sample volumes of 50 μl or more are possible when hybridization is performed at 65°C.

1. In order to minimize the nonspecific binding of probe DNA, prehybridize the microtitration wells with 200 μl of 1× hybridization solution per well at 65°C for at least 30 min. If the Eu-labeled DNA probe is double stranded, denature it by boiling for 10 min, and cool on ice.

2. Add the single-stranded or denatured probe in 25 μl to each 100 μl of the 2× hybridization solution. The optimal concentration of Eu-labeled DNA probes during hybridization is 200–400 ng/ml. Stop the prehybridization by aspirating the wells, and add 125 μl of hybridization solution containing Eu-labeled DNA probe to the wells.
3. Pipette the pretreated samples (75 μl) into the wells. Cover the wells carefully with adhesive tape, and place the microtitration plate in a humidified box at 65°C for overnight hybridization.
4. Wash the wells three times with 250 μl hybridization wash solution at 50°C for 20 min. Finally rinse the wells three times with 250 μl of the same solution at RT.

***c. Fluorescence Measurement***

1. Add 200 μl of Enhancement Solution to the microtitration wells.
2. Shake the wells (position "slow" in the DELFIA Plateshake) at RT for 25 min, and measure fluorescence in a time-resolved fluorometer.

## 2. Solution Hybridization

In solution hybridization assays, one probe is labeled with an affinity label and the other probe, with a detectable label (Syvänen *et al.*, 1986). The target DNA forms a bridge between the two labeled probes (Fig. 4). The affinity-labeled probe facilitates the collection of formed hybrids onto the solid phase, and subsequently the hybridized probe labeled with a detectable marker can be measured. We have used a Eu chelate as the detectable marker, biotin as the affinity label, and streptavidin-coated microtitration wells as the solid phase. Solution hybridization combines the high specificity and easy sample pretreatment of sandwich hybridization with the rapid reaction rate of a solution phase reaction.

***a. Sample Pretreatment***

Crude samples can also be analyzed in solution hybridization assays as in sandwich hybridization (see Section III,B,1).

***b. Hybridization Reaction***

The optimal amount for Eu-labeled probes is around $2 \times 10^9$ molecules and for photobiotinylated probes $2–3 \times 10^9$ molecules per reaction.

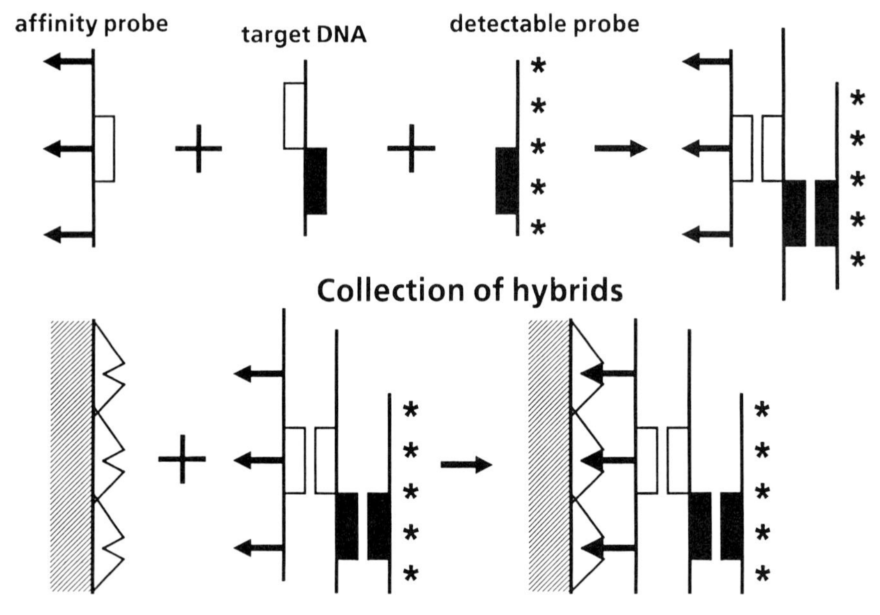


**Fig. 4** Schematic presentation of the solution hybridization. First two different probes, one labeled with an affinity label (e.g., biotin) and the other labeled with a detectable label (e.g., europium), hybridize with the target DNA in solution. In the next step, the formed hybrids are collected onto an affinity (e.g., streptavidin) matrix.

1. Add the single-stranded and denatured probes in 25 $\mu$l to each 100 $\mu$l 2× solution hybridization mixture. Combine in 1.5 ml Eppendorf tubes pretreated samples (75 $\mu$l) and 125 $\mu$l solution hybridization mixture containing the probes.
2. Vortex and hybridize at 65°C for 3 hr. Transfer the content from the hybridization tubes to individual microtitration wells coated with biotinylated IgM and saturated with streptavidin. Incubate at RT or 37°C for 2 to 3 hr. (This two-step hybridization assay can also be performed in one step. Combine the samples and hybridization solution containing the probes in microtitration wells. Cover the wells carefully with adhesive tape, and incubate at 65°C in a humidified box overnight).
3. Wash the wells twice for 5 min at 50°C with 250 $\mu$l of the hybridization wash solution. Additionally, rinse the wells four times with 250 $\mu$l of the same wash solution at RT.

### *c. Fluorescence Measurement*

1. Add 200 $\mu$l Enhancement Solution to microtitration wells and shake for 25 min.

2. Measure the fluorescence in a time-resolved fluorometer. See Table II for typical results achieved in the detection of HPV 16 DNA (Huhtamäki *et al.*, manuscript in preparation).

### 3. Useful Hints

#### *a. Eu Labeling*

The Eu chelate used for DNA labeling is an isothiocyanate analog, which requires that the transaminated DNA to be labeled with the Eu chelate must be absolutely free from other compounds able to react with isothiocyanate (such as Tris and other amino group-containing reagents).

Special care should be taken to avoid contamination of the laboratory with Eu during the labeling procedure. If possible, the Eu-labeling reaction should be done in a laboratory where hybridization tests are not performed.

#### *b. Hybridization Procedures*

The Eu chelate for DNA labeling is nonfluorescent and requires the use of Enhancement Solution in the measurement. In practice, Eu-labeled DNA probes are not suitable for *in situ* and Southern hybridizations. Moreover, Eu-labeled probes are not well suited for dot blot hybridizations on filters, because each dot has to be measured separately in En-

**Table II**
Detection of HPV 16 DNA Using Solution Hybridization[a]

| HPV 16 DNA molecules | Fluorescence (cps) | CV (%) |
|---|---|---|
| 0 | 968 | 16.7 |
| $10^6$ | 3299 | 7.1 |
| $10^7$ | 22403 | 9.5 |
| $10^8$ | 171840 | 7.8 |
| $10^9$ | 458544 | 8.4 |

[a] Three nonoverlapping fragments of HPV 16 DNA were used as probes. Two fragments cloned in M13 DNA were Eu-labeled and one fragment cloned in pBR 322 was biotinylated. Hybridization was performed using the one-step procedure in microtitration wells. The results are mean values based on four determinations.

hancement Solution. Whenever possible, use microtitration plates as the solid-phase matrix.

# IV. ENZYMATIC EUROPIUM LABELING OF DNA PROBES

The first efficient *in vitro* method for labeling DNA was nick translation (Rigby *et al.*, 1977), which requires the action of two enzymes, namely DNA polymerase and deoxyribonuclease. In the presence of all four deoxynucleotide triphosphates, one or more of which carry a label, these enzymes incorporate the labeled nucleotide into double-stranded DNA.

A more recently introduced alternative method for the enzymatic labeling of DNA is random priming (Feinberg and Vogelstein, 1983). A set of random oligonucleotides, usually hexanucleotides, is allowed to anneal with single-stranded DNA. The annealed oligonucleotides serve as primers for the Klenow fragment of DNA polymerase I, which in the presence of all dNTPs, synthesizes a DNA strand complementary to single-stranded DNA. If one or more of the four dNTPs are labeled, the newly synthesized DNA strand becomes labeled. Random priming is the labeling method for single-stranded DNA (e.g., M13 DNA) but also for double-stranded DNA, which has to be denatured before the labeling reaction.

A Eu-labeled 2′-deoxycytidine 5′-triphosphate (Eu-dCTP) (Fig. 5) has been synthesized to be used for the enzymatic labeling of DNA. Eu-dCTP is a substrate for the holoenzyme and the Klenow fragment of DNA polymerase I. The reaction rates of nick translation and random priming are somewhat slower when dCTP is replaced by Eu-dCTP. However, Eu-labeled DNA probes of high specific activity can be obtained with both of the enzymatic procedures. The properties and use of enzymatically Eu-labeled DNA probes are identical to those of chemically labeled ones.

## A. Materials

### 1. Reagents

DNA polymerase I from *Escherichia coli* (several suppliers, e.g., BRL, Pharmacia-LKB Biotechnology)

Deoxyribonuclease I from bovine pancreas (several suppliers, e.g., Pharmacia-LKB Biotechnology)

Eu-dCTP (Wallac Oy)

Oligolabeling kit (Pharmacia-LKB Biotechnology, Cat. No. 27-9250-01)

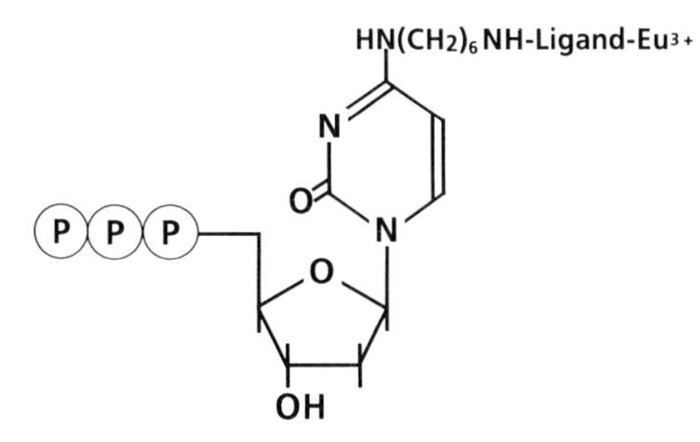


**Fig. 5** Eu-labeled dCTP for enzymatic labeling of DNA.

BSA, DNase free (Pharmacia-LKB Biotechnology, Cat. No. 27-8915-01)

### *a. Solutions*

10× nick translation buffer: 0.5 *M* Tris-HCl, pH 7.5, containing 0.1 *M* $MgCl_2$, 1 m*M* dithiothreitol, and 500 μg/ml DNase-free BSA.

Deoxynucleotide triphosphate solution: dATP, dCTP, dGTP, and dTTP, 125 μ*M* of each in 50 m*M* Tris-HCl, pH 7.5.

Eu-dCTP solution: 0.5 m*M* Eu-dCTP in 50 m*M* Tris-HCl, pH 7.5.

Deoxyribonuclease I stock solution: 1 mg/ml solution of deoxyribonuclease I in 0.15 *M* NaCl, containing 50% glycerol. Divide the solution into small aliquots.

Store all the above solutions at −20°C.

## 2. Component Preparation

### *a. Nick Translation Using Eu-dCTP*

1. Keep all the necessary solutions on ice after thawing. Make a proper dilution of DNase I stock solution (1:1000–1:10,000) in water. The exact amount of DNase I to be added must be estimated for each batch of enzyme. Make the following solution in an Eppendorf tube placed on ice:

| | |
|---|---|
| 10× nick translation buffer | 5 μl |
| dNTP solution | 8 μl |
| Eu-dCTP | 4 μl |
| DNA | 0.5–2μg |
| DNA polymerase I (2.5 U/μl) | 2 μl |
| DNase I | to be estimated |
| Add water to 50 μl | |

2. Vortex and spin down briefly. Incubate the reaction mixture in a water bath at 15°C. Allow the reaction to proceed for 3 hr and purify the Eu-labeled DNA immediately by gel filtration (e.g., Pharmacia NICK column) using 50 m*M* Tris-HCl, pH 8.5, containing 0.1 *M* NaCl as eluent. Do not stop the reaction by adding EDTA, because Eu may dissociate from labeled DNA. Store Eu-labeled DNA probe at −20°C.

***b. Random-Priming Reaction***

Random-priming labeling kits from several manufacturers have been tested with Eu-dCTP. The best results have been achieved with Pharmacia's random priming system (Oligolabeling Kit). The kit contains all components except dCTP and Eu-dCTP.

1. Use 1 nmole dCTP and 2 nmoles Eu-dCTP in a 50 $\mu$l reaction volume. Up to 1 $\mu$g DNA can be labeled, and the optimum reaction time is 1.5–2 hr at 37°C.
2. Purify the labeled probe immediately by gel filtration (e.g., Pharmacia NICK column) using 50 m*M* Tris-HCl, pH 8.5, containing 0.1 *M* NaCl. Do not stop the reaction by adding EDTA because Eu may dissociate from labeled DNA. Store the Eu-labeled DNA at −20°C.

### 3. Instrumentation

See Section II.

## B. Procedures

### 1. Hybridizations

Enzymatically Eu-labeled DNA probes behave like chemically labeled ones. Suitable hybridization procedures for Eu-labeled DNA probes have been described in the section on chemical Eu labeling of DNA probes.

# V. EUROPIUM-LABELED OLIGONUCLEOTIDES

Short synthetic pieces of DNA, oligonucleotides, can be used as specific hybridization probes. Indeed, the use of oligonucleotides as probes has significantly increased over the last few years. This is mainly owing to the breakthrough of *in vitro* DNA amplification methods. In particular the polymerase chain reaction (PCR) has become popular. Target sequences can be amplified up to $10^8$ times or even more before detection by hybridization. This facilitates the use of synthetic oligonucleotide DNA probes, previously limited in use because of their lack of sensitivity. The advantages of oligonucleotides include no need for denaturation, rapid hybridization kinetics, and high specificity. Indeed, short oligonucleotides (shorter than 20 bases) are able to detect even single point mutations (Wallace *et al.,* 1979; Ikuta *et al.,* 1987). In addition, these probes can be produced in large quantities under well-controlled conditions in an oligonucleotide synthesizer.

Several methods for labeling oligonucleotides with Eu chelates have been developed (Sund *et al.,* 1988). The preferred method is to incorporate primary amino groups into the oligonucleotide during synthesis. By using an amino-modified building block, such as the diaminohexane-modified deoxycytidine phosphoramidite, the functionalized base can be introduced at any position during the synthesis. It is possible to introduce up to 50 modified bases to the 5′ end of the oligonucleotide, although a tail of 15 to 25 modified nucleotides is generally used. An Eu chelate containing isothiocyanate as the reactive group is reacted with the primary amino groups after deprotection and purification of the oligonucleotide-containing modified bases. We have successfully labeled a single oligonucleotide with 35 Eu chelates, even though 15 to 20 Eu chelates are sufficient for a very good sensitivity.

The melting temperature ($T_m$) and the hybridization kinetics are not significantly affected by the labeling procedure. The sensitivity that can be achieved in a hybridization test, using oligonucleotides labeled with Eu, is $4 \times 10^6$ to $1 \times 10^7$ molecules of target DNA depending on the hybridization format, degree of labeling, etc. This sensitivity is at least comparable to that achieved with $^{32}P$-labeled probes. An important feature of the Eu-labeled oligonucleotides is the possibility to quantitate the amount of target DNA detected.

Eu-labeled oligonucleotides have been used as hybridization probes for the detection of DNA immobilized on a solid support, as well as in a rapid solution hybridization test (Sund *et al.,* 1988; Dahlén *et al.,* 1991b). In a model study, Eu-labeled oligonucleotides have been applied to the detection of $\lambda$ phage DNA immobilized in microtitration wells. The detection sensitivity in this direct hybridization test was $4 \times 10^6$ target molecules or 200 pg (Sund *et al.,* 1988). In this hybridization format, however, the target DNA is immobilized onto a plastic surface in an overnight coating procedure. This is impractical and time-consuming, and therefore other hybridization formats are needed.

When detecting *in vitro* amplified nucleic acid sequences it is possible to use a rapid *in solution* hybridization test, in which two probes complementary to the amplified DNA are allowed to hybridize simultaneously with the amplified target sequence. One of the probes is labeled with Eu chelates, and the other is labeled with biotin. The formed hybrid is collected onto streptavidin-coated strips employing affinity-based collection. The assay principle is outlined in Fig. 6. The sensitivity of the test is $10^7$ target molecules. This methodology has been successfully applied to a wide range of analytes, human immunodeficiency virus (HIV)-1 (Dahlén *et al.*, 1991b), HIV-2 and human T-cell leukemia virus (HTLV)-I and II (Iitiä *et al.*, submitted). The use of PCR and a time-resolved, fluorescence-based hybridization test makes it possible to detect as few as five copies of proviral target DNA. It is, in addition, possible to quantitate the amount of target DNA originally present in the sample using a 100–1000-fold linear concentration range.

Figure 7 describes schematically the design strategy of the amplifiable/detectable region, primers and probes. The nucleic acid sequences to be

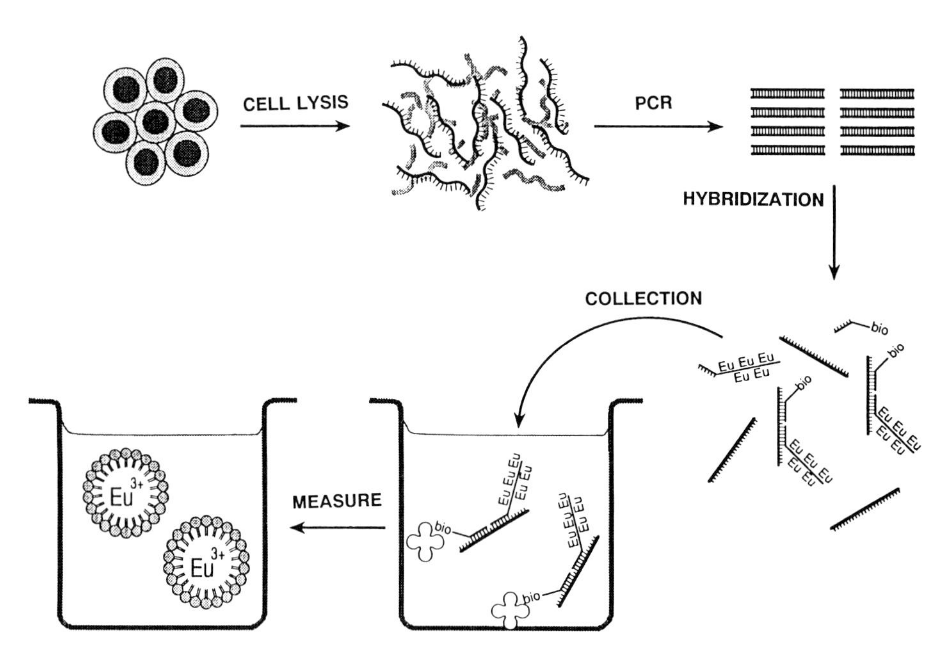


**Fig. 6** Schematic presentation of PCR amplification and a solution hybridization assay employing labeled oligonucleotides. The cells are lysed, and the PCR is performed on the crude sample. Two probes, one biotinylated and the other labeled with Eu complementary to the amplified fragment, are hybridized to the same target. The hybrids are subsequently collected onto streptavidin-coated microtitration wells. The bound europium dissociated with the Enhancement Solution is measured in a time-resolved fluorometer.

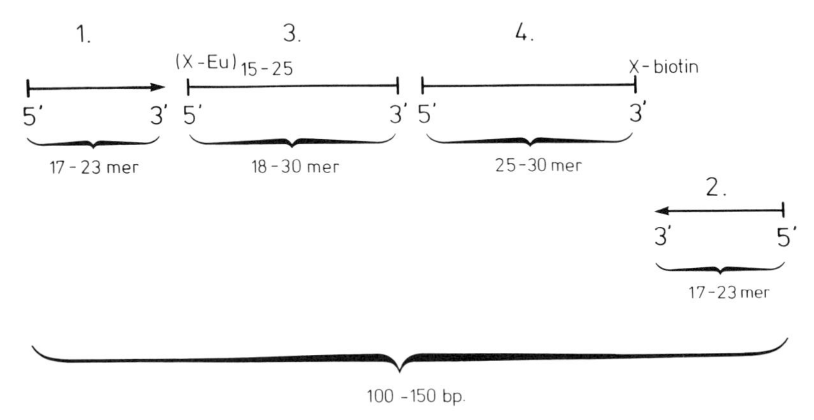


**Fig. 7** Typical design of amplifiable region of primers and probes. 1, sense primer; 2, antisense primer; 3, europium-labeled sense probe; and 4, biotin-labeled sense probe.

amplified should be 100–150 bp long. Generally accepted guidelines for the rationale of choosing the amplifiable sequence, designing the primers, etc., for the PCR should be followed (McInnis *et al.*, 1990). The amplified region must be long enough for two oligonucleotides to hybridize, and the hybridizing sequences should be conserved to ensure proper detection of the target DNA. In order to avoid strand displacement, fragments longer than 200 bp should not be used. The probes should be designed to have approximately the same $T_m$ to ensure maximal hybridization kinetics of both probes.

Europium-labeled oligonucleotides have been applied in the diagnosis of genetic diseases and oncogene detection (to be published). This hybridization assay has been modified for the detection of a single base change in the human genome using allele-specific oligonucleotides (Conner *et al.*, 1983). The Eu-labeled oligonucleotide probe is designed in such a way that the point mutation site is in the middle of the hybridizing sequence. By washing under stringent conditions only the perfectly matched probes will remain hybridized. Thus a simple test can be developed by which specific point mutations can be determined (Liukkonen *et al.*, and Huoponen *et al.*, in preparation).

Moreover, the Eu-labeled oligonucleotides have been used as 5′ primers in a PCR reaction (Dahlén *et al.*, 1991a). The 3′ primer was labeled with biotin, thus allowing direct collection of the amplified double-stranded, Eu-labeled target DNA. Sufficient specificity was achieved by using a nested primer approach. The region of interest was first amplified with a pair of outer unlabeled primers, and then a second amplification round was performed using an inner labeled pair of primers.

# A. Materials

## 1. Reagents

The equipment and reagents necessary for oligonucleotide synthesis using the phosphoramidite chemistry are available from many suppliers. Instructions for use of these reagents should be carefully followed.

## 2. Component Preparation

### *a. Preparation of Eu-Labeled Oligonucleotides: Reagents and Materials*

Oligonucleotide synthesizer, with option to use additional phosphoramidites (for instance the Gene Assembler Plus, Pharmacia-LKB Biotechnology, Cat. No. 80-2081-10).

Standard reagents for oligonucleotide synthesis, including the phosphoramidites (A,T,G,C), tetrazole, acetonitrile, etc. These are available from several suppliers (Pharmacia-LKB Biotechnology, Applied Biosystems, Clontech, Sigma, etc). For a complete list of reagents and suppliers, see the manual of the oligonucleotide synthesizer.

Amino-modified phosphoramidite, 5′-*O*-dimethoxytrityl-*N*4-(trifluoroacetylamidohexyl)-deoxycytidine-3′-*O*-(β-cyanoethyl-diisopropylamino)-phosphoramidite (modC). This compound will become commercially available from Wallac Oy.

Europium chelate of 4-[2-(4-isothiocyanatophenyl)ethyl]-2,6-*bis*[*N,N*-*bis*(carboxymethyl)aminoethyl]pyridine. The chelate will become commercially available from Wallac Oy.

Gel electrophoresis, standard reagents for urea PAGE are available from a number of suppliers (USB, BRL, Hoefer Scientific, etc.). See Sambrook *et al.*, 1989, for complete listing of the necessary reagents.

NAP-5 and NAP-10 columns (Pharmacia-LKB Biotechnology) or a gel filtration column (1 × 50 cm) packed with Sephadex G-50 DNA Grade (Pharmacia-LKB Biotechnology).

*Solutions*

1 *M* $Na_2CO_3$,

10 m*M* Tris-HCl, pH 7.5, containing 50 μ*M* EDTA.

### *b. Procedure for the Preparation of Eu-Labeled Oligonucleotides*

The synthetic route for the preparation of Eu-labeled oligonucleotides is outlined in Fig. 8. The oligonucleotide synthesizer is programmed and used according to the instructions of the manufacturer. Program the machine to synthesize an oligonucleotide starting with the probing sequence at the 3′ end, and continue with 10 to 30 modC at the 5′ end.

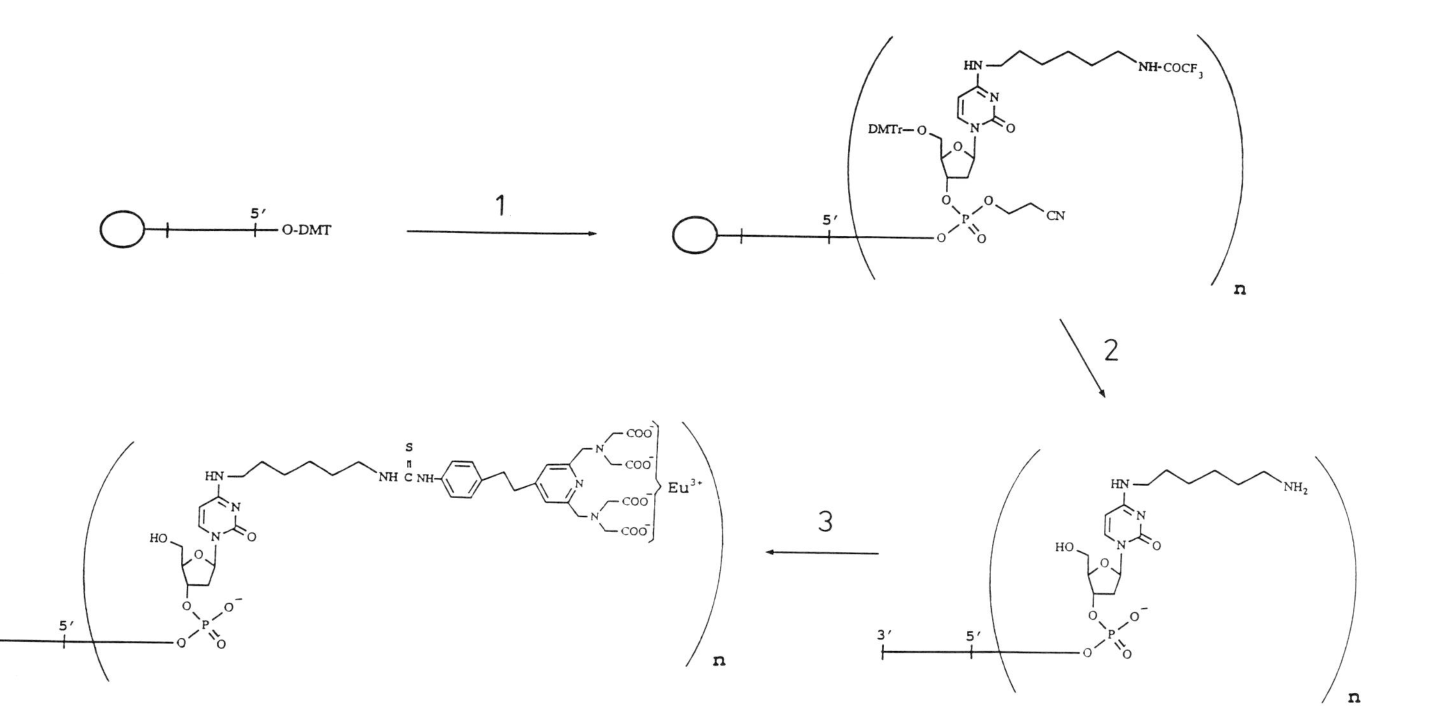


**Fig. 8** Synthesis of Eu-labeled oligonucleotides. (1) The fully protected diaminohexane-modified cytidine (modC) is incorporated at the 5′ end of the oligonucleotide during synthesis. (2) The deprotection and purification is carried out according to standard procedures. (3) The isothiocyanate analog of the Eu-chelate is reacted to the amino groups available on the modified oligonucleotide.

1. Dissolve the modC in dry acetonitrile to a final concentration of 0.1 *M*. Attach the modC solution to the synthesizer at the appropriate site. Upon complete synthesis, the deprotection is done in concentrated ammonia overnight at 55°C according to standard procedures (instruction manual of the synthesizer).
2. Purify the oligonucleotide containing the tail of modC's on the 5′ end by urea polyacrylamide gel electrophoresis (PAGE), and subsequently elute the product according to standard procedures (see Sambrook *et al.*, 1989). Change the buffer of the eluted oligonucleotide to $H_2O$ by using NAP-5/NAP-10 columns. The amount of pure oligonucleotide is determined by absorbance measurement at 260 nm, 1 $OD_{260}$ = 33 μg/ml.
3. Dry 1.5–10 nmole of the oligonucleotide in $H_2O$ in a Speed-Vac centrifuge (Savant). Redissolve the oligonucleotide in 50 μl $H_2O$. Adjust the pH to 9.5 by adding 2.5 μl 1 *M* $Na_2CO_3$.
4. Dissolve about 0.5–4 μmole Eu chelate in 100 μl $H_2O$. Quantitate by making serial dilutions (1:100, 1:10, etc.) into 1 ml Enhancement Solution, and measure the fluorescence in a time-resolved fluorometer against a 1 n*M* Eu standard. The exact concentration of Eu chelate is calculated using the formula:

$$\text{Eu chelate (n}M) = \frac{y \cdot z}{x}$$

where $y$ = fluorescence in cps of Eu chelate containing sample
$x$ = fluorescence in cps of the 1 n*M* Eu standard
$z$ = dilution factor

Add to the oligonucleotide solution a molar excess of the Eu chelate based on the following formula:

$$\text{Eu chelate excess (nmole)} = A \times B \times 10$$

where $A$ = nmoles of oligonucleotide
$B$ = number of modC per oligonucleotide

5. Incubate the reaction mixture overnight at 4°C, and purify the Eu-labeled oligonucleotide by gel filtration. When less than 200 nmole of Eu chelate is used in the labeling reaction (see formula 2), the labeled oligonucleotide can be rapidly purified over a NAP-5 and NAP-10 column, respectively. When larger amounts of oligonucleotide are labeled, a bigger column is required for the purification (1 × 50 cm, Sephadex G-50 column). Elution is carried out in 10 m*M* Tris-HCl, pH 7.5, containing 50 μ*M* EDTA. It is recommended that absorbance be recorded at

260 nm and the Eu profile be determined by diluting 1 $\mu$l aliquots of each fraction into 1 ml Enhancement Solution and measuring Eu content in a time-resolved fluorometer. Identify the first peak based on the UV and Eu profile, and pool the fractions containing the Eu-labeled oligonucleotide. A typical elution profile is shown in Fig. 9.

The final product is analyzed spectrophotometrically to determine the amount of oligonucleotide recovered. The Eu content of the pooled fraction is measured and the Eu/oligonucleotide ratio is obtained, as follows:

Calculate the total amount of Eu chelate (in nmole) in the pooled fractions by the following formula:

$$\mathrm{Eu_{TOT}}\ (\mathrm{nmole}) = [\mathrm{Eu}] \times V \times 10^{-3}$$

where V = the volume of the pooled fractions in ml.

$$[\mathrm{Eu}] = \mathrm{n}M$$

6. Record the absorbance at 260 nm using the elution buffer as reference. Typical UV-spectra are shown in Fig. 10. The Eu chelate absorbs at 260 nm; thus, absorbance measured at 260 nm for the oligonucleotide cannot be used for the estimation of the

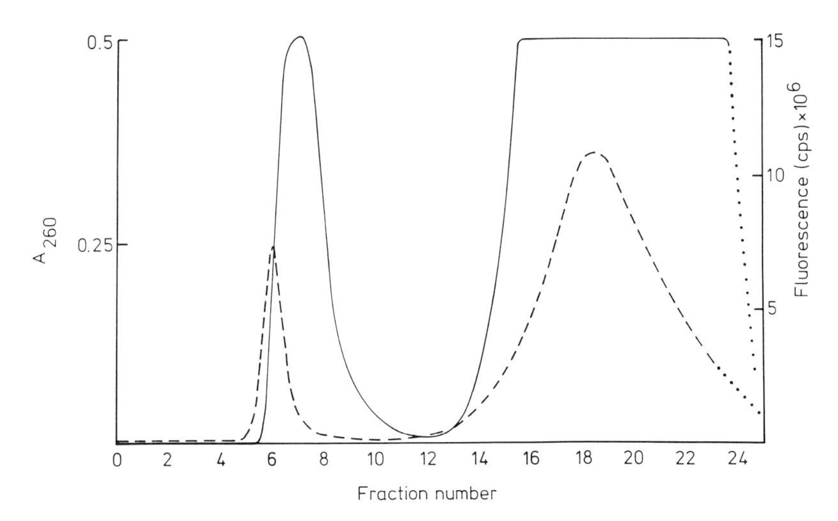


**Fig. 9** Gel filtration profile on the purification of an Eu-labeled oligonucleotide (containing 30 modC's at the 5′ end) from free Eu-chelate using a Sephadex G-50 (1 × 50 cm) column. The absorbance at 260 nm (---) and the Eu content of the fractions (———) was followed. Fractions (6 and 7) containing the Eu-labeled oligonucleotide were pooled.

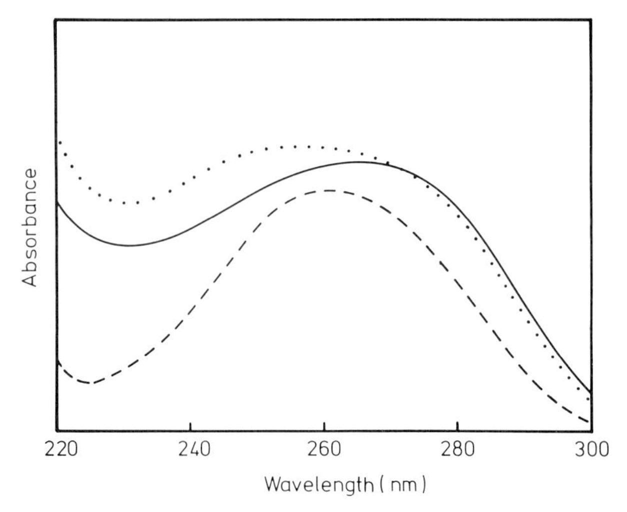


**Fig. 10** Typical absorbance spectra of an oligonucleotide before modification (---), after the addition of 25 modC's (———), and after Eu labeling of the modified oligonucleotide (....).

amount of oligonucleotide recovered. The oligonucleotide concentration is given by:

$$C = \left[ A_{260} - \left( \frac{[Eu]}{50} \times 0.83 \right) \right] \times 33$$

C = concentration of oligonucleotide in $\mu$g/ml.

Eu = $\mu M$

The amount of oligonucleotide (nmole) is calculated according to the following formula:

$$\text{Oligonucleotide}_{TOT} \text{ (nmole)} = \frac{C \times V \times 1000}{MW}$$

where C = concentration of oligonucleotide in $\mu$g/ml
V = volume of the pooled fractions in ml
MW = molecular weight of the oligonucleotide (length × 330 g/mol)

The Eu: oligonucleotide ratio is obtained from the formula:

$$\text{Eu:oligonucleotide} = \frac{\text{Eu}_{TOT} \text{ (nmole)}}{\text{Oligonucleotide}_{TOT} \text{ (nmole)}}$$

7. To obtain a usable concentration of the probe, dilute part of the Eu-labeled oligonucleotide in the elution buffer to a concentration

of 1 $\mu$g/ml and store at 4°C. The solution is stable for 6 months. The stock solution is stored at −20°C and is stable for at least one year.

### *c. Preparation of Biotinylated Oligonucleotides*

*Reagents and materials*

Biotinamidocaproate *N*-hydroxysuccinimide ester, Act-biotin (Sigma Cat No. B 2643) or other equivalent amino reactive biotin derivative. 10 m*M* Tris-HCl, pH 7.5, containing 1 m*M* EDTA. NAP-5 and NAP-10 columns (Pharmacia).

The oligonucleotide synthesis, deprotection, and purification is done as previously, except that in the actual synthesis modC is inserted at the 3′ end as the first nucleotide immediately after the base on the support (see Fig. 7).

1. Five nmole of the oligonucleotide in $H_2O$ containing a single modC at the 3′ end is dried in a Speed-Vac centrifuge. Dissolve the oligonucleotide in 50 $\mu$l $H_2O$ and add 2.5 $\mu$l 1 *M* $Na_2CO_3$. Then a 50 molar excess of act-biotin (0.11 mg) is added dissolved in dry *N,N*-dimethylformamide. Incubate at RT for 3 hr.
2. Purify the biotinylated oligonucleotide by gel filtration on NAP-5 and NAP-10 columns using 10 m*M* Tris-HCl, pH 7.5, containing 1 m*M* EDTA as eluent. Measure the recovery of the oligonucleotide spectrophotometrically at 260 nm, using the elution buffer as the reference (1 $OD_{260}$ = 33 $\mu$g/ml). Dilute part of the biotinylated oligonucleotide to a concentration of 1 $\mu$g/ml in 10 m*M* Tris-HCl, pH 7.5, containing 50 $\mu$*M* EDTA. The diluted probe solution is stable for 6 months when stored at 4°C. The stock solution is stable for 1 year when stored at −20°C.

### *d. PCR Reagents*

Primer-mix: 3$\mu$*M* of each primer in 10 m*M* Tris-HCl, pH 7.5 containing 1 m*M* EDTA (see Fig. 7).

Taq DNA polymerase: a 0.5 U/$\mu$l stock solution (Perkin-Elmer Cetus, Stratagene, Boehringer-Mannheim, Amersham, etc.)

*2× PCR-buffer*

20 m*M* Tris-HCl, pH 8.4, containing 5 m*M* $MgCl_2$, 400 $\mu$*M* dNTP, 0.4 mg/ml gelatin and 100 mM NaCl.

**Note:** The PCR-buffer given here is only an example of a typical buffer composition.

Mineral oil

### *e. Coating of the Solid Phase with Biotinylated BSA*

*Reagents and materials*

DELFIA Assay Buffer (Wallac Oy, Cat. No. 1244-106)
Streptavidin (Porton Products, BRL, Amersham, etc.)
Microtitration Plate (Wallac Oy, Cat. No. 1244-550)
DELFIA Wash Concentrate (Wallac Oy, Cat. No. 1244-500).

*Coating buffer*

50 m*M* $K_2HPO_4$, pH 9.0, containing 0.05% Na-azide and 0.9% NaCl.

---

*Biotinylated BSA*

1. Dissolve 10 mg BSA in 1 ml 50 m*M* $Na_2CO_3$. Add a 50-molar excess of biotin-amidocaproate *N*-hydroxysuccinimide ester in 200 μl dry *N,N*-dimethylformamide. Incubate at RT for 3 hr.
2. Purify the biotinylated BSA twice on PD-10 columns (Pharmacia-LKB Biotechnology) using 20 m*M* HEPES, pH 7.5, containing 0.05% Na-azide and 0.9% NaCl as the eluent.

*Coating procedure*

3. Dilute the biotinylated BSA (bio-BSA) in the coating buffer to a concentration of 500 ng/ml. Dispense 200 μl/well in the microtitration plate. Incubate the plate at RT overnight covered with adhesion tape. Wash the plate with DELFIA Wash Solution three times.
4. Dilute the streptavidin in DELFIA Assay Buffer to a concentration of 2 μg/ml. Dispense 200 μl/well in the microtitration plate coated with bio-BSA. Incubate at RT for 3 hr. Wash the plate with DELFIA Wash Solution four times. The streptavidin-coated plates are stable for at least 6 months when stored at 4°C covered with adhesion tape.

---

### *f. Solution Hybridization Reagents*

Eu-labeled oligonucleotide probe, a 1-μg/ml probe solution (see preceding sections).

Biotinylated oligonucleotide probe, a 1-μg/ml probe solution (see preceding sections).

Streptavidin-coated microtitration plates (see above).

2 × hybridization buffer 100 m*M* HEPES, pH 7.5, containing 0.1% Tween-20, 200 μ*M* EDTA and 1 *M* NaCl.

### 3. Instrumentation

See the section on Indirect labeling.

## B. Procedures

### 1. Detection of Specific DNA Using PCR and Europium-Labeled Oligonucleotide Probes in an *in Solution* Hybridization Assay

#### *a. Sample Pretreatment*

Several rapid sample pretreatment methods intended to be used with PCR protocols have been described in the literature (Kawasaki, 1990; Witt and Erickson, 1989). The sample pretreatment is very dependent on the nature of the sample to be examined. In many cases boiling at 100°C for 10 to 20 min in water is sufficient to lyse the cells and liberate the DNA from the nucleus. If more drastic methods are required, boil in 0.01 to 0.1 *M* NaOH. Ensure proper neutralization on completion, using an equimolar amount of HOAc supplemented with 50 m*M* Tris-HCl, pH 8.4. More sophisticated sample pretreatment methods include use of Proteinase K and/or SDS, etc. However, with these there is a risk of inactivation of the Taq polymerase.

#### *b. PCR*

Many protocols for the PCR have been described for a wide variety of analytes (see McInnes *et al.*, 1990). The PCR solution can be set up according to the following scheme.

1. Mix 10 μl 3 μ*M* primer-mix, 10 μl 0.5 U/μl Taq polymerase, 50 μl 2 × PCR-buffer, and 30 μl sample. Cover the reaction solution with 100 μl of mineral oil.
2. Perform the PCR in an automated heat-block using a suitable temperature profile for 30 cycles or less, depending on the amount of starting template DNA. Always include proper positive and negative controls.

### c. *Hybridization Test*

1. Denature 10 μl of the amplified DNA solution by boiling at 100°C for 10 min. Rapidly cool the denatured samples on ice.
2. Add 100 μl 1 × hybridization buffer containing the Eu-labeled and biotinylated oligonucleotide probes, at 10 ng/ml each. Include as negative controls $H_2O$ and 100 ng of unspecific DNA (such as herring sperm DNA).
3. Hybridize at 40 to 60°C, for 30 to 60 min, depending on the length of the hybridizing sequences of the probes.
4. Transfer 100 μl of the hybridization reaction mixture to individual streptavidin-coated wells in the microtitration plate. Add 100 μl/well of the DELFIA Assay Buffer. Incubate with continuous shaking on a DELFIA Plateshake at RT for 1 hr. Wash the strips 6 times with the DELFIA Platewash using DELFIA Wash Solution.
5. Dispense 200 μl of Enhancement Solution to each well. Incubate with continuous shaking in a DELFIA Plateshake for 25 min at RT. Measure the resulting Eu fluorescence in a time-resolved fluorometer. See Table III for typical results.

**Table III**
Detection of HIV-2 Proviral DNA from MOLT 4 Genomic DNA[a]

| Template DNA[b] molecules | Fluorescence[c] (cps) | CV (%) |
|---|---|---|
| 0 | 959 | 22 |
| 1 | 8098 | 30 |
| 2 | 11363 | 45 |
| 5 | 30540 | 22 |
| 10 | 67613 | 31 |
| 50 | 204492 | 7 |
| 100 | 358491 | 1 |
| 500 | 527811 | 12 |

[a] Primers were used from the LTR region of HIV-2. A 168-bp fragment was PCR amplified for 30 cycles using MOLT 4 genomic DNA as template. The solution hybridization test was performed using a Eu-labeled oligonucleotide.
[b] Purified MOLT 4 DNA was used in the PCR that was run in triplicates.
[c] The hybridization assay was performed in a single reaction, and the signal is the mean of three individual PCR runs.

## 2. Useful Hints

### *a. Preparation of the Probes*

It is not recommended that the tail of modC be more than 30 modC units long, since the solubility of the oligonucleotide containing a longer tail decreases. The yield of a 60-mer is usually good, but longer oligonucleotides are more difficult to produce; the purification of longer oligonucleotides is more complex. Use ordinarily 15–25 modC's incorporated at the 5′end.

The biotinylated probe may be checked by HPLC for the presence of biotin (see Agrawal *et al.*, 1986). The biotinylation reaction should be at least 90–95% efficient. Unreacted act-biotin must be carefully removed by gel filtration. If unreacted act-biotin is present, it will drastically lower the signal level achieved in the hybridization test, since the free biotin will effectively bind to the streptavidin on the solid support.

See Section II,B,3 on chemical Eu labeling of DNA probes.

### *b. PCR*

Avoid extensive cycling of the PCR; 30 cycles is sufficient for detection of a few molecules of target DNA. In most applications 20 to 25 cycles of PCR will provide detectable amounts of amplified DNA. The linearity of the overall assay is strongly dependent on the number of cycles used in the PCR versus the amount of starting material.

Since the PCR amplification produces double-stranded DNA, there will be a competition in the hybridization reaction between the two oligonucleotides and the complementary strand for hybridization to the target sequence. This is reflected in the linearity of the assay. By running an asymmetric PCR, the target DNA will be almost exclusively single stranded (Gyllensten and Erlich, 1988), thus improving the linearity of the assay.

### *c. The Hybridization Test*

Usable concentrations of biotinylated and Eu-labeled oligonucleotide probes are 2 to 20 ng/ml. The biotin-binding capacity of the solid support limits the amount of biotinylated oligonucleotide that can be used in the hybridization reaction. Unspecific binding of Eu-labeled oligonucleotide to the solid support can often be decreased by lowering the concentration of labeled probe.

The time required for reaching maximal hybridization is dependent on the length of the probing sequence of the oligonucleotides. Short oligonucleotides (18-mers) hybridize rapidly (15 min), whereas longer (30-mers) require at least 2 hr for complete hybridization. The procedure can be simplified by combining the hybridization and collection reaction, but longer reaction times (2–6 hr) are required.

A low-salt stringent wash solution (50 m*M* HEPES, pH 7.5, containing 15 m*M* NaCl and 50 $\mu M$ EDTA) and elevated temperature can also be used to increase the specificity. The same wash solution at a temperature close to the $T_m$ of the matched probe is recommended for the detection of single base changes.

## ACKNOWLEDGMENTS

We are grateful to Mrs. Leena Huhtamäki for performing the HPV-16 assays, and to Mr. Veli-Matti Mukkala for synthesizing the modC. The secretarial assistance of Mrs. Teija Ristelä in the preparation of the manuscript is appreciated.

## REFERENCES

Agrawal, S., Christodoulou, C., and Gait, M. J. (1986). Efficient methods for attaching nonradioactive labels to the 5′ ends of synthetic oligodeoxyribonucleotides. *Nucleic Acids Res.* **14,** 6227–6245.

Conner, B. J., Reyes, A. A., Morin, C., Itakura, K., Teplitz, R. L., and Wallace, R. B. (1983). Detection of sickle-cell $\beta^s$-globin allele by hybridization with synthetic oligonucleotides. *Proc. Natl. Acad. Sci. U.S.A.* **80,** 278–282.

Dahlén, P. (1987). Detection of biotinylated DNA probes by using Eu-labeled streptavidin and time-resolved fluorometry. *Anal. Biochem.* **164,** 78–83.

Dahlén, P., Syvänen, A-C., Hurskainen, P., Kwiatkowski, M., Sund, C., Ylikoski, J., Söderlund, H., and Lövgren, T. (1987). Sensitive detection of genes by sandwich hybridization and time-resolved fluorometry. *Mol. Cell. Probes* **1,** 159–168.

Dahlén, P., Hurskainen, P., Lövgren, T., and Hyypiä, T. (1988). Time-resolved fluorometry for the identification of viral DNA in clinical specimens. *J. Clin. Microbiol.* **26,** 2434–2436.

Dahlén, P., Iitiä, A., Mukkala, V-M., Hurskainen, P., and Kwiatkowski, M. (1991a). The use of europium ($Eu^{3+}$)-labeled primers in PCR amplification of specific target DNA. *Mol. Cell. Probes,* **5,** 143–149.

Dahlén, P., Iitiä, A., Skagius, G., Frostell, Å., Nunn, M., and Kwiatkowski, M. (1991b). Detection of HIV-1 using the polymerase chain reaction and a time-resolved fluorescence based hybridization assay. *J. Clin. Microbiol.* **29,** 798–804.

Feinberg. A. P., and Vogelstein, B. (1983). A technique for radiolabeling DNA restriction endonuclease fragments to high specific activity. *Anal. Biochem.* **132,** 6–13.

Forster, A. C., McInnes, J. L., Skingle, D. C., and Symons, R. H. (1985). Nonradioactive hybridization probes prepared by the chemical labeling of DNA and RNA with a novel reagent, photobiotin. *Nucleic Acids Res.* **13,** 745–761.

Gyllensten, U. B., and Erlich, H. A. (1988). Generation of single-stranded DNA by the polymerase chain reaction and its application to direct sequencing of the *HLA-DQA* locus. *Proc. Natl. Acad. Sci. U.S.A.* **85,** 7652–7656.

Hemmilä, I., Dakubu, S., Mukkala, V-M., Siitari, H., and Lövgren, T. (1984). Europium as a label in time-resolved immunofluorometric assays. *Anal. Biochem.* **137,** 335–343.

Hurskainen, P., Dahlén, P., Ylikoski, J., Kwiatkowski, M., Siitari, H., and Lövgren, T. (1991). Preparation of europium-labeled DNA probes and their properties. *Nucleic Acids Res.* **19,** 1057–1061.

Ikuta, S., Takagi, K., Wallace, R. B., and Itakura, K. (1987). Dissociation kinetics of 19 base paired oligonucleotide-DNA duplexes containing different single mismatched base pairs. *Nucleic Acids Res.* **15,** 797–811.

Kawasaki, E. S. (1990). Sample preparation from blood, cells and other fluids. *In* "PCR Protocols. A Guide to Methods and Applications" (M. A. McInnis, D. H. Gelfand, J. J. Sninsky, and T. J. White, eds.), pp. 146–152. Academic Press, San Diego.

Langer, P. R., Waldrop, A. A., and Ward, D. C. (1981). Enzymatic synthesis of biotin-labeled polynucleotides: Novel nucleic acid affinity probes. *Proc. Natl. Acad. Sci. U.S.A.* **78,** 6633–6637.

Leary, J. J., Brigati, D. J., and Ward, D. C. (1983). Rapid and sensitive colorimetric method for visualizing biotin-labeled DNA probes hybridized to DNA or RNA immobilized on nitrocellulose: Bio-blots. *Proc. Natl. Acad. Sci. U.S.A.* **80,** 4045–4049.

Lövgren, T., and Pettersson, K. (1990). Time-resolved fluoroimmunoassay, advantages and limitations. *In* "Luminescence Immunoassay and Molecular Applications" (K. Van Dyke and R. Van Dyke, eds.), pp. 233–250. CRC Press, Boca Raton, Florida.

McInnis, M. A., Gelfand, D. H., Sninsky, J. J., and White, T. J. (eds.) (1990). "PCR Protocols. A Guide to Methods and Applications." Academic Press, San Diego.

Ranki, M., Palva, A., Virtanen, M., Laaksonen, M., and Söderlund, H. (1983). Sandwich hybridization as a convenient method for the detection of nucleic acids in crude samples. *Gene* **21,** 77–85.

Reisfeld, A., Rothenberg, J. M., Bayer, E. A., and Wilchek, M. (1987). Nonradioactive hybridization probes prepared by the reaction of biotin hydrazide with DNA. *Biochem. Biophys. Res. Commun.* **142,** 519–526.

Rigby, P. W. J., Dieckmann, M., Rhodes, C., and Berg, P. (1977). Labeling deoxyribonucleic acid to high specific activity *in vitro* by nick translation with DNA polymerase I. *J. Mol. Biol.* **113,** 237–251.

Sambrook, J., Fritsch, E. F., and Maniatis, T. (eds.) (1989). "Molecular Cloning. A Laboratory Manual." 2nd Ed. Cold Spring Harbor University Press, Cold Spring Harbor, New York.

Shapiro, R., and Weisgras, J. M. (1970). Bisulfite-catalyzed transamination of cytosine and cytidine. *Biochem. Biophys. Res. Commun.* **40,** 839–843.

Sheldon, E. L., Kellogg, D. E., Watson, R., Levenson C. H., and Erlich, H. A. (1986). Use of nonisotopic M13 probes for genetic analysis: Application to HLA class II loci. *Proc. Natl. Acad. Sci. U.S.A.* **83,** 9085–9089.

Soini, E., and Lövgren, T. (1987). Time-resolved fluorescence of lanthanide probes and applications in biotechnology. *CRC Crit. Rev. Anal. Chem.* **18,** 105–154.

Sund, C., Ylikoski, J., Hurskainen, P. and Kwiatkowski, M. (1988). Construction of europium ($Eu^{3+}$)-labeled oligo DNA hybridization probes. *Nucleosides Nucleotides* **7,** 655–659.

Syvänen, A. C., Laaksonen, M., and Söderlund, H. (1986). Fast quantification of nucleic acid hybrids by affinity hybrid collection. *Nucleic Acids Res.* **14,** 5037–5048.

Syvänen, A. C., Tchen, P., Ranki, M., and Söderlund, H. (1986). Time-resolved fluorometry: a sensitive method to quantify DNA-hybrids. *Nucleic Acids Res.* **14,** 1017–1028.

Takahashi, T., Mitsuda, T., and Okuda, K. (1989). An alternative nonradioactive method for labeling DNA using biotin. *Anal. Biochem.* **179,** 77–85.

Viscidi, R. P., Connelly, C. J., and Yolken, R. H. (1986). Novel chemical method for the preparation of nucleic acids for nonisotopic hybridization. *J. Clin. Microbiol.* **23,** 311–317.

Wallace, R. B., Shaffer, J., Murphy, R. F., Bonner, J., Hirose, T., and Itakura, K. (1979). Hybridization of synthetic oligonucleotides to ΦX 174 DNA: The effect of single base pair mismatch. *Nucleic Acids Res.* **6,** 3543–3557.

Witt M., and Erickson, R. P. (1989). A rapid method for detection of Y-chromosomal DNA from dried blood specimens by the polymerase chain reaction. *Hum. Genet.* **82,** 271–274.

# 11 Detection of Lanthanide Chelates and Multiple Labeling Strategies Based on Time-Resolved Fluorescence

Eleftherios P. Diamandis, Department of Clinical Biochemistry, Toronto Western Hospital, Toronto, Ontario, Canada and Department of Clinical Biochemistry, University of Toronto, Toronto, Ontario, Canada

Theodore K. Christopoulos, Department of Clinical Biochemistry, Toronto Western Hospital, Toronto, Ontario, Canada



## I. INTRODUCTION

The diagnosis of genetic, infectious, malignant, and other diseases will, in the future, partially be carried out on a large scale by using molecular biology techniques. This kind of diagnosis is still not very widely used for a number of different reasons, one of which is the (until recently) limited availability of nonisotopic techniques for nucleic acid hybridization. The major detection systems for nucleic acid hybridization are still based on the radionuclides $^{32}P$, $^{35}S$, and $^{3}H$. These radionuclides, although possessing some important favorable characteristics (e.g., very low detectability), have a number of serious disadvantages (i.e., danger of health hazards, limited shelf-life, difficulty in waste disposal, and requirements for special licensed laboratory facilities. These problems associated with

*Nonisotopic DNA Probe Techniques*

the radionuclides make them clearly unsuitable for use in large-scale routine clinical diagnosis (Diamandis, 1990).

In the field of immunological assays, a number of successful nonisotopic methodologies have been introduced (Ngo, 1988) (Diamandis and Christopoulos, 1990). Many of these are now being tried for nucleic acid hybridization and are described in other chapters of this volume. Although many of these methodologies are not so sensitive as those using radionuclides, they still have many applications because often, extreme sensitivity is not needed or, it is possible, using the polymerase chain reaction (PCR), to amplify the target of interest before the final detection step.

Time-resolved fluorometry with lanthanide chelates as labels is now an established technique in nonisotopic immunoassay (Diamandis, 1988; Soini and Lovgren, 1987). During the last 10 years, two different assay designs have evolved and are described in detail elsewhere (Diamandis, 1988). In these assay designs, the label can either be $Eu^{3+}$ or a $Eu^{3+}$ chelator. A fluorescent europium chelate can then be formed by adding either suitable organic chelators (the DELFIA system, LKB-Pharmacia), or $Eu^{3+}$ (the FIAgen system, CyberFluor Inc., Toronto, Ontario, Canada), respectively. The fluorescent europium chelates (and some other lanthanide chelates) possess certain advantages, in comparison to conventional fluorescent labels like fluorescein (i.e., large Stokes shifts, narrow emission bands, and long fluorescence lifetimes). The fluorescence lifetime of most conventional fluorophores is approximately 100 nsec or less; the lifetime of lanthanide chelates, on the other hand, is 100–1000 $\mu$sec. A pulsed light source and a time-gated fluorometer can be used to measure the fluorescence of these compounds in a window of 200 to 600 $\mu$sec after each excitation. This method decreases the background interference from the short-lived fluorescence of natural materials in the sample (cuvettes, optics, etc). Commercially available instruments for time-resolved fluorescence immunoassays have been described (Diamandis, 1988; Soini and Kojola, 1983). The chelates used in the DELFIA system are complexes of the type $Eu(NTA)_3(TOPO)_2$, where NTA is naphthoyltrifluoroacetone and TOPO is trioctylphosphine oxide. The $Eu^{3+}$ label is introduced into antibodies or streptavidin (SA) by using a strong $Eu^{3+}$ chelator of the ethylenediaminetetraacetic acid (EDTA) type. Release of $Eu^{3+}$ and recomplexing with NTA and TOPO can be achieved by lowering the pH to approximately 3.0. DELFIA is well established in the field of nonisotopic immunoassay; it is characterized by high sensitivity and broad dynamic range. It has been criticized because of its vulnerability to $Eu^{3+}$ contamination effects (Diamandis, 1988). The $Eu^{3+}$ label can be measured down to $10^{-13}$ moles/L using time-resolved fluorescence and $10^{-17}$–$10^{-18}$ moles per cuvette ($\sim 6 \times 10^6 - 6 \times 10^5$ molecules) can be detected routinely. Analytes can also be measured down to these levels. The newly developed

FIAgen system uses the europium chelator 4,7-*bis*(chlorosulfophenyl)-1,10-phenanthroline-2,9-dicarboxylic acid (BCPDA) as label (Evangelista *et al.*, 1988); it does not suffer from any $Eu^{3+}$ contamination effects. This system works best when biotinylated antibodies are used with BCPDA-labeled SA. Three SA preparations achieve different detection limits: (1) SA directly labeled with BCPDA ($SA[BCPDA]_{14}$) for assays with detection limits of $10^{-10}$ to $10^{-11}$ moles/L (Diamandis and Morton, 1988); (2) SA covalently linked to thyroglobulin (TG) carrying 160 BCPDA molecules ($SA[TG][BCPDA]_{160}$) for assays with detection limits of $10^{-12}$ (Diamandis *et al.*, 1989); and (3) the preparation under (2) has been activated by an empirical process and is suitable for assays with detection limits of $10^{-12}$ to $10^{-13}$ moles/L or less (Morton and Diamandis, 1990). This reagent will be briefly described later in the text. The best detection limit achieved with the latter reagent for a model $\alpha$-fetoprotein assay was ~300,000 molecules per cuvette (5 $\mu$l sample volume) (Christopoulos *et al.*, 1990).

DNA hybridization assays using the principles of the immunological version of DELFIA have been described (Dahlen, 1987; Sund *et al.*, 1988; Syvanen *et al.*, 1986; Oser *et al.*, 1990). This assay works well in applications in which the spatial distribution of DNA bands on solid supports is not studied (e.g., in dot blot hybridizations). However, it is unsuitable for Southern, Western, and Northern types of analysis, because the spatial distribution of $Eu^{3+}$ in the bands is lost during the extraction procedure necessary for fluorescence enhancement (Soini *et al.*, 1990). Thus, other principles involving stable fluorescent europium or terbium chelates have been investigated (Soini *et al.*, 1990; Hemmila and Batsman, 1988; Oser *et al.*, 1990).

In this chapter, we describe some applications of a newly developed SA-based reagent labeled with the europium chelate of BCPDA (Morton and Diamandis, 1990) for DNA hybridization and other assays. This reagent is highly fluorescent and can stain biotinylated DNA separated on solid supports. Although not described in this chapter, preliminary experiments with other types of assays (Western, Northern, dot–slot blot, etc.) show that the new reagent can find application in any assay design that involves a biotinylated moiety.

# II. MATERIALS

## A. Reagents

Biotinylated DNA molecular weight markers (*Hin*dIII lambda DNA digests) were obtained from Vector Laboratories, Burlingame, California, as a 50-$\mu$g/ml solution. Purified linearized plasmid pBR 328 and its digestion fragments were obtained from Boehringer-Mannheim, Indianapolis, Indiana as part of their DNA Labeling and Detection Kit, Nonradioactive. The Prime-it Random Primer Kit supplemented with Bio-11-dUTP/dTTP mix was obtained from Stratagene, La Jolla, California. Supported nitrocellulose (Hybond-C-Extra) and nylon membranes (Hybond-N) were obtained from Amersham International, Buckinghamshire, United Kingdom. Sonicated salmon sperm DNA (10 mg/ml) was obtained from Sigma Chemical Co., St. Louis, Missouri.

## 1. Streptavidin-Based Macromolecular Complex

The SA-based macromolecular complex (SBMC) was prepared as previously described (Morton and Diamandis, 1990). Because of the involved preparation of this reagent and the commercial unavailability of BCPDA, it is advised that this reagent be purchased from CyberFluor, Inc., as a stock solution of 15 mg/l SA. The SBMC is diluted 50-fold, just before use, in a tracer diluent [also available from CyberFluor, Inc., or prepared as described elsewhere (Morton and Diamandis, 1990)].

A schematic diagram describing the formation and the possible structure of the SBMC is shown in Fig. 1. It is postulated that this is a complex containing (1) SA covalently linked to BCPDA-labeled bovine TG; (2) BCPDA-labeled bovine TG, and (3) $Eu^{3+}$ noncovalently bound to BCPDA, acting as a bridge between components (1) and (2). The molar ratio of components (1), (2), and (3) is approximately 1:2.3:480.

## 2. Solutions

Forty × tris acetate EDTA (TAE) buffer was prepared by dissolving (per liter) 193.6 g *tris*(hydroxymethyl)aminomethane (Tris) base, 108.9 g sodium acetate ·3 $H_2O$, 15.2 g $Na_2$ EDTA, and adjusting the pH to 7.2 with acetic acid. The gel loading buffer was prepared by mixing 5 ml glycerol, 250 $\mu$l 50 × TAE buffer, 1 ml saturated bromophenol blue solution, 1 ml of a 10% suspension of xylene cyanol, and adjusting the volume to 10 ml with water. Tris EDTA (TE) buffer is a 10 m*M* Tris, 0.1 m*M* EDTA solution of pH 7.80. The DNA denaturation buffer (Southern A solution) was prepared by mixing 300 ml 5 *M* NaCl, 50 ml 10 *M* NaOH, and 650 ml $H_2O$. The neutralization buffer was a 5 *M* ammonium acetate solution. Membrane-blocking solution was a 6% (w/v) bovine serum albumin (BSA) solution in a 50 m*M* Tris buffer, pH 7.80. The wash solution was a 10 m*M* Tris buffer, pH 7.80, containing 0.05% Tween-20 and 9 g/liter NaCl. Southern B solution was prepared by dissolving 77 g ammonium acetate and 0.8 g NaOH in 1 liter water.

Twenty × saline sodium phosphate EDTA (SSPE) solution contains

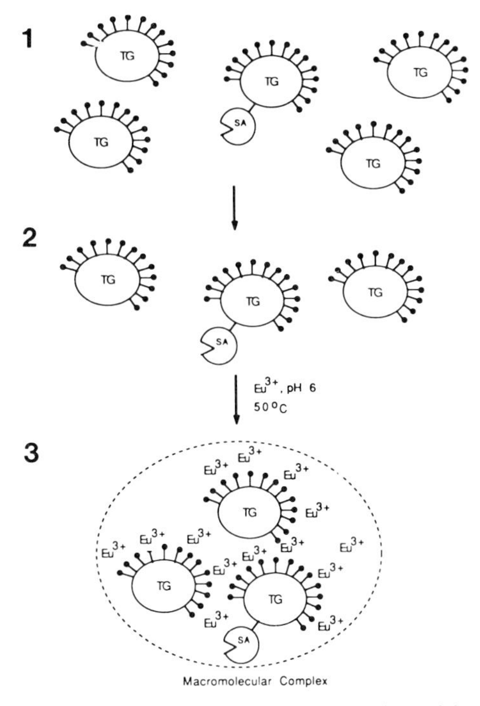


**Fig. 1** Schematic representation of the proposed mechanism of formation of the streptavidin-based macromolecular complex where —●, BCPDA; SA, streptavidin; and TG, thyroglobulin. SA–TG conjugates aggregate with BCPDA-labeled TG in the presence of $Eu^{3+}$, at 50°C, to form the macromolecular complex. For more information, see text and also Morton and Diamandis (1990). [Reprinted with permission from Morton and Diamandis (1990).]

(per liter) 174 g NaCl, 27.6 g $NaH_2PO_4 \cdot H_2O$, 7.4 g EDTA, and the pH adjusted to 7.4 with 10 *M* NaOH. One hundred × Denhardt's solution is 2% polyvinylpyrrolidone, 2% BSA, and 2% Ficoll. Twenty × standard saline citrate (SSC) solution contains (per liter) 175.3 g NaCl, 88.2 g trisodium citrate·$2H_2O$, and the pH was adjusted to 7.0 with 10 *M* HCl.

## B. Instrumentation

For scanning of nitrocellulose or nylon membranes to locate and quantify fluorescent spots or bands, we used the CyberFluor 615 Immunoanalyzer equipped with a special software described elsewhere (Christopoulos and Diamandis, 1991). Instrument and software are available from CyberFluor Inc. (Toronto, Canada). For visual membrane observation, we used a Fotodyne standard UV transilluminator. For regular photography of the membranes under UV illumination, we used a Polaroid camera with instant film, in a manner identical to photography of ethidium bromide-stained gels. Exposure time was 13 sec. Color photography under UV illumination is also possible. All hybridizations were performed in a Robbins Scientific hybridization incubator.

# III. PROCEDURES

## A. Detection of Biotinylated Nucleic Acids

1. Serially dilute twofold at a time in TE buffer, containing 400 $\mu$g/ml salmon sperm DNA. Add in Eppendorf microcentrifuge tubes 10 $\mu$l diluted biotinylated DNA and 10 $\mu$l DNA denaturation buffer and incubate for 10 min at room temperature (RT).
2. Add 20 $\mu$l neutralization buffer, mix and immediately spot 1 $\mu$l on prewetted (50 m*M* Tris, pH 7.40) nitrocellulose or nylon strips. Bake the strips in a vacuum oven for 1 hr at 60°C.
3. Transfer the strips in a 6% BSA blocking solution and incubate with rotary mixing for 1 hr to overnight. Visualize the biotinylated nucleic acids by incubating them for 3 hr at RT with rotary mixing in a 50-fold diluted SA-based macromolecular complex (SBMC) solution.
4. After incubation, wash the strips × 3 with 200 ml wash solution each time and then incubate in the wash solution with rotary mixing for 1 hr or longer. Dry the strips with a hair dryer.
5. Detect or quantify bound DNA as follows:
   (a) UV transillumination (spotted side down): detection limit ≥ 400 pg.
   (b) Photograph with a regular Polaroid camera and instant film in a manner similar to that of photographing ethidium bromide-stained agarose gels. Optimal exposure time is ~13 sec—at least 100 pg DNA can be visualized on 1 $\mu$l spots.
   (c) Scan with the CyberFluor 615 Immunoanalyzer, working in the scanning mode, with use of a special software—~10 pg of DNA can be quantified.

Some results are shown in Fig. 2. Nylon membranes give signals similar to those of nitrocellulose, but the background signal is ~10-fold higher (2000 versus 20,000 arbitrary fluorescence units).

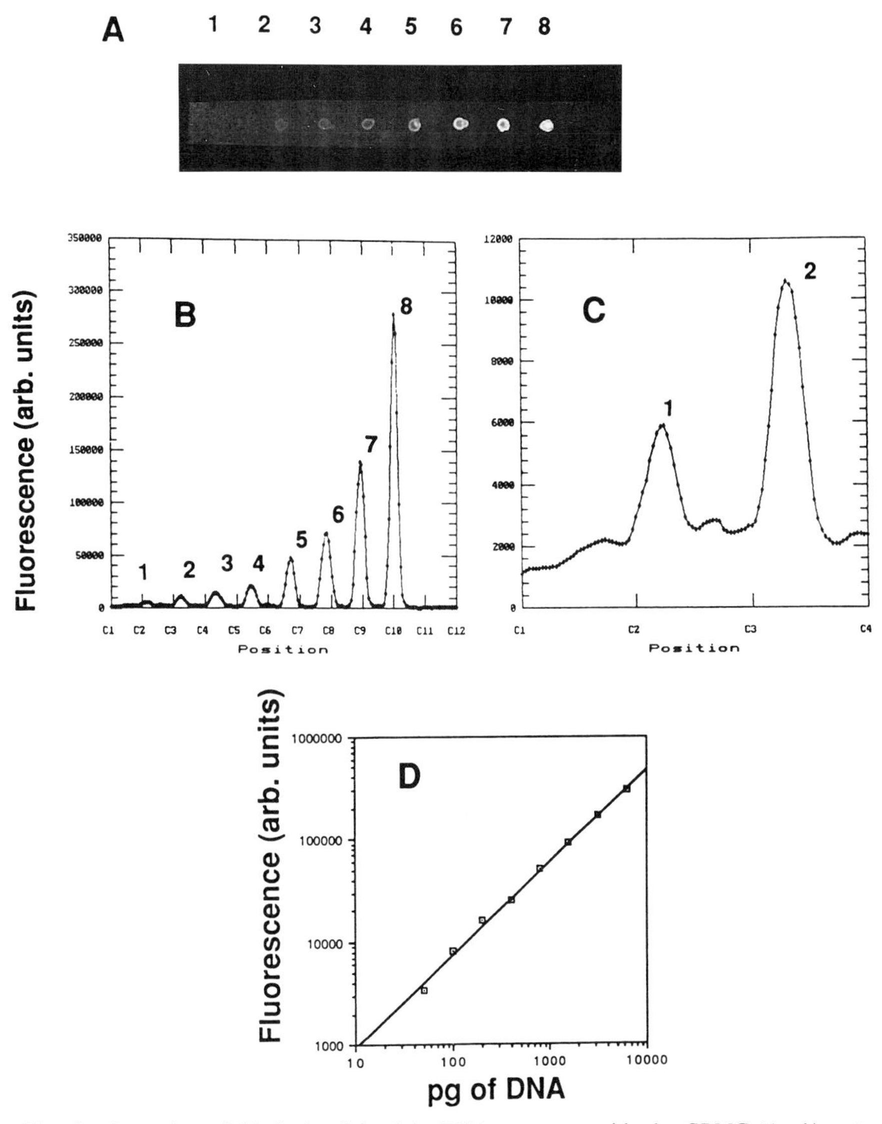


**Fig. 2** Detection of biotinylated lambda DNA on spots with the SBMC (1 μl/spot) (A) photograph of the spots under UV transillumination; (B) scanning of the spots with the modified 615 Immunoanalyzer (200 measurement points); (C) scanning of positions C1–C4 of (B) to improve resolution; and (D) double logarithmic plot of fluorescence versus amount of biotinylated DNA spotted. Biotinylated DNA spotted (pg/1 μl) was 50(1); 100(2); 200(3); 400(4); 800(5); 1600(6); 3200(7); 6400(8).

## B. Southern Transfer of Biotinylated Markers

1. Separate various amounts of biotinylated *Hind*III-digested lambda DNA molecular weight markers on 0.8% agarose minigels using TAE buffer (loadings, 10 $\mu$l sample per lane). Mixed samples 1:1 with gel loading buffer before application. Electrophorese for 2 to 3 hr at 60V.
2. Immerse the gel in 500 ml Southern A solution for 30 min. Repeat

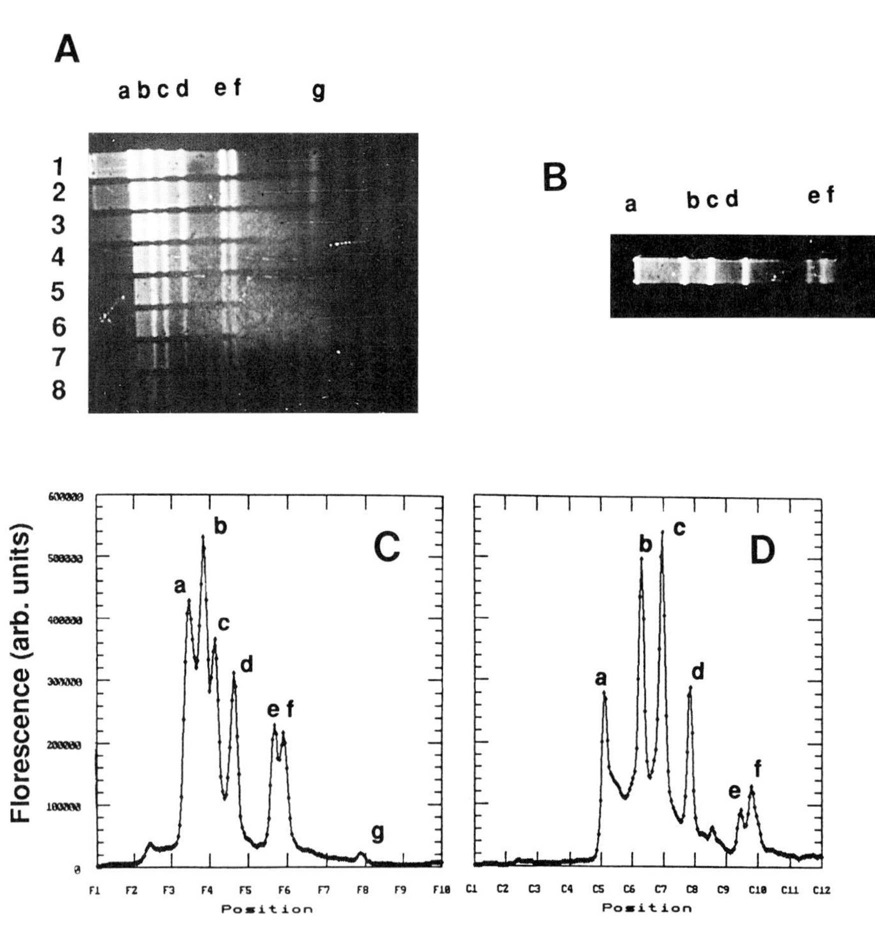


**Fig. 3** Transfer of biotinylated DNA markers from agarose to nitrocellulose and staining with SBMC. (A) Amount of DNA loaded on a 1% agarose gel was (in ng): 500(1); 250(2); 125(3); 62(4); 31(5); 16(6); 8(7); 4(8). (B) Amount of DNA loaded on a 0.7% agarose gel was 250 ng. (C) and (D) represent instrument scans (200 points) of A (250 ng lane) and B, respectively. The length of the fragments (kb) are 23.1 (1); 9.4 (b); 6.7 (c); 4.4 (d); 2.3 (e); 2.0 (f); 0.56 (g).

this step once more. Immerse the gel in Southern B solution for 30 min and repeat the step once more.

3. Perform the Southern transfer of the markers to nitrocellulose exactly as described by Davis *et al.* (1986) using Southern B solution and overnight incubation. Bake the membrane for 2 hr at 80°C, block and stain with the SBMC, and wash exactly as described under detection of biotinylated nucleic acids.

Some typical results are shown in Fig. 3. Four ng or more of total biotinylated DNA markers can be detected by the instrument. The bands obtained are identical to those obtained by using SA-alkaline phosphatase and the colorimetric (5-bromo-4-chloro-3-indolyl phosphate)-(nitroblue tetrazolium) (BCIP-NBT) reagent.

## C. DNA Probe Hybridization Experiments

1. Load the intact, linearized plasmid pBR 328 and its digestion fragments on 0.8% agarose gels and separate by electrophoresis as previously described.
2. Transfer the separated fragments to nitrocellulose by the Southern method and bake as described.
3. Prehybridize the membrane at 65°C in 20 ml prehybridization buffer (5× SSPE, 5× Denhardt's, 0.1% sodium dodecylsulfate (SDS), 100 μg/ml salmon sperm DNA) for 2 hr to overnight, in the hybridization chamber. Use 20 ml of the hybridization buffer (exactly as the prehybridization buffer but with 1× Denhardt's) at 65°C overnight.
4. Biotinylate the probe, 200 ng of intact linearized pBR 328, with the random primer method following the instructions of the Prime-it kit. Boil the biotinylated probe for 5 min before adding it to the hybridization buffer at a concentration of 10 ng/ml.
5. After hybridization, wash the membranes as follows:
   (a) three times with 200 ml each 2× SSC, 0.1% SDS at RT. Soaking time is 5 min with vigorous shaking.
   (b) three times as above, but with 0.2× SSC, 0.1% SDS.
   (c) two times with 0.2× SSC, 0.1% SDS as 65°C, 15 min each.
6. After blocking with 6% BSA, stain the membranes with the SBMC and wash exactly as described under detection of biotinylated nucleic acids. Some typical results are shown in Fig. 4. Sharp hybridization bands at the expected molecular weights are seen even with loadings of 5 ng total DNA.

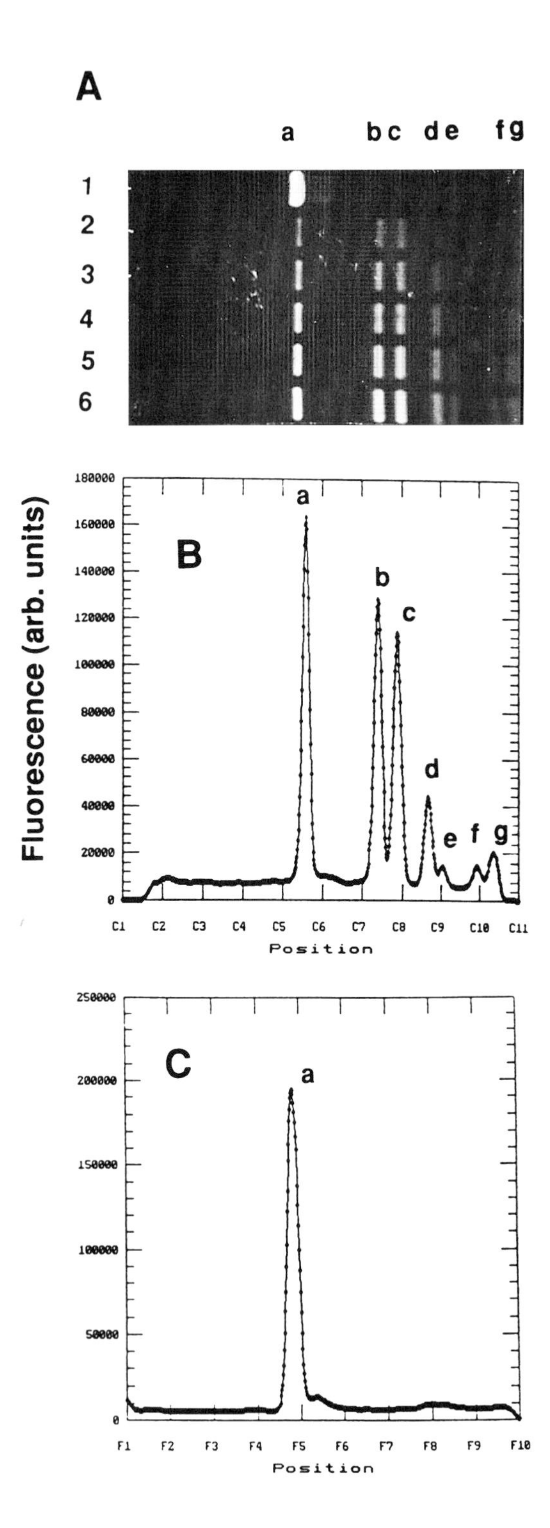
A
a
b c
d e
f g
1
2
3
4
5
6
Fluorescence (arb. units)
180000
160000
140000
120000
100000
80000
60000
40000
20000
0
B
a
b
c
d
e
f
g
C1
C2
C3
C4
C5
C6
C7
C8
C9
C10
C11
Position
250000
200000
150000
100000
50000
0
C
a
F1
F2
F3
F4
F5
F6
F7
F8
F9
F10
Position

## D. Other Applications

The above DNA hybridization experiments are the first reported so far with use of the SBMC (Christopoulos and Diamandis, 1991). Preliminary experiments involving dot immunodetection on nitrocellulose of a variety of antigens ($\alpha$-fetoprotein, ferritin, carcinoembryonic antigen, growth hormone, mouse IgG) gave satisfactory results. It is anticipated that the principle of detecting biotinylated moieties with SBMC and time-resolved fluorometry will also find other applications involving the techniques of Western and Northern blots. Furthermore, the system is now under evaluation for applications involving PCR amplification and DNA sequencing, immunohistochemistry, and flow cytometry.

In comparison to other isotopic and nonisotopic methodologies, the system described has the following advantages and disadvantages. Sensitivity is not so good as that of some other recent nonisotopic methodologies (Bronstein and McGrath, 1989; Pollard-Knight *et al.*, 1990; see also Chapter 5 of this book) and currently, the detection of single-copy genes on Southern blots is not possible. Despite this limitation, the proposed nonisotopic method uses extremely stable reagents (SBMC is stable for over 1 yr), and does not need autoradiography or enzyme substrates. The signal on nitrocellulose is stable for months to years, and is visible on UV illuminators with the possibility of instant photography. With use of the high-resolution scanning time-resolved fluorometer, the system can quantify the fluorescent bands. Additionally, this system is more sensitive and has a much wider range of applications than other time-resolved fluorometric methods based on $Eu^{3+}$ labeling and extraction.

Work is now in progress to improve the sensitivity of this system so that is can detect single-copy genes on Southern blots from 10 $\mu$g of total genomic DNA.

## REFERENCES

Bronstein, I., and McGrath, P. (1989). Chemiluminescence lights up. *Nature (London)* 338, 599–600.

Christopoulos, T. K., and Diamandis, E. P. (1991). Quantification of nucleic acids on nitrocellulose membranes with time-resolved fluorometry. Submitted for publication.

---

**Fig. 4** (A) Hybridization experiments after Southern blotting, of various amounts of enzyme-digested (lanes 2–6) or undigested (lane 1) linearized plasmid pBR 328 from 0.8% agarose gels. The probe was biotinylated linearized pBR 328 plasmid. The total amount of DNA (in ng) in lanes, was 160(1); 5(2); 10(3); 20(4); 40(5); 80(6). The length of each fragment is (in base pairs) 4907 (a); 2176 (b); 1766 (c); 1230 (d); 1033 (e); 653 (f); 517 (g). (B) Time-resolved fluorometric scanning of lane 6 (700 points) indicating the seven bands. (C) as in (B) but for lane 1.

Christopoulos, T. K., Lianidou, E. S., and Diamandis, E. P. (1990). Ultrasensitive time-resolved fluorescence method for α-fetoprotein. *Clin. Chem.* **36,** 1497–1502.

Dahlen, P. (1987). Detection of biotinylated DNA probes by using $Eu^{3+}$-labeled streptavidin and time-resolved fluorometry. *Anal. Biochem.* **164,** 78–83.

Davis, L. G., Dibner, M. D., and Battey, J. F. (1986). "Basic Methods in Molecular Biology." Elsevier, New York.

Diamandis, E. P. (1990). Analytical methodology for immunoassays and DNA hybridization assays—Current status and selected systems—Critical review. *Clin. Chim.* **194,** 19–50.

Diamandis, E. P. (1988). Immunoassays with time-resolved fluorescence spectroscopy: Principles and applications. *Clin. Biochem.* **21,** 139–150.

Diamandis, E. P., and Christopoulos, T. K. (1990). Europium chelate labels in time-resolved fluorescence immunoassays and DNA hybridization assays. *Anal. Chem.* **62,** 1149A–1157A.

Diamandis, E. P., and Morton, R. C. (1988). Time-resolved fluorescence using a europium chelate of 4, 7-*bis* (chlorosulfophenyl)-1, 10 phenanthroline-2,9-dicarboxylic acid (BCPDA). Labeling procedures and applications in immunoassays. *J. Immunol. Methods* **112,** 43–52.

Diamandis, E. P., Morton, R. C., Reichstein, E., and Khosravi, M. J. (1989). Multiple fluorescence labeling with europium chelators. Application to time-resolved fluoroimmunoassays. *Anal. Chem.* **61,** 48–53.

Evangelista, R. A., Pollak, A., Allore, B., Templeton, E. F., Morton, R. C., and Diamandis, E. P. (1988). A new europium chelate for protein labeling and time-resolved fluorometric applications. *Clin. Biochem.* **21,** 173–178.

Hemmila, I., and Batsman, A. (1988). Time-resolved immunofluorometry of hCG. *Clin. Chem.* **34,** 1163–1164.

Morton, R. C., and Diamandis, E. P. (1990). Streptavidin-based macromolecular complex labeled with a europium chelator suitable for time-resolved fluorescence immunoassay applications. *Anal. Chem.* **62,** 1841–1845.

Ngo, T. T. (1988). "Nonisotopic Immunoassay". Plenum Press, New York.

Oser, A., Callasius, M., and Valet, G. (1990). Multiple end labeling of oligonucleotides with terbium chelate-substituted psoralen for time-resolved fluorescence detection. *Anal. Biochem.* **191,** 295–301.

Oser, A., Roth, W. K., and Valet, G. (1988). Sensitive nonradioactive dot-blot hybridization using DNA probes labeled with chelate group-substituted psoralen and quantitative detection by europium ion fluorescence. *Nucleic Acids Res.* **16,** 1181–1196.

Pollard-Knight, D., Simmons, A. C., Schaap, A. P., Akhavan, H., and Brady, M. A. W. (1990). Nonradioactive DNA detection on Southern blots by enzymatically triggered chemiluminescence. *Anal. Biochem.* **185,** 353–358.

Soini, E., Hemmila, I., and Dahlen, P. (1990). Time-resolved fluorescence in biospecific assays. *Ann. Biol. Clin.* **48,** 567–571.

Soini, E., and Kojola, H. (1983). Time-resolved fluorometer for lanthanide chelates—a new generation of nonisotopic immunoassays. *Clin. Chem.* **29,** 65–68.

Soini, E., and Lovgren, T. (1987). Time-resolved fluorescence of lanthanide probes and applications in biotechnology. *CRC Crit. Rev. Anal. Chem.* **18,** 105–154.

Sund, C., Ylikoski, J., Hurskainen, P., and Kwaitkowski, M. (1988). Construction of europium ($Eu^{3+}$)-labeled oligo DNA hybridization probes. *Nucleosides Nucleotides* **7,** 655–659.

Syvanen, A. C., Tehen, P., Ranki, M., and Soderlund H. (1986). Time-resolved fluorometry: A sensitive method to quantify DNA hybrids. *Nucleic Acids Res.* **14,** 1017–1028.

# 12 Detection of Acridinium Esters by Chemiluminescence

Norman C. Nelson
Gen-Probe, Inc.
San Diego, California
Mark A. Reynolds and Lyle J. Arnold, Jr.,
Genta, Inc., San Diego, California



*Nonisotopic DNA Probe Techniques*

# I. INTRODUCTION

## A. Background

Chemiluminescence—the generation of light via a chemical reaction—has gained widespread use as a detection mode over the last several years (Weeks *et al.*, 1986; McCapra and Beheshti, 1985; Hastings, 1987). Advantages of chemiluminescent labels include high sensitivity; ease of use, handling, and disposal; precise control of detection (the reaction chemistry is typically very simple yet very specific); a wide range of detection methods (from photographic film to sophisticated instrumentation); and long shelf-lives. Chemiluminescence detection has been utilized in a variety of formats, including immmunodiagnostics (Berry *et al.*, 1988; Bronstein *et al.*, 1989; Weeks *et al.*, 1983b) and DNA probe-based assays (Arnold *et al.*, 1989; Bronstein *et al.*, 1990; Daly *et al.*, 1991;Dhingra *et al.*, 1991; Granato and Franz, 1989; Granato and Franz, 1990; Harper and Johnson, 1990; Lewis *et al.*, 1990; Ou *et al.*, 1990; Sanchez-Pescador *et al.*, 1988; Schaap *et al.*, 1989; Tenover *et al.*, 1990; Urdea *et al.*, 1988).

Recently the acridinium ester (AE) depicted in Fig. 1 was developed for use as a chemiluminescent label in bioassays (Weeks *et al.*, 1983a). The AE molecule reacts rapidly (typically 1 to 5 sec) with hydrogen peroxide under alkaline conditions (Fig. 2) to produce light at 430 nm (McCapra, 1970, 1973; McCapra and Perring, 1985). These rapid reaction kinetics permit detection over a very short time frame, thereby minimizing background noise and improving overall sensitivity. Detection in a standard luminometer (see Section II,B) exhibits a linear response over an AE concentration range of more than 4 orders of magnitude, with a detection limit of approximately $5 \times 10^{-19}$ mole (Arnold *et al.*, 1989).

Some other important reactions of the AE molecule are shown in Fig. 3. Hydroxide ion can catalyze the hydrolysis of the ester bond, rendering the AE permanently nonchemiluminescent. This property is utilized in an AE assay format, as described in Section I,B. Under the proper conditions hydroxide ion can also react with the carbon in the acridinium ring (C-9) that reacts with hydrogen peroxide (see Fig. 3), resulting in what is referred to as the *carbinol* or *pseudo-base* form of the AE (Weeks *et al.*, 1983a). This form is nonchemiluminescent since the hydroxide ion blocks reaction with peroxide. This inhibition can be easily reversed by incubation in acid, which rapidly reverts the hydroxide adduct, restoring the AE to its chemiluminescent form. Other nucleophiles can form adducts with the AE (Hammond *et al.*, 1991), and they also block chemiluminescence in a reversible manner. Adduct formation can also protect the AE against ester hydrolysis under certain conditions (Hammond *et al.*, 1991).

The AE reagent shown in Fig. 1 can be covalently attached to primary

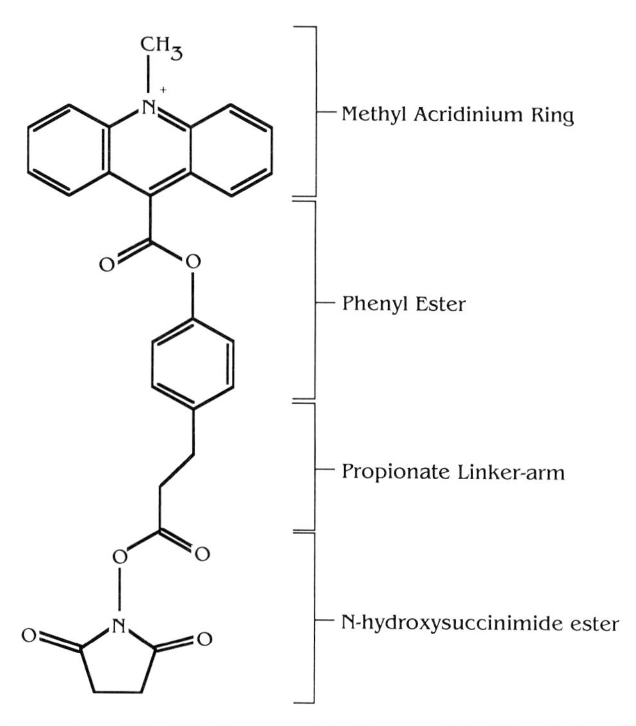


**Fig. 1** The structure of the *N*-hydroxysuccinimide ester of the *N*-methyl acridinium ester.

amine-containing compounds via specific reaction of the *N*-hydroxysuccinimide (NHS) ester of the AE with the primary amine of the compound. DNA probes can therefore be directly labeled with AE by first incorporating a primary amine into the DNA, then reacting this amine with the NHS-AE (see Section III,A). Direct labels of this kind are simpler to use than indirect labels (those attached through an avidin/biotin interaction, for example), since the *capping,* binding of the label, washing, and substrate addition steps associated with indirect labels are unnecessary.

## B. Principle of the Method

First, a chemiluminescent AE is attached directly to a synthetic oligonucleotide probe complementary to the target nucleic acid. An AE-labeled probe (AE-probe) displays the same chemiluminescence characteristics (reaction kinetics, specific activity, and sensitivity of detection) as the free label, demonstrating that the detection properties of the label are not compromised by attachment to oligonucleotides. This is in part owing to

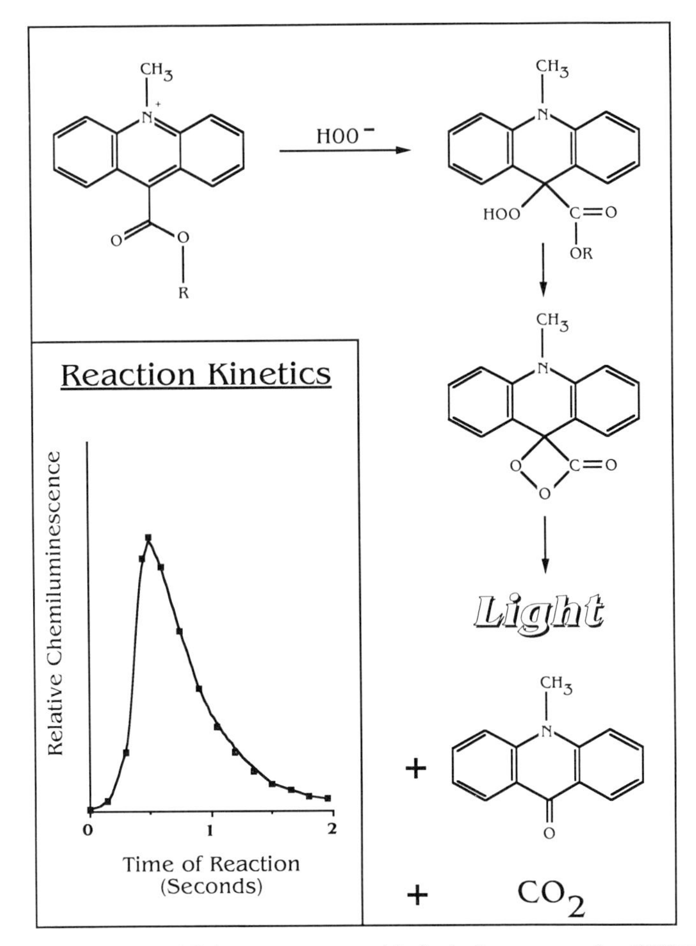


**Fig. 2** The *N*-methyl acridinium ester reacts with the hydroperoxy anion ($H00^-$) through the pathway shown above to produce light. The inset depicts the rapid kinetics of this reaction. The NHS phenyl ester (see Fig. 1) is represented as R.

the cleavage of the light-emitting species (namely, the acridinium ring) from the oligonucleotide during the chemiluminescence reaction (see Fig. 2), thus minimizing intramolecular quenching (Weeks *et al.*, 1983a). Furthermore, an AE-probe displays hybridization characteristics (rate and extent of hybridization, thermal stability, and specificity) essentially equivalent to its $^{32}P$ counterpart, demonstrating that attachment of the AE label also does not compromise hybridization performance of the probe.

The AE-probe can then be used in a variety of simple and rapid assay formats to directly detect target nucleic acid. All these formats utilize in-solution hybridization, which offers significant advantages (see Section

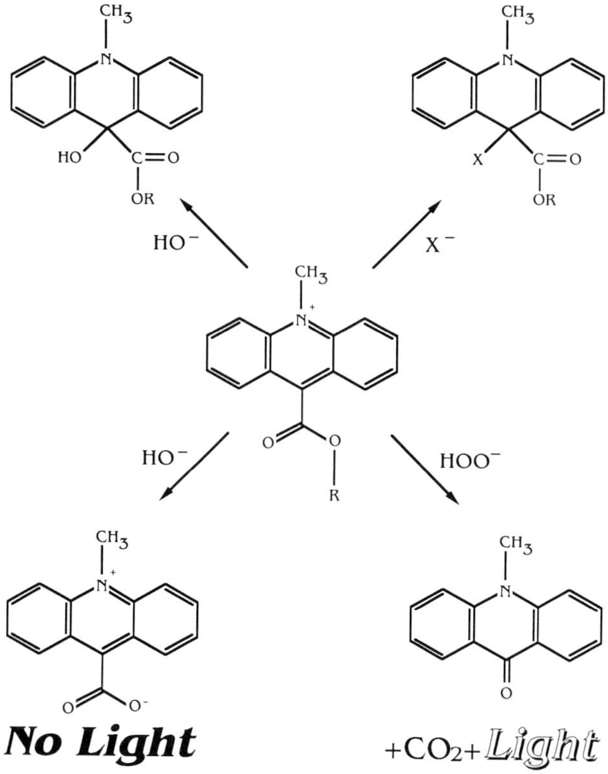

No Light +$CO_2$+ Light

**Fig. 3** The *N*-methyl acridinium ester (AE) reacts with hydroxide ion to form the reversible C-9 *carbinol* adduct (upper left) or to hydrolyze the ester bond, yielding the non-chemiluminescent acridinium carboxylic acid (lower left). Acridinium ester can also react with other nucleophiles ( represented as X-) to form reversible adducts at the C-9 position (upper right). As shown in Fig. 2, reaction with alkaline peroxide produces light (lower right).

I,C) over the more common solid-phase hybridization techniques (Keller and Manak, 1989). Various methods of separating hybridized from unhybridized AE-probe after the in-solution hybridization step have also been developed. In one approach (Arnold *et al.*, 1989), cationic submicron-sized magnetic microspheres are used to specifically capture the hybridized AE-probe from solution. After the spheres are magnetically collected on the side of the tube, the solution, which contains unhybridized AE-probe, is removed, the spheres are washed, and the chemiluminescence associated with hybrid is detected.

In another approach, discrimination between hybridized and unhybridized AE-probe is performed completely in-solution, thus eliminating the need for any physical separation steps. This method is based on chemical hydrolysis of the ester bond of the AE molecule, cleavage of which renders the AE permanently nonchemiluminescent (see Section I,A and Fig. 3). Conditions have been developed under which the hydrolysis of this ester bond is rapid for unhybridized AE-probe but slow for hybridized AE-probe. Therefore, after hybridization of the AE-probe with its target nucleic acid (under conditions that do not promote ester hydrolysis), the reaction conditions are adjusted so that the chemiluminescence associated with unhybridized AE-probe is rapidly reduced to low levels, whereas the chemiluminescence associated with hybridized AE-probe is minimally affected. Following this *differential hydrolysis* process, any remaining chemiluminescence is a direct measure of the amount of target nucleic acid present.

An example of this differential hydrolysis process is shown in Fig. 4, which depicts the loss of chemiluminescence due to ester hydrolysis as a function of time. From linear regression analysis, hydrolysis half-lives were determined to be 50.5 and 0.72 min for hybridized and unhybridized AE-probe, respectively. The theoretical percentage of remaining chemiluminescent label after a given hydrolysis time can be calculated using the equation

$$(0.5)^{n} \times 100 = \text{percentage remaining chemiluminescence} \qquad (1)$$

where n is the ratio of elapsed time to the half-life of chemiluminescence loss (i.e., how many half-lives have transpired during a given incubation time). Using the half-life values given above, the calculated values for percentage remaining chemiluminescence after a 15-min differential hydrolysis step would be 81% for hybridized probe and 0.00005% for unhybridized probe. This represents greater than a one million-fold discrimination between hybridized and unhybridized AE-probe, achieved in 15 min with the addition of a single reagent and without any physical separation. In practical use, background levels of 0.00005% have not yet been achieved because the rate of AE hydrolysis plateaus before reaching these levels. Nevertheless, background levels of 0.002% have been achieved in a homogeneous format, which is still a significant signal-to-noise discrimination.

A homogeneous assay format for the detection of target nucleic acids based on this differential hydrolysis process has been developed. This format, referred to as the hybridization protection assay, or HPA (Arnold *et al.*, 1989; Nelson and Kacian, 1990; Nelson *et al.*, 1990), consists of the following three basic steps (Fig. 5):

1. *Hybridization*. AE-probe is added to a solution containing the target nucleic acid, and the mixture is incubated. The length of time required for this step is dependent on many factors, including temperature, buffer and

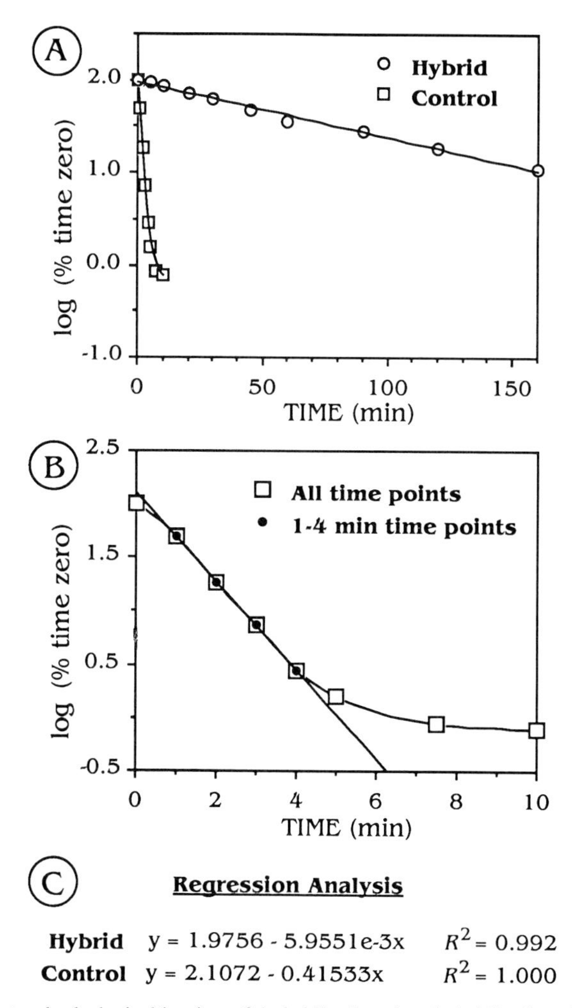


**Fig. 4** (A) Ester hydrolysis kinetics of hybridized and unhybridized acridinium ester-labeled DNA probe (AE-probe) are shown as loss of chemiluminescence versus time. (B) An expanded view of the hydrolysis curve for the unhybridized AE-probe. The hydrolysis rate begins to slow at between 4 and 5 min of incubation, and plateaus at about 10 min. Additionally, the rate of the hydrolysis between 0 and 1 min of incubation is variable since the contents of the tube are going from room temperature to 60°C during that interval. Therefore, the rate of hydrolysis is measured between 1 and 4 min of incubation, during which interval the hydrolysis rate is constant. (C) The equations generated from linear regression analyses for the hybridized and unhybridized AE-probe are given along with the $R^2$ values for each fit.

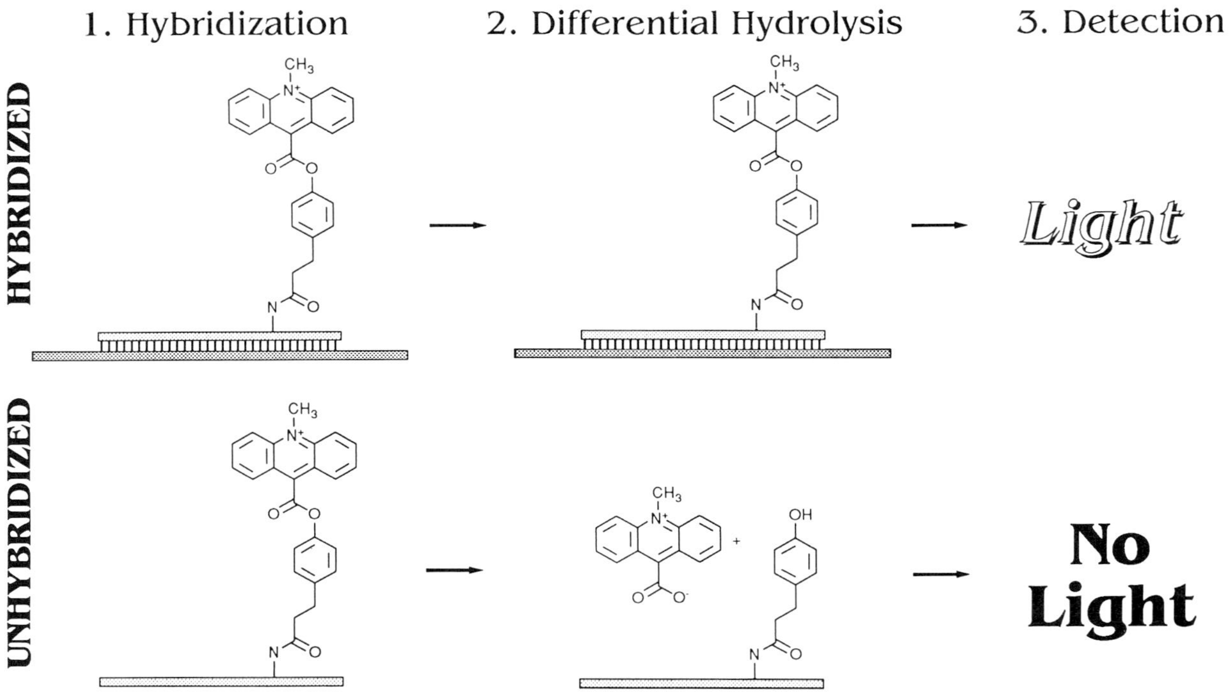

1. Hybridization
2. Differential Hydrolysis
3. Detection
HYBRIDIZED
UNHYBRIDIZED
Light
No Light

**Fig. 5** Schematic representation of the hybridization protection assay (HPA) depicting the three basic steps of the assay. In step 1, acridinium ester-labeled probe (AE-probe) is hybridized with its target nucleic acid. In step 2, acridinium ester (AE) associated with unhybridized AE-probe is rapidly hydrolyzed, whereas AE associated with hybridized AE-probe is protected from hydrolysis. In step 3 (detection), hybridized AE-probe (intact AE) produces chemiluminescence, whereas unhybridized AE-probe (hydrolyzed AE) does not produce chemiluminescence. See text and Figs. 3 and 4 for details.

salt conditions, kinetics of the particular probe used and concentration of the hybridizing nucleic acid strands. At 60°C, hybridization times longer than 1 hr are uncommon, and times of 10 to 15 min are routine.

2. *Differential hydrolysis.* An alkaline differential hydrolysis buffer is added to the hybridization reaction, thereby effecting hydrolysis of AE associated with unhybridized probe. This step is very rapid (typically 6 to 12 min at 60°C).

3. *Detection.* The assay tubes are placed directly into a luminometer, and chemiluminescence is measured automatically. This process takes only 5 to 10 sec per sample.

The limit of sensitivity of the HPA format is approximately $6 \times 10^{-17}$ moles (Arnold *et al.*, 1989), detection is quantitative and linear over a wide range of target nucleic acid concentrations, and the assay is rapid and easy to perform.

In yet another approach, the differential hydrolysis process is combined with separation using cationic magnetic beads, thereby achieving even greater discrimination between hybridized and unhybridized AE-probe. In this format, backgrounds as low as 0.0002% are achieved, and assay sensitivity is approximately $6 \times 10^{-18}$ moles (Arnold *et al.*, 1989). This format is also very rapid and easy to perform because of the simplicity of the magnetic separation step.

Some properties of the AE detection system (both the unconjugated AE and the AE-probe assay formats) are listed in Table I.

## C. Advantages of the Method

There are several advantages to using the chemiluminescent AE as a detection label, including high inherent sensitivity, low quenching, sim-

**Table I**
Selected Properties of the Acridinium-Ester Detection System

| Parameter | AE alone | AE-probe assay formats | |
|---|---|---|---|
| | | HPA | DH + separation |
| Sensitivity | $5 \times 10^{-19}$ moles | $6 \times 10^{-17}$ moles | $6 \times 10^{-18}$ moles |
| Quantitation | Objective | Objective | Objective |
| Specific activity | $1 \times 10^{8}$ RLU/pmol | $1 \times 10^{8}$ RLU/pmol | $1 \times 10^{8}$ RLU/pmol |
| Linear dynamic range | $\geq 10^4$ | $10^4$ | $\geq 10^4$ |
| Time to result | 2–5 sec | 20–75 min | 50–110 min |

plicity, ease of handling and disposal, precise chemical control, and a long shelf-life. Furthermore, these properties are not adversely affected by attachment of the AE to oligonucleotide probes, nor are the hybridization characteristics of the oligonucleotide probe compromised. There are also several advantages to the AE-probe assay formats presented here, including the following:

1. *In-solution approach.* Advantages of the in-solution approach include availability of all the target molecules (in target immobilization procedures, only a portion of the molecules are available for hybridization), faster hybridization kinetics, better quantitation, fewer steps and less complexity, and faster time-to-result.
2. *Sensitivity.* The assays are sensitive. This is in part owing to the high specific activity and inherent low background characteristics of the detection system as well as the low background levels achieved with the assay formats described.
3. *Specificity.* The assays retain the high specificity inherent in DNA probe-based formats. Furthermore, the differential hydrolysis assays described here afford an additional level of specificity not possible with other assay/detection systems. This is because the stability of the AE label in the hybridized form of the AE-probe is very sensitive to duplex structure. Even small perturbations in structure, such as those caused by single base mismatches, can lead to large increases in the AE hydrolysis rate. Therefore, differential hydrolysis characteristics can be used to discriminate between nucleic acids with only small sequence differences (even a single base change), even when overall hybridization characteristics cannot. An example of this has been described previously (Arnold *et al.*, 1989; Nelson *et al.*, 1990); two nucleic acid targets that differed by only a single base were assayed using a single AE-probe perfectly complementary to one of the targets. Using a standard solid-phase separation method, no significant difference in hybridization was seen between the two targets. However, the differential hydrolysis method in an HPA format clearly discriminated between the two targets.
4. *Simplicity.* The assays are very easy to perform (there are only three basic steps; see Section I,B) and the *time-to-result* is very short. Assays utilizing the HPA format are typically complete in 30 to 60 min, of which only about 5 min is *hands-on* time. Also, the entire assay is performed in a single tube, which is simply placed in an automatic luminometer for detection. Furthermore, direct labels are simpler to use than indirect labels since the capping, binding of the label, washing, and substrate addition steps associated with indirect labels are unnecessary.
5. *Quantitation.* The assays are quantitative with a linear response over

several orders of magnitude of target concentration (the limit of the linearity is typically the luminometer and not the assay itself).

6. *Precision*. The assay is reliable and reproducible.

7. *Versatility*. The assay can be used for the detection of both DNA and RNA targets. Several of the formats developed to date target ribosomal RNA (rRNA), which provides the further benefits of natural target amplification (there are approximately 10,000 rRNA copies per bacterial cell) and improved specificity (Enns, 1988; Kohne, 1986; Woese, 1987).

8. *Clinical compatibility*. The AE-probe retains its chemiluminescence and hybridization characterisitics in the presence of relatively large amounts of clinical specimen material, a feature that allows the methods described here to be utilized for direct specimen testing in clinical diagnostic assays.

## D. Limitations of the Method

Limitations of the AE-probe system that should be kept in mind when considering potential applications are as follows:

1. The ester bond of the AE is inherently unstable (this property is the basis for the differential hydrolysis process), and therefore AE-probes can be used only within certain pH and temperature ranges. For example, the AE will not withstand incubation at 95°C and therefore cannot be incorporated *during* polymerase chain reaction (PCR) amplification (although it can be readily used for the detection of the products from *in vitro* amplification procedures; see Sections I,E and III,C). Also, AE-probes cannot be run under standard polyacrylamide gel electrophoresis conditions (although other types or conditions of electrophoresis are feasible).

2. The AE will react with alkaline peroxide to produce light only once. Therefore, samples cannot be re-read for chemiluminescence.

3. Some target nucleic acid samples may contain inherent chemiluminescence. If these levels are high enough, sensitivity of the HPA format will be decreased owing to increased background. In these cases additional steps to eliminate the sample chemiluminescence are necessary.

## E. Applications of the Method

The AE-probe method can be used for the in-solution detection of virtually any target nucleic acid, provided the target possesses a hybridizable probe region (the sequence must be known) and is available in sufficient quantity to be within the sensitivity range of the assay. One of the earliest reported uses of this method (Arnold *et al.*, 1989) was the detection and quantitation of purified ribosomal RNA (rRNA). More recently, applications ranging from single mismatch discrimination (Nelson *et al.*, 1990) and detection of amplified human immunodeficiency virus (HIV) DNA (Ou *et al.*, 1990) to the detection of nucleic acids from clinically relevant pathogens (see the following) have been reported. Potential applications include the detection and identification of a wide variety of pathogens and other microorganisms in medicine, dentistry and veterinary; forensic science; therapeutic drug monitoring; genetic screening (including disease susceptibility) and typing; food science; the petroleum industry; and environmental testing; as well as a wide spectrum of basic research applications involving nucleic acids.

A particularly useful application of the HPA format is the detection of products from target amplification reactions such as the PCR (Mullis and Faloona, 1987), and the transcription amplification system, or TAS (Kwoh *et al.*, 1989). The HPA format retains the distinct advantage of specific probe hybridization to unequivocally identify the amplified sequence. Methods that assay simply for extension of primers, such as gel electrophoresis followed by ethidium bromide staining, are equivocal since nontarget species of similar size can be produced with some frequency (Innis and Gelfand, 1990). Detection of HIV DNA (Kacian *et al.*, 1990; McDonough *et al.*, 1989; Ou *et al.*, 1990), the Philadelphia chromosome (Arnold *et al.*, 1989; Dhingra *et al.*, 1991), hepatitis B virus (HBV) DNA (Kacian *et al.*, 1990), and the genes that code for the rRNA of *Neisseria gonorrhoeae* (Gegg *et al.*, 1990) and *Chlamydia trachomatis* (Kranig-Brown *et al.*, 1990) using this procedure have been reported. Test kits utilizing this detection approach are commercially available (Gen-Probe, San Diego).

Another useful application of the AE-probe system is the detection and identification of pathogenic organisms in the clinical laboratory. Diagnostic assays for a variety of organisms have been reported, including test for *Chlamydia trachomatis* (Harper and Johnson, 1990; LeBar *et al.*, 1987; Woods *et al.*, 1990), *Neisseria gonorrhoeae* (Lewis *et al.*, 1990; Granato and Franz, 1989; Granato and Franz, 1990; Harper and Johnson, 1990), *Escherichia coli* (Watson *et al.*, 1990), *Streptococcus agalactiae, Haemophilus influenzae, Enterococcus* species (Daly *et al.*, 1991), and *Campylobacter* species (Tenover *et al.*, 1990). Kits for the detection of a variety of pathogenic organisms both directly in clinical specimen material and as culture confirmation assays are commercially available (Gen-Probe, San Diego).

# II. Materials

## A. Reagents

### 1. Preparation of Acridinium Ester-Labeled DNA Probe (AE-Probe)

The *N*-hydroxysuccinimide (NHS) ester of AE is synthesized as described by Weeks and co-workers (Weeks *et al.*, 1983a). The dimethylsulfoxide (DMSO) used in the preparation of the stock solution of the AE used in the labeling reaction must be completely free of water (otherwise the NHS ester will be hydrolyzed, rendering the labeling reagent inactive). Therefore, DMSO is purchased as reagent grade (Aldrich, Wisconsin), distilled and stored over molecular sieves (activated, Type 4A, 8–12 mesh; J.T. Baker). There is also DMSO added to the labeling reaction mixture in addition to that added as the AE labeling reagent (see Section III,A). This DMSO does not have to be perfectly dry, since it is mixed with water. Therefore, for this particular use, reagent grade DMSO (Aldrich) is adequate (although it certainly does not hurt if the dry DMSO is used). Ethanol used in nucleic acid precipitations is 200 proof from Quantum Chemical Corporation. For small quantities of DNA (less than about 10 $\mu$g), a carrier should be used in the ethanol precipitation steps to maximize recovery. Glycogen (Sigma) is used as the carrier in the procedures given here.

High-pressure liquid chromatography (HPLC) buffer compositions are as follows: Buffer A, 20 m*M* sodium acetate (pH 5.5), 20% (v/v) acetonitrile; Buffer B, 20 m*M* sodium acetate (pH 5.5), 20% (v/v) acetonitrile, 1*M* LiCl; all reagents used in the preparation of these buffers must be HPLC grade (Fisher Scientific or equivalent). After preparation, both of these buffers must be filtered through a 0.45 $\mu$m Nylon-66 filter (Rainin); a vacuum filtration apparatus designed for this purpose is useful (also available from Rainin). The column used is a Nucleogen-DEAE 60-7 ion-exchange column (also available from Rainin). See Section II,B below for a discussion of HPLC instrumentation. For all other chemicals, reagent grade is adequate (Fisher Scientific or equivalent).

### 2. Determination of Differential Hydrolysis Kinetics of AE-Probe

Lithium succinate, ethylenediaminetetraacetic acid (EDTA), ethylene glycol *bis*($\beta$-aminoethyl ether) *N*,N′-tetraacetic acid (EGTA), sodium tetraborate (or boric acid), lithium lauryl sulfate and Triton X-100 are reagent grade (Fisher Scientific or equivalent). The borate buffer may be made as 0.15 *M* tetraborate or 0.6 *M* boric acid (titrated to the correct pH with NaOH). The solubility of sodium tetraborate is approximately 0.2 *M* at pH 8. Hybridizations are performed in 1.5-ml screw-capped microcentrifuge tubes (Eppendorf); AE hydrolysis and chemiluminescence measurements are performed in 12 × 75 mm polystyrene tubes (Sarstedt, West Germany).

### 3. AE-Probe Assays

The formamide used in the differential hydrolysis + separation protocol is from Fluka; the magnetic microspheres are from Advanced Magnetics, Inc. (Cambridge, Massachusetts). The magnetic separation solution is made by adding a small volume of a concentrated stock solution of magnetic microspheres in water to 0.4 *M* phosphate buffer, pH 6.0 (1 to 50 μl of magnetic spheres per ml of buffer is used, depending on the amount of spheres required for a given application); this solution must be used within 8 hr of preparation. Reagents used in these assays that are also used in the determination of the differential hydrolysis kinetics are described in Section II,A,2; all other reagents used are reagent grade (Fisher Scientific or equivalent).

The magnetic particles can be separated from solution with virtually any magnet of adequate strength. Racks that hold the 12 × 75 mm assay tubes and that have built-in magnets are particularly convenient. Racks that are compatible with the assays described here are commercially available (Corning; Gen-Probe).

### 4. General

Chemiluminescence measurements are performed in 12 × 75 mm polystyrene tubes (Sarstedt, West Germany). Some lots of tubes possess chemiluminescent contamination; several empty tubes of each lot should therefore be read in the luminometer. Tubes should always be blotted with a mildly damp tissue or paper towel before chemiluminescence measurement to remove any static charge from the tubes (this charge can lead to errantly high readings). The tissue or towel should not be *too* wet, as this can potentially damage the interior of the luminometer. Blotting before reading will also remove any contamination that may be present on the outside of the tube, thus protecting the luminometer from contamination and/or damage.

The reagents required for the chemiluminescence reaction of AE are base and peroxide. Reagent grade NaOH and hydrogen peroxide (supplied as a 30% solution; Fisher Scientific) are used for this purpose. Typically, an automatic two-injection protocol is used on the luminometer to inject these reagents (see Section II,B), the first injection containing 0.1% peroxide and the second injection containing 1 *M* NaOH. Hydrogen peroxide is very stable at acidic pH and unstable at basic pH. Therefore if the 0.1% peroxide solution is acidified (1 m*M* nitric acid), it will be stable at room temperature for weeks to months. In some assay modes, the peroxide solution contains a higher concentration of acid (e.g., 0.4 *M* nitric acid) in order to reverse carbinol formation of the AE (see Section III,B). This solution is also stable over extended periods. On occasion, the peroxide and the base are mixed and injected as a single reagent. The peroxide is stable in this basic reagent for only about 10 hr at room temperature; this reagent must therefore be prepared fresh daily.

Standard pipettors (Gilson Pipetman) are used in these protocols; standard vortex mixers are purchased from Fisher Scientific; samples are evaporated in a Savant Speed-Vac rotary/vacuum drying apparatus.

## B. Instrumentation

### 1. Luminometer

The reaction of the AE with alkaline peroxide to produce chemiluminescence is very rapid (see Section I,A and Fig. 2). Therefore a luminometer that automatically injects the alkaline peroxide into the sample and immediately measures the chemiluminescence is required for accurate detection of the AE described here. Furthermore, the assay protocols described here are based on a 12 × 75 mm reaction tube, and the luminometer selected must therefore be able to accommodate such tubes. Luminometers that fulfill both these requirements are commercially available from Gen-Probe, Berthold (Germany), and Turner. A luminometer that is capable of automatically injecting the alkaline peroxide but is based on a microtiter well format is available from Dynatech. All procedures described here were developed using a LEADER I luminometer (Gen-Probe).

### 2. DNA Synthesizer

Any DNA synthesizer capable of standard phosphoramidite solid-phase chemistry is appropriate for synthesis of the DNA probes. All procedures described here were developed using a Biosearch Model 8750 or an ABI Model 380A synthesizer.

### 3. High-Performance Liquid Chromatograph

Any of a number of instruments currently available would be appropriate for the chromatography described herein. The instrument must be capable of performing automatic gradient elution and must be equipped with a detector capable of measurement at 260 nm; a dual-wavelength detector capable of measurement at 260 and 325 nm is preferable. The chromatography described here was performed on a Beckman System Gold HPLC.

# III. PROCEDURES

## A. Preparation of an Acridinium Ester-Labeled DNA Probe (AE-Probe)

An oligomer of virtually any sequence can be labeled, and selection of this sequence is obviously dependent on the target nucleic acid to be detected. The oligonucleotide is synthesized using standard phosphoramidite solid-phase chemistry (Caruthers *et al.*, 1987) on a commercially available automated DNA synthesizer. A primary amine is introduced into the oligomer during synthesis using the nonnucleotide-based linker-arm chemistry described previously (Arnold *et al.*, 1988). This chemistry allows incorporation of the linker-arm at any position in the growing oligonucleotide chain, including the 3′ terminus, the 5′ terminus, or at any position within the oligonucleotide sequence. The resulting primary amine-containing oligomer is purified using standard polyacrylamide gel electrophoresis and desalting techniques (water is the final solvent). Labeling of this oligomer with AE-NHS proceeds as follows:

1. Evaporate to dryness the oligomer to be labeled in a 1.5-ml screw-capped microcentrifuge tube using a Savant Speed-Vac drying apparatus.
2. After drying, add to the tube 8 $\mu$l of 0.125 *M* HEPES buffer (pH 8.0)/50% DMSO and 2 $\mu$l of 25 m*M* AE-NHS in DMSO. Cap the tube, gently mix the contents, and then incubate for 20 min at 37°C.
3. Add an additional 3 $\mu$l 25 m*M* AE-NHS in DMSO, mix gently, then add 2 $\mu$l 0.1 *M* HEPES buffer (pH 8.0). Recap, mix gently, and then incubate an additional 20 min at 37°C.
4. Quench the reaction by adding 5 $\mu$l 0.125 *M* lysine in 0.1 *M* HEPES buffer (pH 8.0)/50% DMSO. Mix gently.

5. Add 30 μl 3 *M* sodium acetate buffer (pH 5.0), 245 μl water, and 5 μl glycogen (40mg/ml); vortex to mix. Add 640 μl chilled (−20°C) absolute ethanol; cap tube, and vortex to mix. Incubate on dry ice for 5 to 10 min, then centrifuge at 15,000 rpm using the standard head in a refrigerated microcentrifuge (Tomy or Eppendorf).
6. Carefully aspirate the supernatant, making sure not to disturb the pellet. Resuspend the pellet in 20 μl 0.1 *M* sodium acetate (pH 5.0) containing 0.1% SDS.

The labeled oligomer (AE-probe) is then purified by gradient elution ion-exchange high-performance liquid chromatography (IE-HPLC) using a 4.0 × 125 mm Nucleogen-DEAE 60-7 column as follows:

1. Pre-equilibrate the IE column in 60% Buffer A, 40% Buffer B at a flow rate of 1 ml/min.
2. To the 20 μl resuspended AE-probe pellet from step 6 of the labeling reaction, add 190 μl of the pre-equilibration buffer (i.e., 60% Buffer A, 40% Buffer B). Vortex to mix.
3. Load 100 μl of the AE-probe sample into a 100-μl injection loop.
4. Inject the AE-probe onto the IE column equilibrated as described in step 1. Elute the AE-probe using a linear gradient from 40 to 70% Buffer B in 30 min at a flow rate of 1 ml/min; monitor absorbance at 260 nm as well as 325 nm (if possible); collect 0.5-ml fractions.
5. If absorbance at 325 nm is not monitored, or the amount of AE-probe being purified is too small to give a discernable absorbance at this wavelength, measure the chemiluminescence of each fraction as follows: Pipette 2 μl of each fraction into corresponding 12 × 75 mm polystyrene tubes containing 100 μl 0.1% SDS; vortex to mix. Measure the chemiluminescence of each tube with the automatic injection of 200 μl 0.4 *M* $HN0_3$, 0.1% $H_2O_2$, then 200 μl 1 *M* NaOH, followed by measurement of signal for 5 sec. Plot chemiluminescence (measured as relative light units, or RLU) versus fraction number. The fractions may need to be diluted if the chemiluminescence contained in 2 μl is too high for the luminometer to measure. If so, dilute in 0.1% sodium dodecyl sulfate (SDS) and vortex thoroughly to ensure a homogeneous sample.
6. Precipitate fractions containing the purified oligomer (see Fig. 6) as follows: Add 5 μl glycogen (40 mg/ml) and 1 ml chilled (−20°C) absolute ethanol; cap the tube, and vortex to mix. Incubate on dry ice for 5 to 10 min, then centrifuge at 15,000 rpm using the standard head in a refrigerated microcentrifuge (Tomy or Eppendorf).

7. Carefully aspirate the supernatant, making sure not to disturb the pellet. Resuspend the pellet in 20 μl 0.1 *M* sodium acetate (pH 5.0) containing 0.1% SDS.

Purification of an AE-labeled 26-mer is shown in Fig. 6. The specific activity of the AE-labeled probe is determined as follows:

1. Perform serial dilutions of the AE-probe stock in 0.1 *M* sodium acetate (pH 5.0) containing 0.1% SDS; the final measurement of chemiluminescence should be in the 50,000 to 250,000 RLU range.
2. Pipet 10 μl of the final dilution of AE-probe into 200 μl 5% Triton X-100 (3 to 5 replicates is recommended). Measure the chemiluminescence of each tube with the automatic injection of 200 μl 0.4 *M* $HNO_3$, 0.1% $H_2O_2$, then 200 μl of 1 *M* NaOH, followed by measurement of signal for 5 sec. Based on the dilution factor from (1), calculate the RLU/μl of the stock AE-probe.
3. Measure the absorbance at 260 nm of AE-probe in 0.1 *M* sodium acetate (pH 5.0) containing 0.1% SDS; the final measurement should be in the 0.2 to 1.0 optical density (OD) range (depending on the amount of probe labeled, dilutions may or may not be necessary to achieve this concentration).
4. Calculate the concentration of DNA probe in the stock solution. Since the AE as well as the oligomer contributes to the OD at

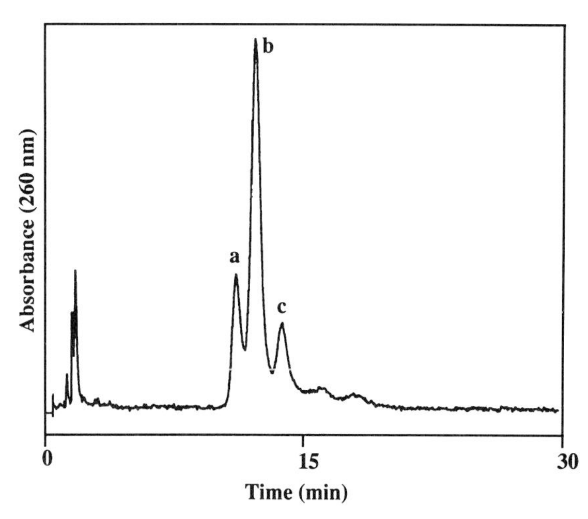


**Fig. 6** A DNA probe was labeled with acridinium ester (AE) and purified used ion-exchange HPLC, as described in the text. The resulting chromatogram is shown. Peak a represents unlabeled probe; peak b represents AE-labeled probe; and peak c represents AE-probe that has undergone ester hydrolysis (see Fig. 3).

260 nm, first calculate the contribution of the oligomer to the OD reading. The extinction coefficient of the oligomer is roughly 10,000 $M^{-1}$ times the number of nucleotides in the oligomer (the exact extinction coefficients for each base can be used if desired). The extinction coefficient of the AE is 78,800 $M^{-1}$. The total extinction coefficient of a 20-mer, for example would be

$$[10{,}000\ M^{-1} \times 20] + 78{,}800\ M^{-1} = 278{,}800 M^{-1}$$

The contribution *from the oligomer* will therefore be

$$200{,}000\ M^{-1}/278{,}800\ M^{-1} = 71.7\%$$

For an AE-labeled 20-mer, the OD reading will have to be multiplied by .717 to calculate the OD reading *of the oligomer.*

To calculate the concentration of the oligomer from the OD, the following two conversions must be used: 1 OD unit = 40 $\mu$g of DNA (approximate value; other values may be used if preferred), and

$$\text{molecular weight of an oligomer} = 330\ \text{g/mol nucleotide} \times \text{the number of nucleotides}$$

For example, if the OD/ml reading of the AE-labeled 20-mer is 1.0, the concentration is calculated as follows:

$$\text{MW of oligo} = 330\ \mu\text{g}/\mu\text{mol nucleotide} \times 20\ \text{nucleotides} = 6600\ \mu\text{g}/\mu\text{mol} \quad (2)$$

$$1\ \text{OD/ml} \times .717 = 0.717\ \text{OD/ml (contribution from the oligomer)} \quad (3)$$

$$0.717\ \text{OD/ml} \times 40\ \mu\text{g/OD} = 28.68\ \mu\text{g/ml} \quad (4)$$

$$28.68\ \mu\text{g/ml} \times 1\ \mu\text{mol}/6600\ \mu\text{g} = 4.35\ \text{nmol/ml} = 4.35\ \text{pmol}/\mu\text{l} \quad (5)$$

[**Note:** this value will have to be multiplied by the dilution factor if any dilutions of the stock AE-probe were made.]

5. Determine the specific activity based on the results of 1 through 4 above:

$$(\text{RLU}/\mu\text{l})/(\text{pmol}/\mu\text{l}) = \text{specific activity in RLU/pmol} \quad (6)$$

Specific activities of AE-labeled probes are typically in the range of 0.5 to 1.0 × $10^8$ RLU/pmol.

---

*Procedural Notes/Troubleshooting*

1. Oligomers as long as 40 bases can be purified using the IE-HPLC protocol given. For oligomers longer than 40 bases, reverse-phase HPLC

and or IE-HPLC using an alternate column is used to achieve purification (data not shown).

2. HPLC resolution can be somewhat improved by extending the duration of the gradient. Additionally, the initial buffer conditions should be slightly adjusted for different length oligomers to optimize elution time—a lower percentage Buffer B for shorter oligomers and a higher percentage B for longer oligomers. The exact conditions can be determined empirically.

3. Absorbance at 325 nm is used to monitor the AE without any interference from the DNA.

4. Various amounts of oligomer can be labeled using this protocol. Approximately 10 nmol can be labeled using the volumes described above; proportionally larger amounts of oligomer can be labeled using proportionally larger amounts of each reagent (the limiting factor is solubility of the AE-probe).

5. An acceptable alternative to the labeling protocol described above is as follows: to the dried oligomer, add 15 $\mu$l of a mixture containing 0.1 *M* HEPES buffer (pH 8.0), 60% DMSO, and 5 m*M* AE; mix gently; incubate 90 min at 30°C.

6. Be extremely careful when using the stock solution of AE not to contaminate your buffers, pipettors, workspace, etc. (a 1 $\mu$l aliquot of the 25 m*M* stock contains approximately $10^{11}$ RLU of chemiluminescence).

7. After precipitation of the AE-probe, be careful to remove as much of the ethanol as possible. However, avoid drying in the Speed-Vac (or other similar drying device) as this can lead to breakdown of the AE on the oligomer.

8. If labeling efficiency is poor, the AE-NHS may be hydrolyzed (this reagent is very water sensitive, and once it is broken down, it will no longer react with the primary amine of the oligomer). Prepare a fresh stock of the AE-NHS in freshly distilled DMSO and repeat the procedure.

9. If peak "c" in the HPLC chromatogram (hydrolyed AE-probe; see Fig. 6) is large, the pH of the reaction mixture may be incorrect. Measure the pH of the existing HEPES buffer as well as a mock reaction cocktail, and/or prepare fresh buffer and repeat the procedure.

10. If the labeling reaction did not proceed to a high extent (peak "a" high and peaks "b" and "c" low; see Fig. 6), the pH of the reaction mixture may be incorrect. Measure the pH of the existing HEPES buffer as well as a mock reaction cocktail, and/or prepare fresh buffer and repeat the procedure. Alternatively, the NHS ester of the AE labeling reagent may be hydrolyzed, in which case the labeling procedure should be repeated using a freshly prepared stock AE-NHS.

11. Mixing HEPES and DMSO produces heat; therefore, these reagents should be pre-mixed and allowed to cool before adding the AE-NHS labeling reagent.

12. See Section III,D, Procedural Notes/Troubleshooting, General, for tips that are applicable to all AE-based procedures.

## B. Determination of the Differential Hydrolysis Kinetics of an AE-Probe

Once an oligomer is labeled and purified, the AE hydrolysis kinetics of the hybridized and unhybridized oligomer are measured using the protocol given. These data are then utilized to evaluate the differential hydrolysis performance characteristics of an AE-probe and to determine optimal hydrolysis times to be used in an assay.

1. In 1.5-ml microcentrifuge tubes, prepare three 30 μl reaction mixtures to contain the following: Hybrid—0.1 *M* lithium succinate buffer, pH 5.2, 2 m*M* EDTA, 2 m*M* (EGTA), 10% (w/v) lithium lauryl sulfate, 0.1 pmol AE-probe, and 0.5–1.0 pmol of the target nucleic acid (i.e., nucleic acid complementary to the AE-probe); Control (i.e., unhybridized AE-probe)—same as Hybrid, except containing no target nucleic acid; Blank—same as Control except containing no AE-probe. Incubate all three reactions in a water bath 30 min at 60°C.
2. Remove reactions from water bath, add 270 μl 0.1*M* lithium succinate buffer, pH 5.2, 2 m*M* EDTA, 2 m*M* EGTA, 10% (w/v) lithium lauryl sulfate to each, and vortex thoroughly.
3. To measure the AE hydrolysis rate for hybridized AE-probe, pipette 15-μl aliquots of the diluted Hybrid reaction into each of seven 12×75mm assay tubes. Add 100 μl 0.15 *M* sodium tetraborate buffer, pH 7.6, 5% (v/v) Triton X-100 to each tube, and vortex each 3 sec. Measure the chemiluminescence of two of the replicate tubes immediately (see detection protocol in step 5); these will be the *time zero* values. Cover the remaining five replicates and place them in a water bath at 60°C; remove a single tube at 5, 10, 15, 20, and 30 min. Immediately after removal from the water bath, place each tube in an ice/water slurry for 15 sec, remove to room temperature and let stand for 30 sec, then measure the chemiluminescence as described in step 5.
4. To measure the AE hydrolysis rate for unhybridized AE-probe, the protocol in step 3 is repeated with the Control reaction from step 1, with the exception that tubes are removed from the water bath at 1, 2, 3, 4, and 5 min. The Blank reaction is also analyzed in this manner (to be run concurrently with the Control analysis).

5. Detect chemiluminescence in a luminometer by the automatic injection of 200 $\mu$l 0.4 $M$ $HNO_3$, 0.1% $H_2O_2$, then 200 $\mu$l of 1 $M$ NaOH, followed by measurement of signal for 5 sec. Remember to blot the tubes with a moist tissue or paper towel before measuring the chemiluminescence (see Section II,A,4).
6. Analyze the data as follows: First, subtract the average Blank value from each Hybrid and Control value. Then plot the data for Hybrid and Control as the log of the percentage of time zero chemiluminescence versus time. Finally, determine the slopes of the linear portions of the hydrolysis curves (and therefore the rates of AE hydrolysis), using standard linear regression analysis.

An example of the data generated using this protocol is given in Fig. 4. From the slope of the line the half-life of hydrolysis can be calculated (half-life in min = 0.3 log units per half-life divided by the slope in log units per min), and the ratio of the hybrid to control half-lives determined. This differential hydrolysis ratio (or DH ratio) is a useful parameter for describing AE-probe performance. Obviously the higher the DH ratio, the higher the discriminating power achieved in an HPA format.

*Procedural Notes/Troubleshooting*

1. Since hydrolysis rate is temperature sensitive, it is very important that the temperature used in this procedure be accurately maintained during the hydrolysis time course. It is also very important that the temperature be very consistent from tube to tube. For these reasons a circulating water bath is recommended for this procedure.

2. It is very important that half-lives of hydrolysis be calculated from the linear portions of the kinetic curves. Small differences in slope can lead to relatively large differences in half-lives and DH ratios. Values for $R^2$ in the regression analysis of 0.98 or greater are recommended. Errant values for half-life of hydrolysis are particularly likely in the Control since hydrolysis rates are much faster in this case. More data points in the early portion of the curve are useful in improving accuracy. In the first 15 to 30 sec, the contents of the tubes are coming to temperature, and therefore data points at 30 sec and beyond are typically more significant.

3. The Blank reaction is necessary to factor out chemiluminescence inherent in the reagents. This is especially important for the later time points of the Control sample. The lower this amount of chemiluminescence, the more accurate the analysis; therefore, care in preparing reagents without introducing contaminating chemiluminescence is very important.

4. Hydrolysis-rate analysis is used to determine optimal differential hydrolysis time in a full assay format. This is described in Section III,C under Procedural Notes/Troubleshooting for the general HPA protocol.

5. Acid reversion is used in this protocol to revert any pseudo-base that formed during the DH step. If acid reversion is not used, the protocol becomes a measurement of the *combined* rates of ester hydrolysis and pseudo-base formation.

6. For the Hybrid sample, the y-intercept of the line determined using regression analysis represents the percentage hybridization of the AE-probe being tested. In most cases the extent of hybridization will be approximately 100% (the reaction is performed in target excess), and the y-intercept will be 2 (i.e., the log of 100%). In some cases, however, the extent will not be 100%. In these cases the hydrolysis curve will be biphasis—an initial rapid phase representing hydrolysis of the unhybridized AE-probe, and a second, slow phase representing the hydrolysis of the remaining hybridized AE-probe. Back extrapolation of the slow phase to the y axis is then used to determine the percentage of AE-probe that is hybridized.

7. The DH characteristics of an AE-probe can also be determined at different pH values using basically the same protocol given. However, hydrolysis rates will be faster at higher pH and slower at lower pH, and the time points must be chosen accordingly. It is difficult to measure accurately ester hydrolysis rates at 60°C for unhybridized AE probes at pH higher than about 8.5 since hydrolysis is so rapid at these pH values. Preheating the hydrolysis buffer and using very short intervals between time points helps the accuracy of the analysis in these cases.

8. If the data points to use for the regression analysis are unclear, or if the $R^2$ value is less than 0.98, the analysis should be repeated.

9. If necessary or desired, more data points can be collected than recommended.

10. A small square of standard laboratory label tape (*Time tape*, VWR Scientific) pressed snugly onto the assay tubes is an ideal covering during incubation at 60°C.

11. See Section III,D, Procedural Notes/Troubleshooting, General, for tips that are applicable to all AE-based procedures.

## C. AE-Probe Assay Formats

The Hybridization Protection Assay can be formatted in a number of ways depending on the exact application. However, each format contains the same three basic steps: hybridization, differential hydrolysis, and detection. There are also quite a large number of ways a sample can be prepared for assay depending on the type of sample and the form and concentration of the target nucleic acid(s). If rRNA is to be assayed from a clinical specimen suspected of containing an infectious microorganism, the organism must be lysed and free rRNA released into solution. If PCR-amplified DNA is to be assayed, the sample must first be denatured since the product of the reaction is double-stranded. If single-stranded nucleic acid is to be quantified after a purification procedure, a sample need simply be pipetted into the assay tube. It is beyond the scope of this chapter to describe all the possible methods by which a sample of nucleic acid can be prepared for assay using the HPA protocol. The general requirement is that the target be available for hybridization in solution at a concentration within the sensitivity limit of the assay. Furthermore, it is preferred that the target nucleic acid be contained in a relatively small volume so the components in the sample mixture will not contribute significantly to the final hybridization conditions.

### 1. General HPA Protocol

1. *Hybridization.* In a 12 × 75 mm assay tube, prepare a 100 μl hybridization reaction containing 0.1 *M* lithium succinate buffer, pH 5.0, 2 m*M* EDTA, 2 m*M* EGTA, 10% (w/v) lithium lauryl sulfate, 0.1 pmol AE-probe and the nucleic acid target to be detected. Also prepare negative controls (all components except the target nucleic acid) and positive controls (all components including a *known* amount of target nucleic acid; this can be a purified and quantitated sample of the target, a synthetic complement to the AE-probe, etc.). Vortex to mix; cover; incubate 5 to 60 min at 60°C.
2. *Differential hydrolysis.* Add 300 μl 0.15 *M* sodium tetraborate buffer, pH 7.6, 5% (v/v) Triton X-100. Vortex 3 sec; cover; incubate at 60°C for 10 min.
3. *Detection.* Immediately after removal from the water bath, place the tubes in an ice/water slurry for 15 sec, remove to room temperature and let stand for 30 sec, then measure the chemiluminescence in a luminometer by the automatic injection of 200 μl 5 m*M* $HNO_3$, 0.1% $H_2O_2$, then 200 μl of 1 *M* NaOH, followed by measurement of signal for 5 sec. Remember to blot the tubes with a moist tissue or paper towel before measuring the chemiluminescence (see Section II,A,4).

As mentioned in Section I, this procedure has been used for the detection of a wide variety of target nucleic acids. As little as $6 \times 10^{-17}$ moles of target can be detected, and chemiluminescence response is linear over three to four orders of magnitude. This basic format can be applied to the assay of almost any nucleic acid if, as mentioned above, the target is available for hybridization in solution at a concentration within the sensitivity limit of the assay. This general procedure can also be varied somewhat.

*Procedural Notes/Troubleshooting*

1. Incubation time for step 1 is dependent on the hybridization kinetics of the particular probe used. These kinetics can be measured and the hybridization time determined accordingly. This can most easily be done in the assay format described previously by setting up replicate hybridization reactions, incubating them all at 60°C, and assaying tubes (steps 2 and 3) at various time points. We have found that, under the conditions described above, hybridization is essentially complete for most AE-probes in 10 to 15 min.
2. The amount of probe used in an assay can be varied; a higher probe concentration will accelerate hybridization, but it will also increase background signal at the end of the assay.
3. The pH of hybridization is mildly acidic to protect the AE from hydrolysis. A range of pH 4.7 to pH 5.5 is acceptable, with pH 4.8 to pH 5.0 optimal.
4. The lithium lauryl sulfate contained in the hybridization mixture can range from 2 to 10% (10% is best).
5. Differential hydrolysis (DH) time is a critical factor in assay design. The optimal time for the DH step can be determined by measuring the AE hydrolysis kinetics of the AE-probe used in the assay as described in Section III,B. The point at which the AE hydrolysis rate of the unhybridized probe reaches a plateau (see Fig. 4) is the key factor in determining the DH time. For example, the optimal DH time for the AE-probe used for the hydrolysis rate study depicted in Fig. 4 is 10 min (under the conditions used in that experiment). After the ester hydrolysis of unhybridized AE-probe plateaus, reduction of signal from unhybridized AE-probe (i.e., background) is minimal, whereas signal from hybridized AE-probe is slowly but continually declining. Therefore assay sensitivity will slowly decrease beyond this point. It should be kept in mind, however, that the AE hydrolysis rate and therefore the DH time is condition dependent (pH, temperature, etc.) and must therefore be measured for each condition used. Ideally, optimal DH time should be determined in the exact assay format to be used.
6. An assay temperature of 60°C was chosen to give optimal performance of the system in regard to multiple factors, including hybridization kinetics, specificity of reaction, stability of the AE, and differential hydrolysis characteristics. However, other temperatures can be used if the particular application warrants the change. At temperatures lower than

60°C, hybridization times as well as differential hydrolysis times will have to be increased (if the pH is constant). Hybridization kinetics and differential hydrolysis rates should be measured, and the appropriate incubation times, selected accordingly.

7. Since the ester hydrolysis rate is directly proportional to the pH of the solution, the differential hydrolysis time can be manipulated by the pH of the differential hydrolysis buffer. A range of pH from 7.5 to 8.5 yields convenient differential hydrolysis times (6 to 12 min) at 60°C. Other pH values can be used if desired, especially if alternate temperatures are also used (for example, a higher pH may be used at a lower temperature in order to keep the duration of the differential hydrolysis step relatively short).

Maintaining an accurate pH during the differential hydrolysis step is obviously very important. The composition of the hydrolysis buffer is designed to maintain an accurate pH, but the samples must be thoroughly mixed after addition of this buffer to assume proper and consistent performance.

8. Since this assay is completely homogeneous, all of the liquid that is put into the tube during the course of the assay remains in the tube during the detection step. Since there is a volume limitation in the luminometer (beyond a certain volume efficiency of *capture* of chemiluminescent signal rapidly diminishes), there is also a volume limitation for the assay itself. For the 12 × 75 mm tube format, a typical volume maximum is 800 μl (this depends on the exact luminometer being used). A typical injection volume for a luminometer is 200 μl. Therefore, if a two-injection regimen is being used for chemiluminescence detection, the volume maximum for all the steps of the assay before detection is 400 μl. If the peroxide and base are added in the same injection, the assay volume can be increased to 600 μl. One disadvantage of this approach is that the peroxide is stable in the base only for about 10 hr. A single-injection regimen also precludes an acid reversion step as described in Section II,B, although acid reversion is not *typically* used in the HPA format (see following sections).

9. Acid reversion is typically not used in the HPA format. This is because a portion of the background component in the assay is in the pseudo-base form, and acid reversion would revert this component and thus raise backgrounds. Hybridized AE-probe, on the other hand, does not require acid reversion to recover all available chemiluminescence. Therefore maximal signal-to-noise ratios can be achieved using a detection regimen without acid reversion.

10. Double-stranded target nucleic acid must be denatured before hybridization with the AE-probe. This is typically accomplished by a short incubation period at 95°C (see HPA Protocol for the Detection of Products from *in vitro* Amplification Techniques).

11. A small square of standard laboratory label tape (Time tape, VWR Scientific) pressed snugly onto the assay tubes is an ideal covering during incubation at 60°C.

12. Possible explanations of high background signals include the following:

- The pH of the differential hydrolysis step may be too low. This can result if the pH of the differential hydrolysis buffer is errantly low, but can also result if the sample being assayed contains buffer components or is acidic such that the borate in the DH buffer is insufficient to maintain the proper pH. If this is the case, lower sample volumes should be used, or the sample should be neutralized, semipurified, other buffer concentrations selected, etc., before being assayed.
- The sample may not be mixed thoroughly after addition of the DH buffer (the solution is fairly viscous because of the high detergent concentrations), leading to an inconsistent pH throughout the mixture.
- The temperature of hydrolysis may be incorrect. For example, when racks of tubes are assayed, the centrally located tubes may not come to the proper temperature during the incubation period.
- The AE-probe may have been splashed up the sides of the tube during one of the steps before the hydrolysis step. This material may be high enough that it does not come into contact with the hydrolysis buffer and therefore remains unhydrolyzed, but low enough that it comes into contact with the detection reagents and therefore contributes to the final chemiluminescent signal. Special care in pipetting and mixing will help to eliminate this problem.
- The sample may contain components that reduce the rate of AE hydrolysis of the unhybridized probe. The most likely candidates in this regard are strongly nucleophilic compounds that can form adducts with the AE, thus protecting it from hydrolysis (see Fig. 3 and Section I,A). For example, certain buffers and reducing agents, such as Tris and dithiothreitol (DTT), will form adducts at millimolar concentrations at basic pH. Solutions to this problem include lower sample volumes per assay, prepurification of the target nucleic acid, and avoiding use of such components in sample preparation. Another approach is to react such components with compounds that render them inert, as described by Hammond and co-workers for the reversion of adduct formation in AE (Hammond *et al.*, 1991).
- Samples to be tested, especially those that are biological in nature, may exhibit inherent chemiluminescence, which can lead to elevated background values. The best way to examine this parameter for a given sample type is to perform the assay normally with the exception that AE-probe is omitted.

13. Possible explanations of low hybridization signals include the following:

• The pH of the differential hydrolysis step may be too high. This can result if the pH of the differential hydrolysis buffer is errantly high, but can also result if the sample being assayed contains buffer components or is basic such that the borate in the DH buffer is insufficient to maintain the proper pH. If this is the case, lower sample volumes should be used, or the sample should be neutralized, semipurified, other buffer concentrations selected, etc., before being assayed.

• The AE-probe may not have hybridized to the target nucleic acid completely. If this is suspected, retest the hybridization kinetics under the exact assay conditions that resulted in low signals.

• The temperature of hydrolysis may be incorrect. Low signal is indicative of too high a temperature.

• The melting temperature ($T_m$) of the AE-probe may be too near the operating temperature. The $T_m$ of the AE-probe should be determined under the exact assay conditions (probe concentration, salt concentration, etc.).

• The target nucleic acid may not be available for hybridization with the AE-probe. The sample may need to be denatured before hybridization (see HPA Protocol for the Detection of Products from *In Vitro Amplification Techniques*).

• The sample may contain components that alter the normal structure of duplex DNA. This can have a negative effect on the protection of the AE since this protection is very structure dependent. Examples of such compounds include polyamines and strong intercalating compounds. Solutions to this problem include lower sample volumes per assay, prepurification of the target nucleic acid, and avoiding use of such components in sample preparation.

## 2. HPA Protocol for the Detection of Products from *in Vitro* Amplification Techniques

A typical protocol follows for the assay of a target amplification reaction such as PCR and TAS (see Section I,E). A denaturation step is often required, and a convenient sample size at this step is 50 $\mu$l. However, it is not typically necessary to assay 50 $\mu$l directly from the amplification reaction, and in fact 50 $\mu$l may be too much to assay (see Procedural Notes/Troubleshooting for the general HPA protocol described). In this case the sample (5 to 10 $\mu$l of the amplification reaction, for example) should be diluted with the appropriate amount of water to bring the volume to 50 $\mu$l. This dilution will also help prevent the strands from reannealing after denaturation. For convenience the AE-probe is typically premixed into a 2X hybridization buffer such that 50 μl contains the proper amount of AE-probe to be used per assay. Therefore addition of 50 μl of this AE-probe reagent to 50 μl of the sample yields an ideal hybridization mixture.

1. *Denaturation.* Pipette 50 μL of sample to be tested into a 12 × 75 mm assay tube; cover. Incubate at 95°C for 5 min to denature the target nucleic acid (this step can be omitted if the target is single stranded). After 5 min immediately place the tubes into an ice/water slurry.
2. *Hybridization.* Cool the tubes in the ice/water slurry for 1 min, then pipet 50 μl prepared AE-probe reagent [0.2 *M* lithium succinate buffer, pH 5.0, 4 m*M* EDTA, 4 m*M* EGTA, 17% (w/v) lithium lauryl sulfate and 0.1 pmol AE-probe] into each tube. Vortex to mix; cover; incubate the tubes in a water bath at 60°C for 10 min.
3. *Differential hydrolysis.* Remove the tubes from the water bath, add 300 μl 0.15 *M* sodium tetraborate buffer, pH 8.5, 5% (v/v) Triton X-100. Vortex 3 sec; cover; incubate at 60°C for 6 to 7 min.
4. *Detection.* Immediately after removal from the water bath, place the tubes in an ice/water slurry for 15 sec, remove to room temperature and let stand for 30 sec, then measure the chemiluminescence in a luminometer by the automatic injection of 200 μl 5 m*M* $HNO_3$, 0.1% $H_2O_2$, then 200 μl of 1 *M* NaOH, followed by measurement of signal for 5 sec. Remember to blot the tubes with a moist tissue or paper towel before measuring the chemiluminescence (see Section II,A,4).

As mentioned in Section I,E, this protocol can be used for the detection of a variety of amplification products. In one application an overlapping pair of AE-probes is used to detect specific sequences within the *gag* gene of human immunodeficiency virus-1 (HIV-1) amplified by PCR. Ou and co-workers (Ou *et al.*, 1990) have reported the detection of HIV-1 proviral DNA amplified in the presence of cell lysate. They were able to detect as little as 4 copies of HIV DNA, and the assay was linearly quantitative over about 3 orders of magnitude. They also compared the HPA format with an isotopic method utilizing a $^{32}P$-labeled probe and found the HPA method to be at least as sensitive as the $^{32}P$ method. (The isotopic method consisted of solution hybridization, restriction enzyme digestion, gel electrophoresis, and autoradiography.) In another application the HPA format is used to detect the presence of the Philadelphia chromosome (Adams, 1985), the chromosomal translocation product associated with chronic myelogenous leukemia (CML). In one report (Dhingra *et al.*, 1991), chimeric mRNA was amplified using PCR and detected using the HPA format described. Samples from 60 different patients were analyzed, and the HPA format gave equivalent results to standard Southern analysis of the PCR reactions, in a fraction of the time. The HPA format also gave complete correlation with the results of karyotype or Southern blot analysis of genomic DNA. The HPA format has also been used to detect amplified nucleic acid from hepatitis B virus (Kacian *et al.*, 1990), the genes that code for the rRNA of *Neisseria gonorrhoeae* (Gegg *et al.*, 1990) and *Chlamydia trachomatis* (Kranig-Brown *et al.*, 1990), the 7.5 kb cryptic plasmid of *Chlamydia trachomatis* known as pCHL1, the HIV-1 envelope region, and the HIV-2 viral protein x and LTR regions.

*Procedural Notes/Troubleshooting*

1. The procedural notes and troubleshooting guidelines given for the general HPA protocol apply to this protocol as well.
2. Never include the AE-probe in the denaturation step, as it will be hydrolyzed. Make sure the sample is cooled to 70°C or below before adding the AE-probe (using the protocol described, the temperature will be below well 70°C).
3. The denaturation protocol can also be used for double-stranded nucleic acid targets other than those from amplification reactions.
4. See Section III,D, Procedural Notes/Troubleshooting, General, for tips that are applicable to all AE-based procedures.

### 3. Differential Hydrolysis + Magnetic Separation Protocol

Following is a protocol that combines the differential hydrolysis process with a magnetic separation step to further reduce backgrounds. The protocol given is one variation on this theme, and represents only one particular way this general procedure can be performed.

---

1. *Hybridization*. In a 12 × 75 mm assay tube, prepare a 200 $\mu$l hybridization reaction containing 0.1 *M* lithium succinate buffer, pH 5.0, 2 m*M* EDTA, 2 m*M* EGTA, 10% (w/v) lithium lauryl sulfate, 0.1 pmol AE-probe, and the nucleic acid target to be detected. Vortex to mix; cover; incubate 10 to 60 min at 60°C.
2. *Differential hydrolysis*. Add 1 ml 0.15 *M* sodium tetraborate buf-

fer, pH 7.6, 5% (v/v) Triton X-100. Vortex 3 sec; cover; incubate at 60°C for 10 min.

3. *Magnetic separation*. Add 1 ml 0.4 *M* sodium phosphate buffer, pH 6.0. containing 0.1 to 5 mg of magnetic amine microspheres (see Section II,A,3). Vortex thoroughly; cover; incubate 5 min at 60°C. Place tubes in a magnetic separation unit (see Section II,A,3) at room temperature and allow the microspheres to be separated from the solution (about 5 min). With the tubes still in the magnetic separation unit, decant the supernatant and blot the tops of the tubes. Wash the spheres one to three times by removing the tubes from the rack, adding 1 ml 0.2 *M* phosphate buffer, pH 6.0, vortexing, placing the tubes back into the magnetic rack for 5 min and decanting the supernatant.
4. *Elution of AE-probe from magnetic microspheres*. After the last wash from step 3 is decanted and the tubes are carefully blotted, remove the tubes from the magnetic separation unit and add 300 $\mu$l of a solution containing 0.2 *M* sodium phosphate buffer, pH 6.0, and 50% (v/v) formamide to each tube. Vortex thoroughly; cover. Incubate the tubes at 60°C for 5 min, place tubes in magnetic separation unit (see Section II,A,3) at room temperature, and allow the microspheres to be separated from the solution (about 5 min). Transfer the supernatant (which now contains the AE-probe) to a new 12 × 75 mm assay tube.
5. *Detection*. Detect chemiluminescence in a luminometer by the automatic injection of 200 $\mu$l 0.4*M* $HNO_3$, 0.1% $H_2O_2$, then 200 $\mu$l 1 *M* NaOH, followed by measurement of signal for 5 sec. Remember to blot the tubes with a moist tissue or paper towel before measuring the chemiluminescence (see Section II,A,4).

---

The limit of sensitivity of this procedure is as low as $6 \times 10^{-18}$ moles of target, and chemiluminescence response is linear over four or more orders of magnitude (the limiting factor in range of linearity is usually the luminometer and not the assay itself). As mentioned, there are alternate protocols to perform a differential hydrolysis + separation assay. For example, a clinical diagnostic assay (*PACE* 2, Gen-Probe) utilizes a differential hydrolysis + separation protocol that includes only one wash step and omits the elution step (chemiluminescence is measured in the presence of the magnetic microspheres).

*Procedural Notes/Troubleshooting*

1. The procedural notes and troubleshooting guidelines given for the general HPA protocol apply to this protocol as well.

2. In high enough concentrations, the magnetic spheres will quench the chemiluminescence from the AE-probe. That is why the AE-probe is eluted from the spheres before the detection step.

3. The elution step melts the hybrid and elutes the free AE-probe from the spheres. Since it is no longer protected by the duplex, the AE can now once again hydrolyze fairly rapidly. Therefore the 60°C portion of the elution step should be no longer than the recommended 5 min, and the samples should be measured for chemiluminescence as soon as they are transferred to new assay tubes.

4. Acid reversion is used in the detection step in this particular protocol since the AE is detected as unhybridized AE-probe and not hybridized AE-probe.

5. Freezing of the magnetic microspheres can lead to clumping of the spheres, which leads to low capture efficiency and errantly low assay readings. Therefore do not freeze this reagent.

6. See Section III,D, Procedural Notes/Troubleshooting, General, for tips that are applicable to all AE-based procedures.

## D. General Procedural Notes/Troubleshooting

Following are some procedural notes and troubleshooting tips that are generally applicable to handling AE and AE-labeled DNA probes. Some of these suggestions have been repeated in selected sections for emphasis.

- The AE is a hydrophobic molecule, and as such has a propensity to stick to surfaces such as test tube walls and pipet tips. For this reason the AE must always be handled in solutions that contain agents that disrupt hydrophobic interactions and keep the AE in solution. Detergents and organic solvents fulfill this criterion very well. For example, laurel sulfate, DMSO, or acetonitrile are commonly used to keep the AE in solution. Never try to transfer AE from water, as you will invariably leave some behind.
- For optimal storage stability, AE-probes should be stored in a pH 5 buffer (sodium acetate, lithium succinate, etc.) with a small amount of detergent (e.g., 0.1% SDS) present to keep the AE in solution (see 1 above). In this form the AE-probe is stable for months at −20°C and years at −70°C. However, multiple freeze–thawing cycles will reduce the shelf-life of AE-probes, which should therefore be stored in small aliquots.
- When thawing aliquots of AE-probe, the solution must be heated to 45 to 60°C for about 1 min and then vortexed to thoroughly resolubilize the AE. If this is not done, poor recoveries of AE-probe from the stock tube(s)

can result. Solutions with high concentrations of detergents may have to be heated for 2 to 3 min in order to fully solubilize all components. For example, AE-probe in 2X hybridization buffer (see Section III,C) can form a gel when stored in the cold. Using the reagent like this can result in improper delivery of AE-probe and hybridization buffer, leading to errantly low hybridization values. This gel can be easily redissolved by heating at 60°C for 2 min, followed by thorough vortexing.

• Due to the "sticky" nature of the AE and the relatively high viscosity of many of the solutions used in the AE protocols, all pipetting steps should be done very slowly and carefully. Pipet tips should just barely pierce the surface of the liquid that is being pipetted from or to, since a relatively large amount of AE will otherwise stick to the outside of the tip. All AE solutions should be delivered as close to the bottom of the tube into which you are pipetting as possible. A fresh disposable pipette tip should be used for *each* sample and positive and negative controls to avoid carryover. Repeating pipettors may be used for the addition of the AE-probe in hybridization buffer, differential hydrolysis buffer, and the separation and wash solutions in the differential hydrolysis + separation format. When using these devices, make sure the liquid is dispensed with a smooth, gentle delivery to avoid splashing of reagent and sample already present onto the sides of the tube.

• Chemiluminescence is typically expressed as relative light unit (RLU).

• Be careful not to contaminate reagents, pipettors, workspace, etc., as any contaminating chemiluminescence can lead to errant results. Be especially careful when using the stock solution of AE, since a 1-$\mu$l aliquot of the 25 m*M* stock contains approximately $10^{11}$ RLU of chemiluminescence. The use of disposable polystyrene or polypropylene labware is strongly recommended for reagent preparation.

• Before insertion into the luminometer, tubes should be gently blotted with a lightly dampened tissue or paper towel. This removes any residue from the outside of the tubes (thus keeping the luminometer free of contamination and protecting against instrument damage due to moisture build-up in the measurement chamber, etc.) and also reduces static charge (a build-up of this charge can lead to errant readings).

• Any AE-probe that becomes deposited on the sides or the rim of an assay tube may not come into proper contact with the hydrolysis buffer, which will lead to high background values. This can be avoided by always pipetting to the bottom of the tube and avoiding touching the sides and the rim of the tube. Thorough vortexing of the tubes after the addition of the hydrolysis buffer will also minimize this problem.

• Temperature is a very important parameter in AE-probe assays. Specificity of hybridization is temperature dependent, and AE hydrolysis rate is very temperature dependent. Therefore hybridization and differential hydrolysis temperatures should be accurately maintained to within a degree of the desired operating temperature. A consistent temperature from tube to tube and a rapid equilibration of the sample tubes to the correct temperature is also important. For these reasons, incubation in a circulating water bath is the recommended procedure, although a noncirculating water bath or a heating block can be used if they fulfil the requirements just described.

- Nucleophilic compounds can form adducts with the AE (Hammond *et al.*, 1991; see also Fig. 3 and Sections I,A and III,C). If formed during the differential hydrolysis step, the AE hydrolysis rate of unhybridized AE-probe may be reduced, leading to elevated negative control values.

## REFERENCES


Adams, J. M. (1985). Oncogene activation by fusion of chromosomes in leukemia. *Nature (London)* **315,** 542–543.

Arnold, L. J., Jr., Bhatt, R. S., and Reynolds, M. A. (1988). Nonnucleotide linking reagents for nucleotide probes. *WO Patent* 8803173.

Arnold, L. J., Jr., Hammond, P. W., Wiese, W. A., and Nelson, N. C. (1989). Assay formats involving acridinium-ester-labeled DNA probes. *Clin. Chem.* **35,** 1588–1594.

Berry, D. J., Clark, P. M. S., and Price, C. P. (1988). A laboratory and clinical evalution of an immunochemiluminometric assay of thyrotropin in serum. *Clin. Chem.* **34,** 2087–2090.

Bronstein, I., Voyta, J. C., Thorpe, G. H. G., Kricka, L. J., and Armstrong, G. (1989). Chemiluminescent assay for alkaline phosphotase: Application in an ultrasensitive enzyme immunoassay for thyrotropin. *Clin. Chem.* **35,** 1441–1446.

Bronstein, I., Voyta, J. C., Lazzari, K. G., Murphy, O., Edwards, B., and Kricka, L. J. (1990). Rapid and sensitive detection of DNA in Southern blots with chemiluminescence *BioTechniques* **8,** 310–314.

Caruthers, M. H., Barone, A. D., Beaucage, S. L., Dodds, D. R., Fisher, E. F., McBride, L. J., Matteuci, M., Stabinsky, Z. and Tang, J.-Y. (1987). Chemical synthesis of deoxyoligonucleotides by the phosphoramidite method. *Methods Enzymol* **154,** 287–313.

Daly, J. A., Clifton, N. L., Seskin, K. C., and Gooch, W. M., III (1991). Use of rapid, nonradioactive DNA probes in culture confirmation tests to detect *Streptococcus agalactiae, Haemophilus influenzae,* and *Enterococcus* ssp. from pediatric patients with significant infections. *J. Clin. Microbiol.* **29,** 80–82.

Dhingra, K., Talpaz, M., Riggs, M. G., Eastman, P. S., Zipf, T., Ku, S., and Kurzrock, R. (1991). Hybridiziation protection assay: A rapid, sensitive, and specific method for detection of Philadelphia chromosome–positive leukemias. *Blood* **77,** 238–242.

Enns, R. K. (1988). DNA probes: An overview and comparison with current methods *Lab. Med.* **19,** 295–300.

Gegg, C., Kranig-Brown, D., Hussain, J., McDonough, S., Kohne, D., and Shaw, S. (1990). A clinical evaluation of a new DNA probe test for *Neisseria gonorrhoeae* that included analysis of specimens producing results discrepant with culture. *Abstracts of the 90th Annual Meeting of the American Chemical Society for Microbiology, Anaheim, California,* p. 356.

Granato, P. A., and Franz, M. R. (1989). Evaluation of a prototype DNA probe test for the nonculture diagnosis of Gonorrhea. *J. Clin. Microbiol.* **27,** 632–635.

Granato, P. A., and Franz, M. R. (1990). Use of the Gen-Probe PACE system for the detection of *Neisseria gonorrhoeae* in urogenital samples. *Diagn. Microbiol. Infect. Dis.* **13,** 217–221.

Hammond, P. W., Wiese, W. A., Waldrop, A. A., III, Nelson, N. C., and Arnold, L. J., Jr. (1991). Nucleophilic addition to the 9 position of 9-phenylcarboxylate-10-methylacridinium protects against hydrolysis of the ester. *J. Biolumin. Chemilumin.* **6,** 35–43.

Harper, M., and Johnson, R. (1990). The predictive value of culture for the diagnosis of gonorrhea and chlamydia infections. *Clin. Microbiol. Newsletter* **12,** 54–56.

Hastings, J. G. M. (1987). Luminescence in clinical microbiology. *In* "Bioluminescence and Chemiluminescence, New Perspectives" (J. Scholmerich, R. Andreesen, A. Kapp, M. Ernst, and W. G. Woods, eds.), pp. 453–462. John Wiley and Sons, Chichester, England.

Innis, M. A., and Gelfand, D. H. (1990). Optimization of PCRs. *In* "PCR Protocols" (M. A. Innes, D. H. Gelfand, J. J. Sninsky, and T. J. White, eds.), pp. 3–12. Academic Press, San Diego, California.

Kacian, D. L., Lawrence, T., Sanders, M., Putnam, J., Majlessi, M., McDonough, S. H. and Ryder, T. (1990). A rapid and sensitive chemiluminescent DNA probe system (HPA) for detection of amplified HIV and HBV DNA. *Abstracts of Biochemische Analytik 90,* Munich, Germany.

Keller, G. H., and Manak, M. M. (1989). "DNA Probes." Stockton Press, New York.

Kohne, D. E. (1986). Application of DNA probe tests to the diagnosis of infectious disease. *Am. Clin. Prod. Rev.* November, 20–29.

Kranig-Brown, D., Gegg, C., Hussain, J., Johnson, R., and Shaw, S. (1990). Use of PCR amplification and unlabeled probe competition for analysis of specimens producing results discrepant with culture in an evaluation of a DNA probe test for *Chlamydia trachomatis. Abstracts of the 90th Annual Meeting of the American Chemical Society for Microbiology,* Anaheim, California, p. 356.

Kwoh, D. Y., Davis, G. R., Whitfield, K. M., Chappelle, H. L., DiMichele, L. J., and Gingeras, T. R. (1989). Transcription-based amplification system and detection of amplified human immunodeficiency virus type 1 with a bead-based sandwich hybridization format. *Proc Natl. Acad. Sci. U.S.A.* **86,** 1173–1177.

LeBar, W., Herschman, B., Jemal, C., and Pierzchala, J. (1989). Comparison of a DNA probe, monoclonal antibody enzyme immunoassay, and cell culture for the detection of *Chlamydia trachomatis. J. Clin. Microbiol.* **27,** 826–828.

Lewis, J. S., Kranig-Brown, D., and Trainor, D. A. (1990). DNA probe confirmatory test for *Neisseria gonorrhoeae. J. Clin. Microbiol* **28,** 2349–2350.

McCapra, F. (1970). Chemiluminescence of organic compounds. *Pure Appl. Chem.* **24,** 611–629.

McCapra, F. (1973). Chemiluminescent reactions of acridines. *Chem. Heterocycl Compounds* **9,** 615–630.

McCapra, F., and Beheshti, I. (1985). Selected chemical reactions that produce light. *In* "Bioluminescence and Chemiluminescence: Instruments and Applications;; (K. van Dyke, ed.), Vol. I, pp. 9–42. CRC Press, Boca Raton, Florida.

McCapra, F., and Perring, K. D. (1985). General organic chemiluminescence. *In* "Chemiluminescence and Bioluminescence" (J. G. Burr, ed.), pp. 259–320. Marcel Dekker, New York.

McDonough, S. H., Ryder, T. B., Cabanas, D. K., Yang, Y. Y., Majlessi, M., Harper, M. E., and Kacian, D. L. (1989). Rapid and sensitive detection of amplified DNA by a chemiluminescent DNA probe system (HPA). *Abstracts of the Twenty-Ninth Interscience Conference on Antimicrobial Agents and Chemotherapy,* Houston, Texas, p. 301.

Mullis, K. B., and Faloona, F. A. (1987). Specific synthesis of DNA *in vitro* via a polymerase catalyzed chain reaction. *Methods Enzymol.* **155,** 335–350.

Nelson, N. C., Hammond, P. W., Wiese, W. A., and Arnold, L. J., Jr. (1990). Homogeneous and heterogeneous chemiluminescent DNA probe-based assay formats for the rapid and sensitive detection of target nucleic acids. *In* "Luminescence Immunoassay and Molecular Applications" (K. van Dyke, ed.), pp. 293–309. CRC Press, Boca Raton, Florida.

Nelson, N. C., and Kacian, D. L. (1990). Chemiluminescent DNA probes: A comparison of the acridinium ester and dioxetane detection systems and their use in clinical diagnostic assays. *Clin. Chim. Acta.* **194,** 73–90.

Ou, C., McDonough, S. H., Cabanas, D., Ryder, T., Harper, M., Moore, J., and Schochetman, G. (1990). Rapid and quantitative detection of enzymatically amplified HIV-1 DNA using chemiluminescent oligonucleotide probes. *AIDS Res. Hum. Retroviruses* **6,** 1323–1329.

Sanchez-Pescador, R., Stempien, M. D., and Urdea, M. S. (1988). Rapid chemiluminescent nucleic acid assays for detection of TEM-1 β-lactamase-mediated penicillin resistance in *Neisseria gonorrhoeae* and other bacteria. *J. Clin. Microbiol.* **26,** 1934–1938.

Schaap, A. P., Akhavan, H., and Romano, L. J. (1989). Chemiluminescent substrates for alkaline phosphatase: Application to ultrasensitive enzyme-linked immunoassays and DNA probes. *Clin. Chem.* **35,** 1863–1864.

Tenover, F. C., Carlson, L., Barbagallo, S., and Nachamkin, I. (1990). DNA probe culture confirmation assay for identification of thermophilic *Campylobacter* species. *J. Clin. Microbiol.* **28,** 1284–1287.

Urdea, M. S., Warner, B. D., Running, J. A., Stempien, M., Clyne, J., and Horn, T. (1988). A comparison on nonradioisotopic hybridization assay methods using fluorescent, chemiluminescent and enzyme-labeled synthetic oligodeoxyribonucleotide probes. *Nucleic Acids Res.* **16,** 4937–4956.

Watson, M., Zepeda-Bakan, M., Smith, K., Alden, M., Stolzenbach, F., Johnson, R. (1990). Evaluation of a DNA probe assay for screening urinary tract infections. *Abstracts of the 90th Annual Meeting of the American Chemical Society for Microbiology,* Anaheim, California, p. 392.

Weeks, I., Beheshti, I., McCapra, F., Campbell, A. K., and Woodhead, J. S. (1983a). Acridinium esters as high-specific-activity labels in immunoassay. *Clin. Chem.* **29,** 1474–1479.

Weeks, I., Campbell, A. K., and Woodhead, J. S. (1983b). Two-site immunochemical assay for human alpha-fetoprotein. *Clin. Chem.* **29,** 1480–1483.

Weeks, I., Sturgess, M., Brown, R. C., and Woodhead, J. S. (1986). Immunoassays using acridinium esters. *Methods Enzymol.* **133,** 366–386.

Woese, C. R. (1987). Bacterial evolution. *Microbiol. Rev.* **51,** 221–271.

Woods, G. L., Young, A., Scott, J. C., Jr., Blair, T. M. H., and Johnson, A. M. (1990). Evaluation of a nonisotopic probe for detection of *Chlamydia trachomatis* in endocervical specimens. *J. Clin. Microbiol.* **28,** 370–372.

# 13 Detection of Energy Transfer and Fluorescence Quenching

Larry E. Morrison
Amoco Technology Company
Naperville, Illinois



*Nonisotopic DNA Probe Techniques*

# I. INTRODUCTION

## A. DNA Assay Formats Using Energy Transfer and Fluorescence Quenching

Methods of detecting DNA[1] by energy transfer and fluorescence quenching were the first homogeneous sequence-specific methods described, and subsequently few additional homogeneous approaches have been developed. The energy transfer and fluorescence-quenching methods rely on the distance dependence of intermolecular quenching interactions between a luminescent label attached to one DNA probe and a second label attached to a second DNA probe. Hybridization of complementary DNA strands serves to bring the two probes together, whereupon interaction between the two labels results in a detectable change in the luminescence properties of one or both labels. The presence of the DNA analyte is thereby detected without the need to physically separate probes bound to the analyte from probes free in the solution. Like its predecessor, the energy transfer-based immunoassay (Ullman *et al.*, 1976), the hybridization assay provides a route to rapid and simple bioanalyses.

The three basic hybridization formats that employ energy transfer or fluorescence quenching are shown in Fig. 1. The DNA probes are shown as short DNA strands to which the labels F and Q are attached. The DNA analyte, otherwise called the target DNA, is shown as the longer unlabeled DNA strand. Part A of Fig. 1 depicts the format first described (Heller *et al.*, 1983), in which the sequences of two DNA probes are selected, such that they will hybridize to contiguous regions of the target DNA. The probe hybridizing toward the 5′-terminus of the target DNA is labeled near its 5′-terminus, whereas the other probe is labeled near its 3′-terminus.

[1] While the work here describes probes composed of DNA used to detect DNA analytes, in principle, the probes could be composed of RNA, and either probe used to detect either RNA or DNA.

**Fig. 1** Schematic representations of label interactions in energy transfer and fluorescence-quenching hybridization assay formats. Parts A, B, and C show the various species present in the noncompetitive-assay format, the competitive-assay format, and the intercalator-assay format, respectively. F represents the label excited by light absorption or chemical reaction. Q represents the label that interacts with F at short separation distances to quench the emission of F. Arrows indicate light emission from individual labels or energy transfer between labels. Arrows are labeled to indicate energy characteristic of F emission ($h\nu_1$) or Q emission ($h\nu_2$). Q may be emissive or nonemissive and may facilitate the loss of energy from excited F labels without energy transfer to Q.

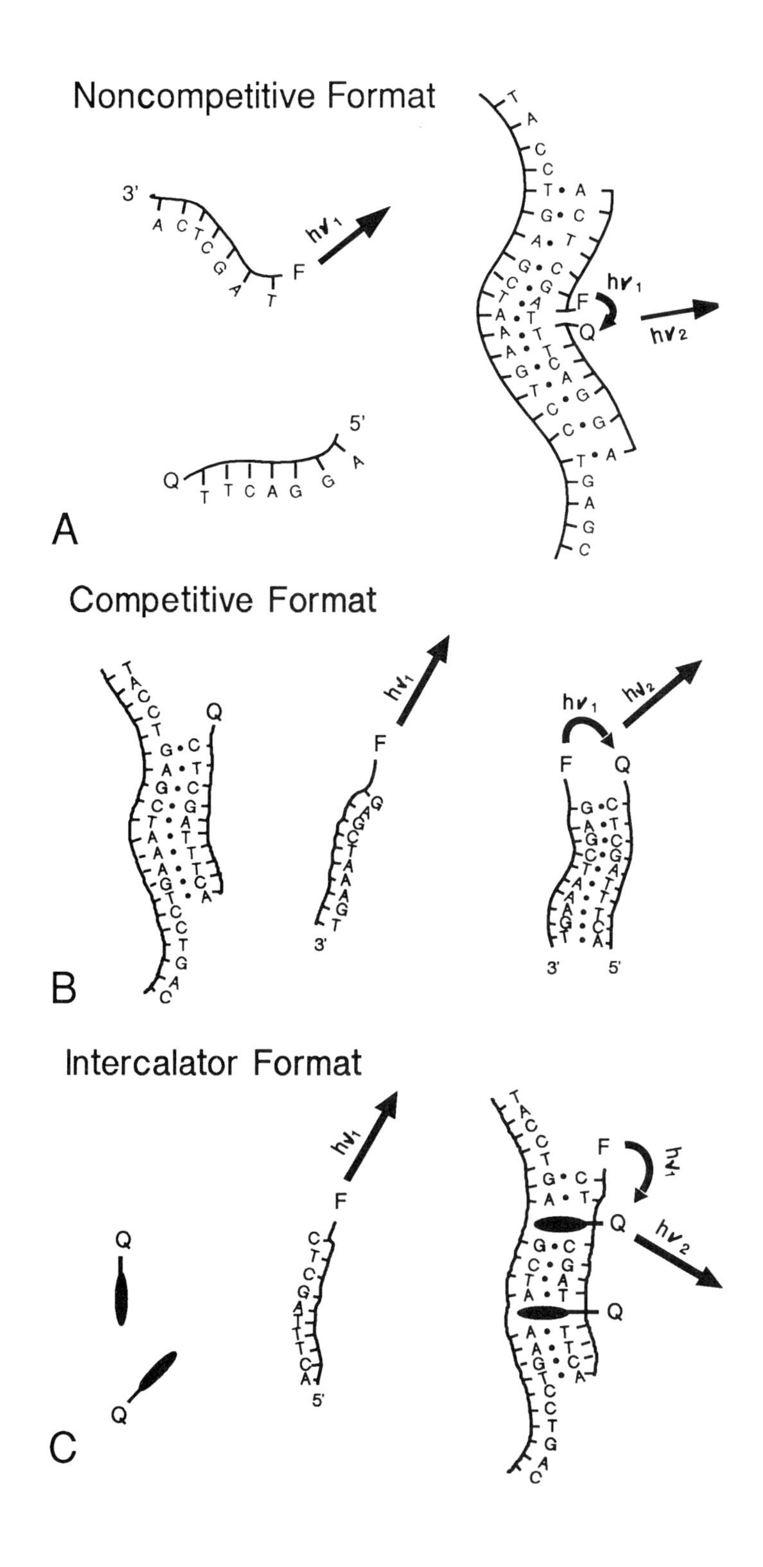
Noncompetitive Format
A
Competitive Format
B
Intercalator Format
C

This arrangement provides for close placement of the two labels on hybridization to the target DNA.

The label F is a luminescent compound that can be excited either by chemical reaction, producing chemiluminescence, or by light absorption, producing fluorescence. The label Q can interact with F to alter its light emission, usually resulting in the decreased emission efficiency of F. This phenomenon is called quenching. A special type of quenching involves the transfer of the excitation energy from F to Q. This special case is referred to loosely as energy transfer. If Q is also fluorescent, light emission from Q can accompany energy transfer. The distance dependence of the quenching of F by Q depends on the particular quenching mechanism involved, but in general, the probes are present at low enough concentrations that the average distance of separation in solution is far greater than that required to observe detectable quenching. Once the probes are hybridized to contiguous regions on the target DNA, however, the labels are close enough to produce appreciable interaction. The presence of target DNA, therefore, is detected as the quenching of F, which is sometimes accompanied by enhanced light emission from Q.

The energy-transfer concept was first demonstrated using two oligomers of thymidine, one containing a 5′-isoluminol label and the other containing a 3′-tetramethylrhodamine label (Heller and Morrison, 1985). Polydeoxyadenylic acid served as the target DNA, and initiation of the chemiluminescent isoluminol reaction resulted in a 24% increase in emission at the red end of the spectrum, where rhodamine fluoresces, relative to red emission in the absence of the polydeoxyadenylic acid. A later demonstration used DNA oligomers with terminal labels of fluorescein and tetramethylrhodamine (Cardullo, *et al.*, 1988). Fluorescence of the fluorescein label was reduced 71% on addition of the target DNA oligomer.

The second assay format is depicted in Fig. 1B and employs complementary DNA probes matching the nucleotide sequence of a selected region within the target DNA (Morrison, 1987). This type of assay is referred to as a competitive assay since probe strands compete with target strands for hybridization. As in the noncompetitive format described in Fig. 1A, one probe is labeled on its 3′-terminus, and the other probe is labeled on its 5′-terminus. When the probes hybridize to one another, the quenching interaction is favored, whereas an insignificant amount of interaction occurs when the probes are separated. The detection of target DNA is achieved by denaturing target and probe DNA together and then allowing the DNA strands to rehybridize. The more target strands present, the larger the number of probe strands that hybridize to target strands, and the smaller the number of F and Q labels placed next to one another by probe-to-probe hybridization. The presence of target DNA, therefore, is detected as increased emission from F, due to reduced quenching, and decreased emission from Q, due to decreased energy transfer.

The competitive assay was demonstrated using complementary probes labeled with a variety of fluorescent labels that interact on hybridization to provide fluorescence quenching or energy transfer (Morrison, 1987; Morrison *et al.,* 1989). DNA oligomers as well as restriction fragments of natural DNA were labeled to serve as probe strands. DNA hybridization assays were performed on samples containing targets of synthetic DNA oligomers or plasmid DNA. In addition, an amplified version of the assay was demonstrated using the polymerase chain reaction, (PCR) (Morrison *et al.,* 1989).

A third assay format, depicted in Fig. 1C, uses only one labeled probe and a dye that binds preferentially to double-stranded DNA (Heller and Morrison, 1985). This dye may intercalate between the base pairs or may bind to the outside of the helix and serves as the quencher Q. In the absence of hybridization Q does not bind to the single-stranded probe, and F is unaffected by Q. In the presence of the target DNA, the probe hybridizes to the target and Q binds to the resulting double helical region, thereby placing Q near F and allowing energy transfer or fluorescence quenching.

This concept was first demonstrated using an isoluminol-labeled oligomer of adenosine and the double-strand-specific intercalator, ethidium bromide (Heller and Morrison, 1985). Energy transfer between the chemiluminescent product of isoluminol oxidation and intercalated ethidium bromide was observed, in the presence of polyuridylic acid, as increased emission at 600 nm, where the ethidium bromide fluoresces. The correspondence between changes in the fluorescence intensity at 600 nm and the absorbance hypochromicity of the RNA, as the RNA was denatured by heating, confirmed that energy transfer was a direct result of hybridization. A similar experiment was reported later in which the DNA-binding dye, acridine orange, was used as a fluorescent energy donor to a tetramethylrhodamine label attached to a DNA oligomer (Cardullo *et al.,* 1988). Again, absorbance hypochromicity measurements were correlated with energy transfer, which was observed this time as both quenching of the acridine orange fluorescence and enhancement of the rhodamine emission.

The intercalator assay format is the least sensitive of the three formats. One reason for this is that any intercalator that binds nonspecifically to single-stranded DNA probe produces label interactions falsely indicative of hybridization between probe and target. In addition, problems result from the relatively high concentration of DNA-binding dye that must be used in order to ensure sufficient binding of the dye to the polynucleotide. For example, if the DNA-binding dye serves as F, emission from the free dye raises the background level of light emission. If the DNA-binding dye is Q, and F is fluorescent, then unbound Q will absorb some of the light intended to excite F and raise the background light level. For these reasons, the noncompetitive and competitive assay formats have been studied to a greater extent, and further disscussions of the energy-transfer and fluorescence-quenching methods will concentrate on these two formats.

## B. A Molecular View of Fluorescence Quenching and Energy Transfer

Fluorescence quenching and energy transfer involve processes that remove energy from an electronically excited luminescent molecule, thereby returning that molecule to its ground electronic state without the emission of light. These processes may best be understood with reference to the simplified energy level diagram shown in Fig. 2. Horizontal lines represent energy levels, and the vertical distance is a measure of the relative energy of each level. Hypothetical energy levels for F and Q are shown in the left and right portions of the figure, respectively. Energy levels are labeled as either singlet states, S, or triplet states, T, with subscripts numbered 0, 1, or 2, representing the ground state, first excited electronic state, or second excited electronic state. Normally a molecule resides in its lowest, or ground, electronic state, $S_0$. However, the molecule may be excited to one of its higher energy levels by any of a number of processes, including light absorption and chemical reaction. The excitation of F is represented in Fig. 2 by the arrows pointing upward from the ground state. Once excited, the molecule has a number of routes available to lose the excess energy and return to its ground state. These routes are marked by arrows pointing downward from the excited states $S_1$ and $S_2$. Molecules excited into their second or higher excited states are rapidly de-excited by nonradiative internal conversion processes to the first excited state, $S_1$. Internal conversion from $S_1$ to $S_0$, labeled in Fig. 2 with the first order rate constant, $k_{ic}$, can also be rapid but in luminescent molecules is slow enough that de-excitation by light emission is competitive. Light emission from $S_1$ is called fluorescence and is governed by the first-order rate constant, $k_f$, used to label this path in Fig. 2. Typically, $k_f$ is on the order of $10^8$ to $10^9$ $sec^{-1}$. Fluorescence competes not only with internal conversion but also with intersystem crossing to the triplet state, labeled $k_{is}$. Light emission from the triplet state, called phosphorescence, is possible; however, the first-order rate constant governing this process, $k_p$, is small (typically $1-10^6$ $sec^{-1}$), and other nonradiative processes generally predominate in solutions at room temperature.

The efficiency of fluorescent light emission is denoted by the quantum yield, $\Phi$, which is the number of fluorescent photons emitted divided by the number of molecules excited, either by light absorption or chemical reaction. In the absence of other molecular species, $\Phi$ is determined by the

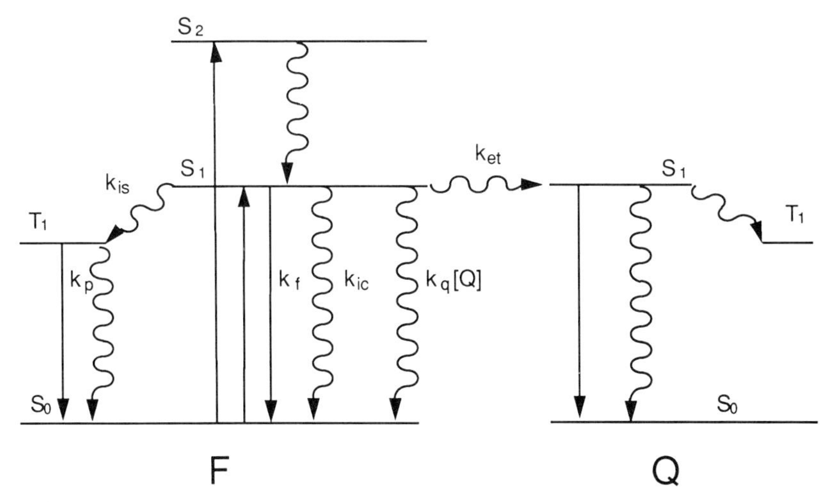


**Fig. 2** Energy level diagrams for F and Q labels describing the possible paths for quenching of F or energy transfer from F to Q. Horizontal lines represent energy levels that increase in energy in the vertical direction. Vertical lines represent transitions between energy levels, where straight lines represent transitions involving photons and curving lines represent nonradiative transitions. Only electronic energy levels are included for simplicity. Further explanation is provided in the text.

rates of the various processes involved in removing energy from $S_1$, according to Eq. 1:

$$\Phi = k_f/(k_f + k_{ic} + k_{is}) \tag{1}$$

The quantum yield approaches the maximum value of 1 when the rates of intersystem crossing and internal conversion are small compared to the rate of fluorescence. If additional molecular species are introduced that may collide with the luminescent molecule, new avenues for de-excitation are introduced. These molecules function as quenchers of the luminescent molecule if the product of quencher concentration, [Q], and the bimolecular quenching rate constant, $k_q$, is significant relative to the sum of the de-excitation rate constants in the absence of quencher. The quantum yield is then determined according to Eq. 2:

$$\Phi_q = k_f/(k_f + k_{ic} + k_{is} + [Q]k_q) \tag{2}$$

where $\Phi_q$ refers to the quantum yield in the presence of quencher. This form of quenching is referred to as dynamic quenching. Collisions between molecules can result in the conversion of the electronic energy into vibrational and kinetic energy, and distortions of the electronic levels due to collision can enhance intersystem crossing to weakly emitting or nonemitting triplet states, particularly when heavy atoms are involved. Examples of collisional quenchers are molecular oxygen and molecules containing bromine and iodine.

Another form of quenching, static quenching, results from the formation of non-emissive complexes between the luminescent molecule and the quencher. As such, static quenching does not compete for de-excitation of $S_1$, but instead reduces the fluorescence intensity by reducing the concentration of the free luminescent molecule. Dynamic and static quenching can be distinguished by the effect of increasing viscosity on the rate of $S_1$ de-excitation, which remains unchanged for static quenching and is decreased for dynamic quenching owing to the reduced rate of diffusion. Another process that removes excited states is chemical reaction of the excited state. An example of this is electron transfer between a molecule in its excited state and a second molecule in its ground state. More detailed information on fluorescence quenching can be found elsewhere (Eftink and Camillo, 1981; Lakowicz, 1983).

One of the better understood quenching processes is long-range energy transfer, also referred to as dipole-coupled and Förster energy transfer. Long-range energy transfer does not require contact between F and Q, but instead can occur at separation distances greater than 100 Å. In this process, de-excitation of F, the energy donor, occurs simultaneous with excitation of Q, the energy acceptor. Photons of light are not involved, and the energy transfer may be compared to that between a radio transmitter and a receiving antenna. The primary requirement for such a transfer is that the energy lost by de-excitation of the energy donor, $S_1$–$S_0$, be matched by the energy required for going from the ground state to an excited state of the energy acceptor. Another way of saying this is that the absorption spectrum of Q must overlap the emission spectrum of F. An example of spectral overlap is shown in Fig. 3, where the fluorescence excitation and emission spectra of fluorescein-labeled DNA and tetramethylrhodamine-labeled DNA are plotted together. The emission spectrum of the energy donor fluorescein and the excitation spectrum of the energy acceptor tetramethylrhodamine are marked with diagonal lines and shading, respectively, to emphasize the relevant region of spectral overlap at their intersection.

Returning to Fig. 2, energy transfer is depicted as the additional de-excitation pathway leading from the $S_1$ level of F to the $S_1$ level of Q, marked with the first-order rate constant for energy transfer, $k_{et}$. Actually the energy transfer can excite Q to $S_2$ or higher levels as long as the energy difference between the higher level and $S_0$ of Q is the same as that between $S_1$ and $S_0$ of F. Once excited, the acceptor can undergo de-excitation by the emissive and nonemissive processes described above for F. With a knowledge of the emission and excitation spectra, and several other spectral properties, $k_{et}$ can be calculated using the relationship derived by

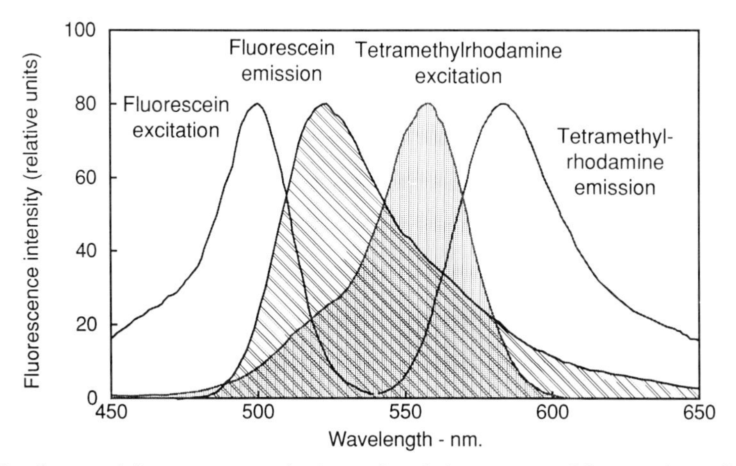


**Fig. 3** Corrected fluorescence excitation and emission spectra of fluorescein- and tetramethylrhodamine-labeled DNA probes indicating the region of spectral overlap (intersection of diagonal lines and shading) between the emission spectrum of fluorescein and the excitation spectrum of tetramethylrhodamine.

Förster (Förster, 1959; Conrad and Brand, 1968):

$$k_{et} = 8.79 \times 10^{-25}\kappa^2\Phi_D k_D \eta^{-4} R^{-6} J \tag{3}$$

where $\kappa$ is an orientation factor, $\Phi_D$ is the donor fluorescence quantum yield, $k_D$ is the first-order rate constant for de-excitation of the donor in the absence of energy transfer, $\eta$ is the refractive index of the medium, R is the distance separating the energy donor and acceptor in centimeters, and J is an integral expression describing the degree of spectral overlap and given by Eq. 4:

$$J = \int_0^\infty F_D(\lambda)\varepsilon_A(\lambda)\lambda^4 d\lambda \tag{4}$$

$F_D(\lambda)$ is the normalized emission spectrum of the energy donor (integrated intensity = 1) and $\epsilon_A(\lambda)$ is the spectrum of the acceptor's molar extinction coefficients for absorption, both expressed as functions of the wavelength, $\lambda$. The orientation factor relates to the relative orientation of the donor emission dipole and acceptor absorption dipole, and can be difficult to determine. In the absence of other information, the value of $\kappa^2 = 2/3$ for randomly oriented donor and acceptor molecules in solution is often assumed.

The important feature to note in Eq. 3 is the strong dependence of the energy-transfer rate on the separation distance. The inverse sixth power relationship dictates that small changes in the donor–acceptor separation distance will make large changes in the rate of energy transfer. The critical separation distance, $R_0$, is defined as the distance at which the rate of energy transfer is equal to the rate of de-excitation in the absence of energy transfer, and is calculated according to Eq. 5:

$$R_0 = (8.79 \times 10^{-25} \kappa^2 \Phi_D \eta^{-4} J)^{1/6} \quad (5)$$

To observe significant amounts of energy transfer, the distance separating the donor and acceptor molecules should be on the order of $R_0$ or smaller. For good donor–acceptor pairs, $R_0$ values of 30 to 60 Å are common. For example, the value of $R_0$ calculated for the fluorescein/tetramethylrhodamine donor–acceptor pair is about 50 Å (Ullman *et al.*, 1976).

According to the above discussion, a homogeneous binding assay based on energy transfer must bring the labeled portions of the reagents within about 30 to 60 Å of one another in response to the analyte. This is achieved in immunoassays by random labeling of antibody reagents via the amino groups of lysines on the antibody surfaces. Since the antibodies are large, about 115 Å in length, the degree of acceptor labeling must be high enough to guarantee that some labels are within 30 to 60 Å of the antibody-binding site, or labeled portions of a neighboring antibody bound to the same polyvalent antigen. A number of acceptor labels placed at a relatively long distance from the donor label can provide a result similar to that of one acceptor placed close to the donor label. The situation is different for DNA hybridization assays because there are no convenient functionalities, such as amino groups, on which to attach luminescent labels. Also, random multiple labeling would serve only to produce large luminescence backgrounds in the noncompetitive hybridization format, shown in Fig. 1A, since the linear nature of the hybrids permits label interactions only near the adjacent termini of the two probes. Random labeling might be effective in the competitive hybridization format, pictured in Fig. 1B, but the high labeling level required to ensure a high degree of label interaction would alter the stability of the hybridized complex.

Specific terminal labeling of probes allows the labels to be placed at locations that maximize the interactions between the labels. In fact, assays based on dynamic and static quenching are feasible since the labels may be brought close enough to allow collisional contact. For common luminescent compounds, contact requires center-to-center separations of about 5 to 10 Å. The length of the DNA double helix is 34 Å per single turn of 10 base pairs, in the most common B form of DNA. Labels placed within 9 to 18 base pairs, therefore, should show appreciable energy transfer, whereas labels placed within 1 to 3 base pairs of each other can additionally collide with one another. The length and rigidity of the chemical linkage between the DNA and the label also affect the separation distance and the ability of the labels to contact one another.

Because energy-transfer and contact-quenching processes have different distance criteria, considerable control of the quenching process is possible. For example, it was shown that energy transfer is maximized when fluorescein and Texas Red labels, attacted to two different nucleotides on the same strand of DNA, are separated by five intervening nucleotides in double-stranded hybrids (Heller and Jablonski, 1987). As the separation distance was increased beyond five intervening nucleotides, energy transfer decreased, as expected; however, energy transfer also decreased as the separation distance was decreased. The latter decrease can be explained by the dominance of static and dynamic forms of quenching when the separation distances become small enough to allow contact between the labels. This is supported by the fact that opposing terminal labels on complementary DNA strands usually show strong quenching of one label with no corresponding increase in the emission of the second label (Morrison, 1987; Morrison *et al.,* 1989). Even label pairs of fluorescein and tetramethylrhodamine displayed strong quenching of fluorescein emission with little or no increase in rhodamine emission. Energy transfer was favored, though, for one label pair examined: a 5′-pyrenesulfonate label opposed by a 3′-fluorescein label. The optimal separation distance for energy transfer, therefore, will depend on the particular label pair, since the type and strength of the short-range quenching interactions will be different. These short-range interactions are difficult to predict and must be determined empirically. In general, labels should be adjacent to one another or oppose one another to maximize dynamic and static quenching, and be separated by several nucleotides to optimize energy transfer.

## C. Comparison of Assay Formats

The noncompetitive and competitive assay formats pictured in Figs. 1A and B are alike in their dependence on label interactions, but differ in several fundamental aspects. The noncompetitive format directly measures the presence of target DNA through the hybridization of both probes to the target. The extent of energy transfer or contact quenching is proportional to the target concentration with one interaction occurring per target. As the target concentration approaches that of the probes, however, the increase becomes less than expected and even begins to decrease as the target concentration exceeds that of the probes. Excess target serves to separate probes as the likelihood of probes hybridizing to different target strands increases. This produces a peak in the response of the signal to the target concentration, similar to those observed for sandwich-type immunoassays. Performing the assay at two different sample dilutions identifies whether the target concentration is above or below the maximal response.

The competitive assay format indirectly detects the target since the label interaction is determined instead by probe-to-probe association. Four different equilibria are coupled together as probe-to-probe hybridization competes with probe-to-target hybridization:

$$T_1 + T_2 \xrightarrow{K_1} T_1 \cdot T_2 \qquad P_1 + P_2 \xrightarrow{K_2} P_1 \cdot P_2$$
$$T_1 + P_2 \xrightarrow{K_3} T_1 \cdot P_2 \qquad T_2 + P_1 \xrightarrow{K_4} T_2 \cdot P_1$$

where $T_1$ and $T_2$ are the complementary target strands, $P_1$ and $P_2$ are the complementary probe strands, and $K_1$ through $K_4$ are equilibrium constants. The four competing equilibria provide for a nonlinear relationship between label interaction and target concentration according to Eq. 6 (Morrison *et al.*, 1989):

$$[T_1 \cdot P_2]/[P_1 \cdot P_2]_0 = 1 - [P_1 \cdot P_2]/[P_1 \cdot P_2]_0 = [T_1 \cdot T_2]_0/([P_1 \cdot P_2]_0 + [T_1 \cdot T_2]_0) \quad (6)$$

where $[T_1 \cdot T_2]_0$ and $[P_1 \cdot P_2]_0$ are the initial concentrations of target and probe, respectively, before their denaturation at the beginning of the assay. If the label F is attached to $P_2$ and the label Q, attached to $P_1$, then the emission from F is greatest in the complex $T_1 \cdot P_2$ where Q is not present, and least in the complex $P_1 \cdot P_2$. An expression similar to Eq. 6 holds when the target is a single-stranded polynucleotide, such as RNA (Morrison *et al.*, 1989). Strictly speaking, Eq 6 only holds when the equilibria constants $K_1$ through $K_4$ are large and $K_1K_2/K_3K_4 = 1$. The first condition is generally true, but even though $K_1K_2/K_3K_4$ is rarely 1, the relationship accurately describes the experimental observations.

Figure 4 shows typical fluorescence changes associated with assays at two different probe concentrations and a number of different target concentrations. The solid curves were calculated using Eq. 6 and fit the data well. It may be noted that the useful assay range covers about a two-order-of-magnitude change in target concentration with a midpoint equal to the initial probe concentration. Lower detection limits are obtained by reducing the amount of probe in the assay. Although the competitive assay format is nonlinear with target concentration, the relationship between the detected signal and target concentration is well defined and the signal does not decrease at high target concentration.

Some examples of label quenching and enhancement observed in both hybridization formats are listed in Table I. It may be seen that large emission changes can be detected using either format with several different label combinations. The largest signal changes so far reported were

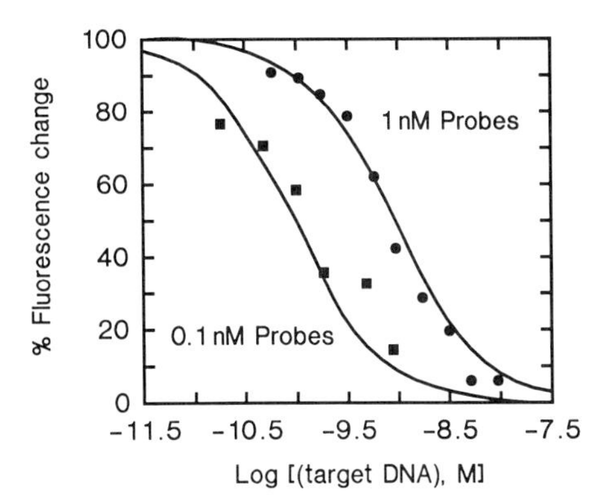


**Fig. 4** Competitive hybridizations of samples containing either 1 n*M* (circles) or 0.1 n*M* (squares) complementary DNA probes and various concentrations of double-stranded target DNA in 1 *M* NaCl, 10 m*M* tris(hydroxymethyl)aminomethane, pH 8.0, containing 10 $\mu$g $\lambda$ DNA/ml at 40°C. The probes were fluorescein-(ethylenediamine)d(AGATTAGCAGGTTT-CCCACC) and d(GGTGGGAAACCTGCTAATCT)(aminohexylaminoadenosine)-pyrene-butyrate and the target DNA was equimolar amounts of d(AGATTAGCAGGTTTCCCACC) and d(GGTGGGAAACCTGCTAATCT). Solid lines represent the theoretical relationships derived from Eq. 6. From Morrison *et al.* (1989) (Fig. 5, p. 240).

achieved using the competitive format with labels placed at the 5′-terminus of one probe and the 3′-terminus of the other probe. In this situation de-excitation of F is usually dominated by static and dynamic quenching. The last entry in Table I shows the exception, in which energy transfer was detected with labels attached to the neighboring terminal positions. Significant emission changes were also reported when labels were placed on the 5′-termini of complementary 8 nucleotide long oligomers (seventh entry in table, Cardullo *et al.*, 1988). With 5′-labeling of both probes, however, interaction between labels is greatly reduced as the lengths of the probes increase, making this method of probe attachment ineffective for practical hybridization assays.

It may be noted that energy transfer has one advantage over other quenching mechanisms in that energy transfer provides a check for unexpected interactions between sample impurities and labeled probes. Quenching of the donor label must be accompanied by a corresponding increase in emission from the acceptor label if no other quenching processes are involved. If only quenching is monitored in an assay, then false positives or false negatives in noncompetitive or competitive assay formats, respectively, result from the presence of unforeseen sample impurities that quench the donor emission. The presence of such impurities would lead to nonproportionate donor and acceptor emission changes if energy transfer were monitored, and the assay result could be discounted.

**Table I**
Emission Changes Due to Label Interactions in Competitive and Noncompetitive Assay Formats

| Primary label (F) | | Quenching label (Q) | | |
|---|---|---|---|---|
| Label[a] | Quenching (%)[b] | Label[a] | Enhancement (%)[b] | Reference |
| Noncompetitive format | | | | |
| 5′-ABEI | NR | 3′-TMR | 24 | Heller and Morrison (1985) |
| 5′-FITC | 71 | 3′-TRITC | NR | Cardullo *et al.* (1988) |
| Competitive format | | | | |
| 5′-FITC | 81 | 3′-Tx Rd | 0 | Morrison (1987) |
| 5′-FITC | 85 | 3′-PYB | 0 | Morrison (1987) |
| 5′-FITC | 83 | 3′-EITC | 0 | Morrison (1987) |
| 5′-FITC | 62 | 3′-EITC | 0 | Morrison (1987) |
| 5′-FITC | 63 | 5′-TRITC | NR | Cardullo *et al.* (1988) |
| 5′-FITC | 84 | 3′-PYB | 0 | Morrison *et al.* (1989) |
| 5′-FITC | 76 | 3′-Tx Rd | 0 | Morrison *et al.* (1989) |
| 5′-PYS | 78 | 3′-FITC | 270 | Morrison *et al.* (1989) |

[a] Abbreviations for fluorophore labels are FITC, fluorescein isothiocyanate; ABEI, *N*-(4-aminobutyl)-*N*-ethylisoluminol; PYS, *N*-hydroxysuccinimidyl 1-pyrenesulfonate; TMR, tetramethylrhodamine; TRITC, tetramethylrhodamine isothiocyanate; Tx Rd, Texas Red (trademark of Molecular Probes Inc.); PYB, *N*-hydroxysuccinimidyl 1-pyrenebutyrate; and EITC, eosin isothiocyanate. The DNA terminus to which the label was attached is indicated by 5′- and 3′-.
[b] NR signifies that quenching or enhancement was indicated, but no quantitative data were provided.

On the other side, it is often difficult to measure increases in acceptor emission because of background acceptor emission resulting from the direct absorption of external light intended for excitation of the fluorescent donor label. A method of overcoming the problem of direct acceptor excitation was demonstrated in an immunoassay using time-resolved detection and the combination of a short-lifetime acceptor label and a long-lifetime donor label (Morrison, 1988). If it were established that quenching impurities would not be present, then assays based on any of the quenching mechanisms would be suitable.

A better method for removing background acceptor emission is to use a chemiluminescent donor. This eliminates the external excitation light source, leaving energy transfer as the sole means to excite the acceptor label. Applications of a chemiluminescent donor in DNA hybridization assays were only reported once (Heller and Morrison, 1985). The relatively small amount of attention paid to chemical excitation is attributable in part to the greater difficulty in controlling chemiluminescent reactions to provide reproducible light intensities, and in part to the limited number of suitable chemiluminescent compounds that are amenable to conjugation. The conjugation of chemiluminescence catalysts, such as peroxidase, to probes would greatly increase the light intensity and provide another avenue for chemical attachment. However, if the chemiluminescent intermediate produced by the catalyst does not remain bound to the catalyst, then diffusion away from the probe before light emission destroys the distance relationship required for energy transfer. A catalyst such as luciferase should not suffer from this problem, and the newer highly purified preparations of luciferase may be suitable for energy transfer assays.

## D. Applications and Limitations

The noncompetitive and competitive fluorescence-quenching and energy-transfer formats share a number of advantages and disadvantages that are fundamental to homogeneous assays. The primary advantage of homogeneous assays is their simplicity, which generally leads to rapid assays using fewer reagents and requiring fewer manipulations. Since the assay is performed in a single phase, no immobilized reagents are required, and no washing of solid immobilization supports or formation and separation of precipitates is necessary. Homogeneous assays also eliminate background problems due to nonspecific adsorption of labeled reagents to solid immobilization supports.

Homogeneous assays, however, are generally less sensitive than heterogeneous assays. Even though nonspecific adsorption is eliminated in homogeneous assays, background signals arise from the fact that the measurable properties of the labels are only partially modified as a result of the specific binding reaction, whether it is antibody–antigen binding or DNA hybridization. This leads to the limitation that a large excess of reagents over analyte cannot be used. Taken together with the fact that the probe concentrations must be kept high enough to allow binding between reagents and analytes within a reasonable incubation time, the minimal amount of detectable target is relatively high.

For example, in the competitive hybridization assay format, if 90% quenching can be achieved when the two probes are hybridized to one another in the absence of target DNA, then the concentration of target DNA required to increase light emission by an amount equal to that of the background level is about one tenth that of the probe concentration. If 99% quenching could be achieved in the absence of target DNA, then target at 1/100 of the probe concentrations would provide signal equivalent to the background unquenched emission. Reducing the detection limit, therefore, requires lowering the probe concentration and/or finding labels that provide more efficient quenching. Reduction of the probe concentration can be taken only so far without drastically increasing the assay time. This can be seen in the times required for 50% hybridization, $t_{1/2}$, for the first pair of complementary oligomers listed in Table II. The concentration dependence of the hybridization time is consistent with that of a bimolecular reaction. Consequently, the $t_{1/2}$ is expected to be 2 hr for a 10-p$M$ probe concentration, and about 1 d for a 1-p$M$ probe concentration.

Heterogeneous assays can use a large excess of probe over target since the excess probe is removed by washing. The high probe concentrations counteract the fact that reaction rates involving surface-immobilized species are inherently slower than purely solution-phase reactions. Nonspecific adsorption increases at higher reagent concentrations, but if time is taken to wash the solid immobilization supports stringently, nonspecific adsorption can be reduced to low levels. The extra manipulations and time required for stringent washing of immobilization supports can be offset by the homogeneous techniques only if hybridization times are kept short, thereby limiting probe concentrations to about 0.1 n$M$ or higher and limiting detectable concentrations of target DNA to about 10 p$M$, or about 1 fmol target in a 100-$\mu$l sample.

A 10-p$M$ level of detection would be satisfactory for many applications in immunoassays. Therapeutic drug monitoring, for example, requires only the detection of micromolar amounts of material. The situation for DNA analyses is different, and there are few instances where 1 fmol of

**Table II**
Hybridization Characteristics of Labeled Oligomers in 1 $M$ NaCl, pH 8.0

| | | | Hybridization $t_{1/2}$ | |
|---|---|---|---|---|
| Oligomer[a] | Oligomer concentration (n$M$) | $T_m$ (°C) | $t_{1/2}$ (s) | Temperature (°C) |
| F-AGATTAGCAGGTTTCCCACC | | | | |
| P-TCTAATCGTCCAAAGGGTGG | 10 | | 12 | 40 |
| | 1 | | 74 | 40 |
| | 0.1 | | 850 | 40 |
| F-TTGGTGATCC | | | | |
| S-AACCACTAGG | 10 | 34 | 18 | 16 |
| | | | 10 | 25 |
| | 1 | 28 | | |
| | 0.1 | 22 | | |
| F-AGATTAGCAGGTTTCCCACC | | | | |
| S-TCTAATCGTCCAAAGGGTGG | 10 | 64 | 11 | 34 |
| | | | 8 | 44 |
| | 1 | 62 | | |
| | 0.1 | 59 | | |

[a] The sequence of the first oligomer in each pair is listed 5′-to-3′. The complementary oligomer sequence is listed in reverse. F, fluorescein isothiocyanate; S, sulforhodamine 101 (Texas Red); P, pyrenebutyrate.

DNA is of diagnostic value. For clinical analyses, detection requirements below 1 amol DNA are common. Fortunately there are polynucleotide amplification systems available that multiply the amount of target DNA using polymerase enzymes. The PCR is the best known and most studied of the amplification systems and is capable of producing well over a million copies of each target DNA strand originally present in a sample (Mullis and Faloona, 1987). More important to the present discussion, PCR is a homogeneous technique that still requires a means of detecting the amplified polynucleotide. Reverting to a heterogeneous technique for detecting the PCR-amplified DNA would introduce additional manipulations and defeat the simplicity inherent in the PCR technique. Coupling PCR with a homogeneous detection technique allows for a completely homogeneous assay having the advantages of both high sensitivity and simplicity. This was demonstrated by the combination of a competitive fluorescence-quenching assay with PCR amplification to detect a region within λ-DNA (Morrison *et al.*, 1989). Following 25 cycles of PCR, the midpoint of the quenching response corresponded to 96,000 molecules of original λ-DNA, thereby providing subattomole detection. The total assay time was less than 4 hr, which includes the 3.5 hr required for the PCR amplification. The amplified fluorescence-quenching assay is schematically represented in Fig. 5.

In summary, homogeneous assays using fluorescence quenching and energy transfer offer the advantages of speed and simplicity but provide inadequate detection levels for most clinical analyses. In light of these properties, the quenching assays appear particularly well suited for coupling with homogeneous amplification techniques that increase the amount of target sequences but still require a means for convenient detection. The following sections provide detailed experimental methods and procedures for the preparation, testing, and use of labeled probes in energy transfer and fluorescence-quenching DNA hybridization assays.

# II. MATERIALS

## A. Reagents

Both the noncompetitive and competitive fluorescence-quenching and energy-transfer hybridization assays require two labeled reagents. Suitable labeled reagents have been prepared by several methods, all of which provide either 3′- or 5′-terminal labels. The 3′-terminal labeling methods include (1) periodate oxidation of a 3′-terminal ribonucleotide, followed by reaction with an amine-containing label (Heller and Morrison, 1985); (2) enzymatic addition of a 3′-aliphatic amine-containing nucleotide using

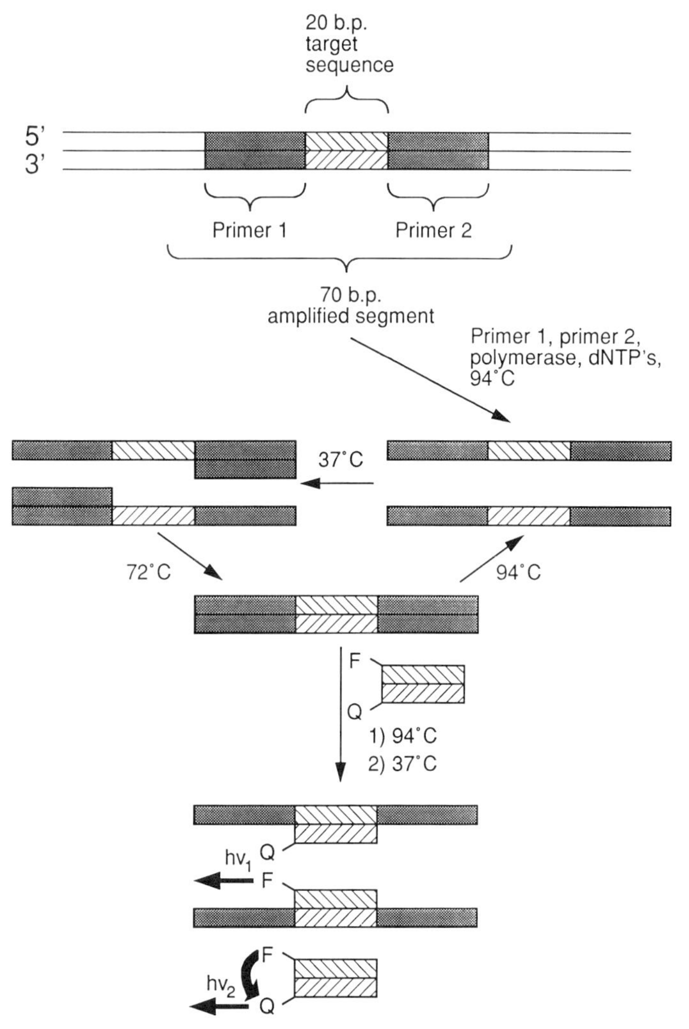


**Fig. 5** Schematic representation of the PCR-amplified energy transfer or fluorescence-quenching assay. Cross-hatching indicates the region within the target DNA that matches the sequence of the complementary labeled probes used for detection. Shading indicates the regions to which the two PCR primers hybridize during the PCR cycles of target amplification. In this example, the probes are 20 nucleotides long, and each primer is 25 nucleotides long. PCR is initiated by mixing the sample containing the target DNA with both primers, heat-stable DNA polymerase, and the four deoxynucleotide triphosphates, and heating the solution to 94°C to denature the target DNA. Repeated cycles of cooling and heating produce double-stranded copies of the target DNA, which include the central 20-base pair target sequence and the two flanking primer sequences (70 base pairs total). For simplicity, only the 70-base pair amplified section of the target DNA is shown in the amplification cycle since this predominates after several amplification cycles. Polymerase-catalyzed extension of the primers, during the 72°C incubation in the amplification cycle, produces two of these 70-base

terminal deoxynucleotidyl transferase, followed by reaction with an amine-reactive label (Morrison, 1987; Morrison *et al.*, 1989); and (3) periodate oxidation of a 3′-ribonucleotide (incorporated using a solid-phase DNA synthesizer), followed by reaction with 1,6-hexanediamine to provide a 3′-terminal aliphatic amine, followed by reaction with an amine-reactive label (Cardullo *et al.*, 1988). The 5′-terminal labeling methods include (1) periodate oxidation of a 5′-to-5′-coupled ribonucleotide, followed by reaction with an amine-containing label (Heller and Morrison, 1985); (2) condensation of ethylenediamine with 5′-phosphorylated polynucleotide, followed by reaction with an amine-reactive label (Morrison, 1987; Morrison *et al.*, 1989); and (3) introduction of an aliphatic amine substituent using an aminohexyl phosphite reagent in solid-phase DNA synthesis, followed by reaction with an amine-reactive label (Cardullo *et al.*, 1988). In addition to these methods, labels can be placed at specific locations within synthetic DNA oligomers using special aliphatic amine-containing nucleotide phosphoramidite reagents (Heller and Jablonski, 1987; Telser *et al.*, 1989a; Telser *et al.*, 1989b). Some common labels compatible with the aforementioned labeling procedures are pictured in Fig. 6. The chemiluminescent labeling compound aminobutylethylisoluminol (ABEI; Sigma Chemical Company, St. Louis, Missouri) contains an aliphatic amine that will couple with the periodate-oxidized ribonucleotides. The chemiluminescent properties of this and other amine-containing luminol derivatives have been compared (Schroeder and Yeager, 1978). The other labels pictured in Fig. 6 are fluorescent compounds, which can be coupled with DNA-containing aliphatic amine substituents. Common amine-reactive functionalities include isothiocyanates, *N*-hydroxysuccinimidyl esters, and sulfonic acid chlorides. Amine-containing luminol derivatives have also been converted to amine-reactive isothiocyanates (Patel *et al.*, 1983).

The following protocols describe 5′- and 3′-labeling of DNA using the condensation of ethylenediamine with 5′-phosphorylated DNA and the enzymatic incorporation of a 3′-amine-containing nucleotide. These reactions were selected since both can be performed easily in any biochemistry laboratory without the use of a DNA synthesizer, and both can be applied to synthetic DNA as well as DNA isolated from natural sources. Each terminal labeling procedure is performed in two parts, as depicted in Fig. 7,

---

pair sections for every one present at the beginning of the cycle (only one is shown). Specific detection of the amplified DNA is achieved by adding the complementary labeled probe DNA, heating to 94°C to denature both probe and target DNA, cooling to promote competitive hybridization, and measuring the amount of quenching and/or enhancement of the labels F and Q, respectively.

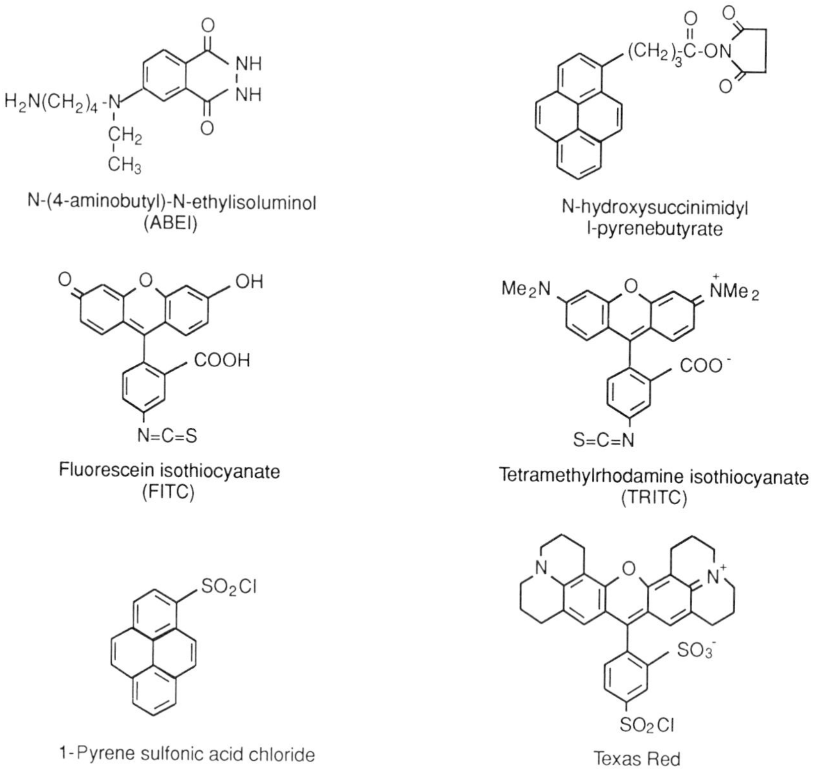

N-(4-aminobutyl)-N-ethylisoluminol (ABEI)

N-hydroxysuccinimidyl l-pyrenebutyrate

Fluorescein isothiocyanate (FITC)

Tetramethylrhodamine isothiocyanate (TRITC)

1-Pyrene sulfonic acid chloride

Texas Red

**Fig. 6** Chemical structures of some common chemiluminescent and fluorescent labeling compounds used in energy-transfer and fluorescence-quenching assays (Texas Red is the trademark of Molecular Probes, Inc. for the sulfonic acid chloride derivative of sulforhodamine 101).

involving (1) modification of the DNA to provide aliphatic amine substituents, and (2) reaction of the amine-modified DNA with amine-reactive fluorophores. The procedure for 5′-labeling also includes steps for phosphorylation of the DNA, if needed.

### 1. 5′-Terminal-Labeled DNA

#### *a. 5′-Phosphorylation*

1. Combine the following in a microcentrifuge tube: 10 nmol DNA strands (based upon 5′-OH termini)[2], 100 μl of 5X kinase buffer

[2] Ten nmol DNA strands corresponds to 3.5 n μg of DNA, where n is the number of nucleotides per strand. Ten nmol DNA strands also corresponds to 0.1 n OD units of single-stranded DNA or 0.07 n OD units of double-stranded DNA.

5' –labeling

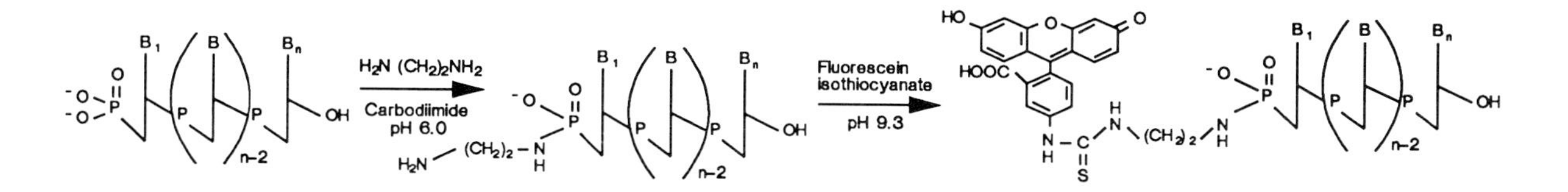


3' –labeling

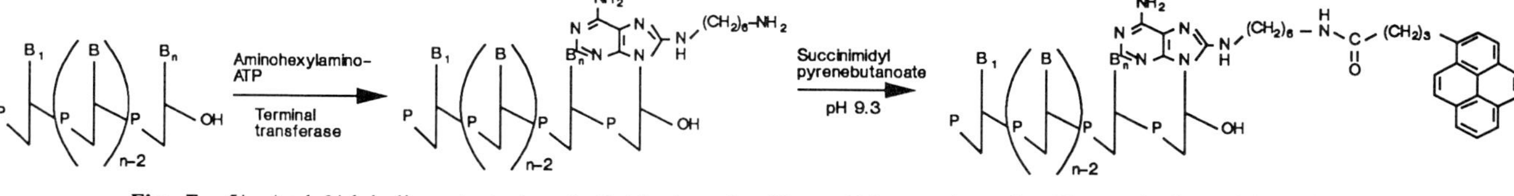


**Fig. 7** 5′- And 3′-labeling strategies. Individual nucleotides within a polynucleotide are indicated by stick structures showing the phosphates (P), bases ($B_n$), and deoxyribose ring (vertical line). Examples of amine-reactive labeling compounds shown are fluorescein isothiocyanate (5′-labeling strategy) and *N*-hydroxysuccinimidyl pyrenebutyrate (3′-labeling strategy).

(5X kinase buffer = 300 m*M* Tris(hydroxymethyl)aminomethane (Sigma), 75 m*M* $\beta$-mercaptoethanol, 50 m*M* $MgCl_2$, pH 7.8), 20 $\mu$l 5 m*M* adenosine triphosphate (ATP) (100 nmol; Pharmacia LKB Biotechnology, Inc., Piscataway, New Jersey), 100 units T4 polynucleotide kinase (New England Biolabs, Inc., Beverly, Massachusettes), and demineralized water to a final volume of 500 $\mu$l.

2. Place the tube in a 37°C water bath.
3. Add an additional 20 $\mu$l 5 m*M* ATP every 10 min for 40 min, and add an addition 100 units T4 polynucleotide kinase after 30 min of reaction.
4. Isolate the phosphorylated DNA by gel permeation chromatography using a prepacked Sephadex G-25 column, such as a NAP 10 column (Pharmacia LKB Biotechnology, Inc.), preequilibrated in demineralized water. Apply the reaction mixture to the column and elute with demineralized water. Collect the phosphorylated DNA in the column void volume.
5. Reduce the volume of the phosphorylated DNA to 500 $\mu$l in a centrifugal evaporator (Savant Instruments, Farmingdale, New York).

***b. 5′-Amination (adapted from Chu* et al., *1983)***

1. Prepare the ethylenediamine/carbodiimide solution just before use as follows: add 0.98 g 2[*N*-morpholino]ethanesulfonic acid (MES; Sigma.) and 830 $\mu$l ethylenediamine (Aldrich Chemical Co., Milwaukee, Wisconsin) to 15 ml demineralized water, titrate the solution to pH 6.0 with 6 *M* HCl, dilute the solution to a final volume of 25 ml, and add 1 g 1-ethyl-3-(dimethylaminopropyl) carbodiimide (Sigma).
2. Add 500 $\mu$l of the ethylenediamine/carbodiimide solution to the 500 $\mu$l phosphorylated DNA.
3. Allow the solution to react overnight (approximately 18 hr), with stirring, at room temperature.
4. Isolate the 5′-aminated DNA using gel permeation chromatography according to step 4 in the 5′-phosphorylation procedure.
5. Dry the 5′-aminated DNA sample in a centrifugal evaporator.

***c. Fluorophore Conjugation***

An amine-reactive label is attached by adding the reactive label, dissolved in an organic solvent, to the aminated DNA in an aqueous buffer suited to the coupling reaction. Good results can be obtained by following several general guidelines. For isothiocyanate and sulfonic acid chloride label derivatives, a suitable buffer is 50 m$M$ $Na_2B_4O_7$ at pH 9.3. For $N$-hydroxysuccinimide esters, a suitable buffer is 0.2 $M$ 3-[$N$-morpholino]propanesulfonic acid (MOPS; Sigma) at pH 7.4. $N$-Hydroxysuccinimidyl pyrenebutyrate, however, reacts well in the $Na_2B_4O_7$ buffer. Organic solvents capable of solubilizing different reactive fluorophores are acetone for $N$-hydroxysuccinimidyl pyrenebutyrate, dimethyl sulfoxide for fluorescein isothiocyanate, and dimethylformamide for Texas Red. Isothiocyanates are added in 50-fold molar excess to amine-modified termini, and sulfonic acid chloride and $N$-hydroxysuccinimidyl ester derivatives are added in a 100- to 200-fold excess to amine-modified termini. The amount of aqueous buffer present should be high enough that the organic solvent does not constitute more than 20% of the final reaction volume. Labeling reagents and labeled probes are sensitive to photodestruction and should be protected from prolonged exposure to light by covering containers with aluminum foil and reducing the room lights when covering is not possible. The fluorophores shown in Fig. 6 and those used to label the probes listed in Tables I through III were obtained from Molecular Probes, Inc., Eugene, Oregon.

1. Dissolve the 10 nmol aminated DNA in 0.6 to 1.0 ml of the labeling buffer (see preceding sections).
2. Dissolve the amine-reactive labeling compound in the solubilizing solvent at a concentration of 0.01 $M$.
3. For isothiocyanate-derivatized labels, add 50 $\mu$l 0.01 $M$ fluorophore solution to the aminated DNA solution. For $N$-hydroxysuccinimidyl and sulfonic acid chloride-derivatized labels, add 150 $\mu$l of the fluorophore solution to the aminated DNA solution.
4. Allow the reaction to stir overnight (approximately 15 hr) at room temperature.
5. Filter the reaction solution through a small syringe-loaded filter with 0.2- to 0.45-$\mu$m pore size (e.g., Acrodiscs from Gelman Sciences, Ann Arbor, Missouri).
6. Isolate the labeled DNA from the bulk of the labeling compound using gel permeation chromatography as described in step 4 of the 5′-phosphorylation procedure. The sample should be dried in a centrifugal evaporator and stored at −20°C.

***d. Purification***

At this point the labeled DNA is sometimes pure enough to use in the quenching assays. Often, however, aggregates of the labeling material elute with the void volume of the small pre-packed columns, and further purification is advised. Complete removal of the unreacted labeling compound can be achieved by performing gel permeation chromatography using a larger column (1 cm diameter × 28 cm length) packed with Sephadex G-25. If medium-or high-pressure chromatography is used (i.e., FPLC and HPLC), the unlabeled DNA can be separated from the labeled DNA as well. Good separations are obtained on an FPLC system (Pharmacia LKB, Inc.) using a linear acetonitrile gradient between 0 and 50% acetonitrile (starting buffer, 10 m*M* tetraethylammonium acetate (TEAA); final buffer, 50:50 acetonitrile:10 m*M* TEAA). The flow rate is 2 ml/min, and the gradient is increased at a rate between 0.5 and 1% acetonitrile/min. Material purified using these elution buffers is obtained free of the volatile buffer salts on drying. Store the dried DNA at −20°C.

### 2. 3′-Terminal-Labeled DNA

#### *a. 3′-Amination*

The following procedure uses terminal deoxynucleotidyl transferase-catalyzed addition of 8-(6-aminohexyl)aminoadenosine to the 3′-termini of DNA strands. Only a small excess of ribonucleotide triphosphate to 3′-termini is used in order to avoid the addition of two or three ribonucleotides. The same procedure also works well with amine-modified dideoxynucleotide triphosphates, in which case the addition of only one nucleotide is assured.

1. Combine the following in a microcentrifuge tube: 10 nmol DNA strands (based upon 3′-termini), 4.7 μl 3.2 m*M* 8-(6-aminohexyl)aminoadenosine triphosphate (Sigma Chemical Co.; 1.5 mol triphosphate/mol 3′-termini), 18 μl 1 *M* potassium cacodylate at pH 7.0, 15 μl 10 m*M* $CoCl_2$, 30 μl of a 500 μg bovine serum albumin/ml solution (nuclease free, Bethesda Research Laboratories, Gaithersburg, Maryland), 6 μl 15 m*M* dithiothreitol, 620 units terminal deoxynucleotidyl transferase (Supertechs, Bethesda, Maryland), and demineralized water to a final volume of 150 μl.
2. Incubate the tube in a 37°C water bath for 5 hr.
3. Isolate the DNA as described in steps 4 and 5 of the 5′-amination procedure.

At this point the fluorophore attachment and purification procedures are identical to those for 5′-labeling.

## B. Instrumentation

The fluorescence-quenching and energy-transfer hybridization assays require instrumentation for detecting the light emitted by one or both luminescent labels. Any fluorometer can be used to perform assays using fluorescent labels, ranging from simple filter fluorometers to photon-counting spectrofluorometers. For chemical excitation, a luminometer or scintillation counter is used to detect the light emission. The complexity of the instrumentation increases if automated reagent delivery and temperature control are incorporated, and if monitoring of the light emission throughout hybridization is desired.

### 1. Fluorometers

Florometers can be obtained in a variety of configurations, any of which is suitable for monitoring hybridization assays based on fluorescence quenching. Filter fluorometers rely on optical filters to select the wavelengths of light used to excite the fluorophores and to isolate the wavelengths of light emitted by the fluorophore. The simple design of a filter fluorometer allows high efficiency of light collection and low instrument cost. Filter fluorometers can be constructed from basic building blocks of a light source, sample compartment, and detector with associated light collection and focusing lenses. Spectrofluorometers use monochromators for selecting excitation and emitted light. The added expense of monochromators is not necessary for performing fluorescence-quenching assays; however, they are necessary when spectral characterization of the labeled reagents is desired.

Energy-transfer assays can be monitored by using a single-channel instrument (one-emission detector), by measuring emission from only one of the labels, or by changing optical filters or monochromator settings to sequentially measure the emissions from both the donor and acceptor labels. Single-channel fluorometers are available from several companies, including Perkin-Elmer (Norwalk, Connecticut), Spex Instruments (Edison, New Jersey), and SLM-AMINCO Instruments (Urbana, Illinois). Energy-transfer analyses are faster if a dual-channel instrument is used, in which two detectors are placed each at 90° to the excitation light path. Fluorometers and sample compartments of this design are available from SLM-AMINCO Instruments, and I.S.S. (Urbana, Illinois).

Temperature control, automatic reagent injection, and time-based measurement capabilities are attractive features if kinetic or thermodynamic analyses are desired. Kinetic measurements can be used to determine the equilibration times required for complete hybridization of probes and target. They also provide the initial and final intensities of label emission from which the net fluorescence change is calculated. The net fluorescence change is less sensitive to fluorescent impurities in a sample, since fluorescence that is unchanged over the course of the hybridization is subtracted. A thermostated sample holder is required for kinetic analyses, because the rates of hybridization are temperature dependent, and the optimal rates are obtained at different temperatures for different length strands of DNA. Another advantage the thermostated sample holder provides is the ability to record fluorescence melting curves. Melting curves recorded on complementary labeled probes or between probes and target DNA are useful for demonstrating that the labeled probes are performing properly and for determining the temperature range in which the assays should be performed. It is difficult to find all of the above features in a single commercial fluorometer, and modifications may be necessary. I have modified an SLM-AMINCO Instruments model 4800 spectrofluorometer for the purposes of conducting automated fluorescence-quenching and energy-transfer assays, in either a static or kinetic mode (Morrison *et al.*, 1989), and for performing thermodynamic and kinetic analyses of labeled probes (Morrison and Stols, 1989). The fluorometer layout is pictured in Fig. 8. The basic SLM-AMINCO sample compartment (A in Fig. 8) has two detection channels in a T format (180° to each other and 90° from each to the excitation beam) and a third channel for monitoring the intensity of the excitation light. To increase sensitivity, the standard analog detectors were replaced with photon-counting detectors, consisting of two Hamamatsu (Middleton, New Jersey) R928 photomultiplier tubes in thermoelectric cooled housings (Model TE-177RF, Products for Research, Inc. Danvers, Massachusettes; H and I in Fig. 8). Two channels of photon-counting electronics consist of EG & G ORTEC (Oak Ridge, Tennessee) preamplifiers, amplifier/discriminators, and counter/timers. Photons emitted from the fluorophores are collected, filtered, and focused onto the photomultiplier tubes, where they are counted and the tallies are transferred to a Hewlett Packard (Palo Alto, California) Model 9836 computer over an IEEE 488 interface. The computer also records the relative intensity of the excitation beam detected at the reference photomultiplier tube (J) and normalizes the detected fluorescence by the reference value, in order to correct for fluctuations in the intensity of the excitation lamp (P). A digital buffer (home-built), analog-to-digital converter, and digital-to-analog converter interfaced to the computer allow the computer to input the reference signal, control a Hamilton Micro Lab P Syringe Dispenser (Reno, Nevada) for injection of probes, and control the temperature of a Haake (Saddlebrook, New Jersey) Model A81 refrigerated water bath, which circulates water through the thermostated sample holder within the sample compartment. The actual temperature within the sample cell is measured with a Yellow Springs Instruments (Yellow Springs, Ohio) pre-

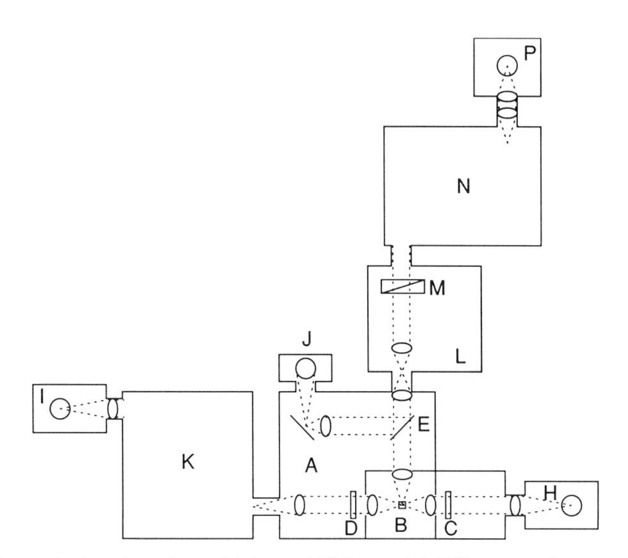


**Fig. 8** Schematic drawing of a multichannel SLM model 4800 spectrofluorometer modified for photon counting. Key to the simultaneous detection of donor and acceptor emission is the SLM T-format sample compartment (A) providing detection of light emitted from a sample (B) in two directions normal to the path of the excitation light. The compartment also includes a beam splitter (E) for diverting a fraction of the excitation light to a detector (J) enabling correction of fluorescence intensities for variations in the intensity of the excitation light source (P). Other components include the excitation monochromator (N), the emission monochromator (K), the emission photomultiplier detectors, housed in thermoelectric cooled housings (H and I), emissions filters (C and D) used in the absence of emission monochromators, and an optics module (L) housing optional equipment such as an electronically driven mechanical shutter (M) for protecting the sample from extended exposure between automated fluorescence measurements and allowing the alternate recording of light and dark measurements.

cision thermistor and Keithly (Cleveland, Ohio) Model 179-20A Digital Multimeter equipped with an IEEE 488 interface.

### 2. Luminometers

A number of commercial luminometers are available that contain a single photomultiplier detector and are suitable for monitoring chemiluminescence in a quenching-based hybridization assay. Such luminometers are available, for example, from Los Alamos Diagnostics, Inc. (Los Alamos, New Mexico), Berthold Analytical, Inc. (Nashua, New Hampshire), Pharmacia LKB, Inc., and Turner Designs (Mountain View, California). Some luminometer models accept optical filters, providing the potential for measuring the emission from two labels sequentially. Only one commercial luminometer (Turner Designs) has two detection channels, which allows the simultaneous monitoring of donor and acceptor label emission. The dual-channel luminometer accepts filters and uses a signal-chopping technique, which approaches the sensitivity of photon counting.

Luminometers are less complicated than fluorometers and can be constructed fairly easily. The luminometer is essentially a light-tight enclosure containing a sample holder and a detector, and may optionally contain focusing optics. Two designs the author has built provide dual-channel detection with filtering capabilities and temperature control. The first design uses two lenses to collect light from the sample in each of two directions and to focus the light onto two photomultiplier detectors. The lenses allow the detectors and sample cell to be separated from each other, providing room for filter holders and large auxiliary components that might be used from time to time in specialized applications. Reagents are injected through a light-tight septum in the sample compartment lid or through a small-diameter teflon tube passing through the side of the sample compartment. Because the light is focused on the detector, either the large cathode end-on photomultiplier tubes or the smaller cathode side-on tubes may be used. Cooled and shuttered photomultiplier housings and photon-counting electronics are identical to those used on the photon-counting fluorometer.

The second design is more compact, owing to the removal of the focusing optics, and is pictured in Fig. 9. The small thermostated sample compartment (A) is required for high-efficiency light collection, as are the large cathode photomultiplier detectors (F). The optical filters are sandwiched between the sample compartment and detectors, fitting within impressions (E) in either side of the sample cell holder (B). The sample holder is designed so that raising the holder is required to open the sample compartment lid (C). Raising the holder in this manner serves to shutter the detectors, providing a convenient means for their protection. Reagents are injected though a rubber septum (H) in the sample compartment lid positioned directly above the sample.

Both luminometers have been automated with computer-controlled reagent dispensers and data-acquisition electronics, similar to the modified fluorometer. Since chemiluminescence does not provide an instantaneous response, like fluorescence, automated recording of melting curves and hybridization kinetics is not possible. However, light emission from a chemiluminescent reaction can be complex, and important information is obtained by monitoring the rate of light emission over the course of the chemiluminescence reaction. For example, inhibitors in the samples can cause the peak of the chemiluminescence to occur at different times relative to time of reagent injection. Acquiring data continuously following

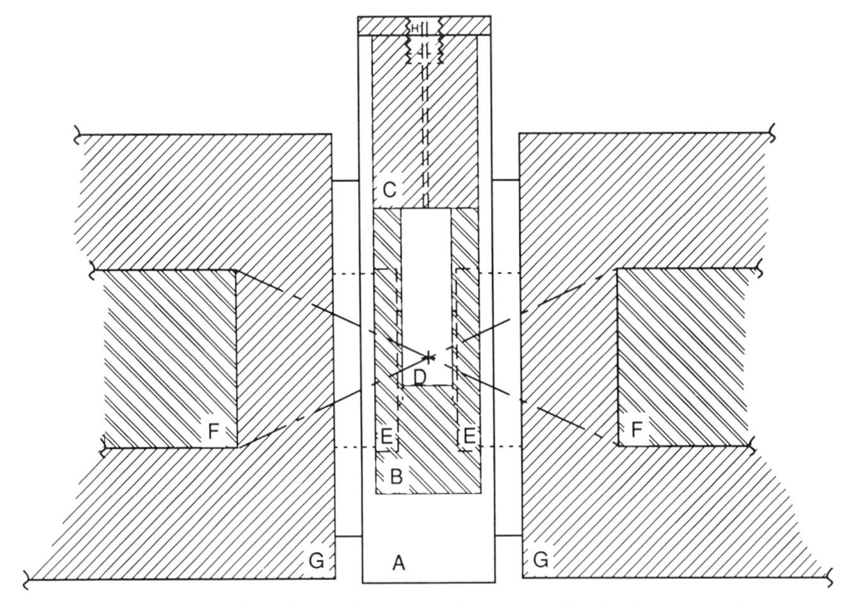


**Fig. 9** Schematic drawing of a dual-channel photon-counting luminometer. Components include a sample compartment (A), 3.6 cm in width, a sample holder (B) with impressions to hold two emission filters (E) and a septum port (H) to permit light-tight reagent delivery to the sample, and photomultiplier detectors (F) housed within thermoelectric cooled housings (G) with the photocathodes placed 5 cm from the center of the sample cuvette (D). The sample holder also serves as a shutter to protect the photomultiplier tubes. In order to access the sample cuvette within the sample compartment, the sample holder must be raised high enough to allow rotation of the sample holder lid (C). In this position, light entering the sample holder is prevented from reaching the photomultiplier tubes.

reagent injection will determine whether such inhibitors are present and provide the peak emission intensity regardless of the delay.

## III. PROCEDURES

Energy-transfer and fluorescence-quenching techniques are best applied to homogeneous solutions and, therefore, will probably find little application to the more familiar heterogeneous methods of DNA analysis in use today. These common methods, including the various blotting techniques, affinity-based separations of DNA, and DNA-sequencing methods, employ solid immobilization supports or gel matrices for separation of DNA species. When solid supports are used, a single labeled probe is sufficient and is easily removed by washing procedures. Also, fluorescence measurements are made more difficult on solid supports, such as nitrocellulose filters, owing to the large amount of light scattering and impurity fluorescence at the surface of the solid support. An exception is the detection of specific DNA sequences within gel bands. Light scattering and impurities would be lower than those on membrane filters, and detection would be rapid since the unbound probe would not have to diffuse out of the gel before measurements could be made. This application, however, is untested at this time.

The following procedures describe how energy-transfer and fluorescence-quenching methods are used in place of blots and other heterogeneous methods to detect DNA free in solution without the need for immobilization. Procedures are included for the direct detection of DNA as well as for the detection of enzymatically amplified DNA. Procedures for characterizing the labeled reagents and for performing the light-emission measurements are also included.

## A. Light-Emission Measurements

Light-intensity measurements are common to all forms of the energy-transfer and fluorescence-quenching assays. One of the most important variables in measuring light is the selection of optical filters and/or monochromator settings. The only instance in which light filtering is not required is when the label L is chemically excited and Q is either nonemissive or emits at wavelengths beyond the sensitivity of the detector being used. Table III provides the excitation and emission maxima for a number

**Table III**

Spectral Characteristics of Labels Commonly Used in Energy-Transfer and Fluorescence-Quenching Hybridization Assays

| Label[a] | Excitation maximum[b] (nm) | Emission maximum[b] (nm) | $\varepsilon_{\lambda max,\ label}$[b] ($M^{-1}cm^{-1}$) | $\varepsilon_{260,\ label}$[c] ($M^{-1}cm^{-1}$) |
|---|---|---|---|---|
| FITC | 494 | 520 | 68,000 | 13,000 |
| Tx Rd | 596 | 606 | 110,000 | 21,000 |
| PYB | 350 | 378 | 27,000 | 10,000 |
| TRITC | 558 | 581 | 88,000 | 33,000 |
| ABEI | (chemical) | 420 | 54,000 | 12,000 |

[a] See Table I for key to label abbreviations.

[b] Except for ABEI, the label spectral values were determined using labels conjugated to DNA.

[c] Label extinction coefficients at 260 nm were estimated by first determining the ratio of the unconjugated label absorbance at 260 nm to its absorbance at the long wavelength maximum. The ratio was then multiplied by $\varepsilon_{\lambda max,\ label}$.

of fluorescent labels, along with other pertinent spectral information. The literature should be consulted for labels not included in this list. It is good practice to record the excitation and emission spectra of new labels after attachment to DNA probes since spectral shifts may occur relative to the unconjugated dyes. Usually the spectral shifts will amount to only several nanometers; however, large changes in the quantum yield of a label are not uncommon on conjugation. Excitation monochromator settings or filters are selected to pass as much light within the excitation region of F as possible without passing light in the wavelength range transmitted by the emission monochromators or filters. The emission monochromator settings or filters are selected to pass as much light within the emission band of F (quenching assays) or Q (energy transfer assays) while excluding the excitation light.

### 1. Static Fluorescence Measurements

1. Select appropriate monochromator settings and/or filters from the values given in Table III.
2. Place the sample in a quartz cuvette (glass is acceptable if all wavelengths involved are in the visible portion of the spectrum), and place the cuvette in the fluorometer. Allow the sample to equilibrate to the temperature of the thermostated sample holder. Room temperature is acceptable where this does not interfere with hybridization.
3. Record the signal produced by the emission detector(s), $S_{em}$. Simultaneously record the signal produced by the reference detector, $S_{ref}$, if the fluorometer is equipped with a reference detector that monitors the intensity of the excitation light.
4. Repeat steps 2 and 3 for a similar sample for which the probe containing the label F has been deleted. This provides the background fluorescence, $S_{bkg}$, and the associated reference signal, $S_{ref,\ bkg}$.
5. Calculate the net fluorescence signal for each emission band monitored, corrected for lamp intensity, according to Eq. 7:

$$S_{em,\ net} = S_{em}/S_{ref} - S_{bkg}/S_{ref,\ bkg} \quad (7)$$

Set the reference signals equal to 1 in the above equation if the fluorometer does not have a reference detector.

Quenching or enhancement of fluorescence can be calculated as the percentage decrease or increase of the maximum fluorescence change, $\%\Delta f_{max}$, according to Eq. 8:

$$\%\Delta f_{max} = 100(S^0_{em,\ net} - S_{em,\ net})/(S^0_{em,\ net} - S^\infty_{em,\ net}) \quad (8)$$

This requires measuring the fluorescence of samples in which the two extremes in fluorescence are obtained, that is, in one sample in which the interaction between the labels is absent, providing $S^0_{em,\ net}$, and in a second sample in which the interaction is maximal, providing $S^\infty_{em,\ net}$. The samples used to determine $S^0_{em,\ net}$ and $S^\infty_{em,\ net}$ are different, depending on the assay format. For the competitive assay format, $S^0_{em,\ net}$ is measured using a sample in which both probes are present together with a 100-fold or larger target excess. $S^\infty_{em,\ net}$ is measured using a sample containing both probes and no target. For the noncompetitive assay format, $S^0_{em,\ net}$ is measured as in the competitive assay format, but can additionally be measured using a sample containing both probes and no target. $S^\infty_{em,\ net}$ is measured on samples containing both probes and a target concentration equal to that of the probe containing the label F.

## 2. Kinetic Fluorescence Measurements

Kinetic fluorescence measurements are made in the same manner as static measurements except that the fluorescence is monitored continuously after initiation of the hybridization. Often hybridization is initiated by injecting the denatured DNA solution into a cuvette in the fluorometer sample holder, which is maintained at some temperature below the melting temperatures of target/probe and probe/probe double-stranded complexes. Continuous fluorescence measurements are obtained by several methods, including (1) manually recording $S_{em}$ at regular intervals; (2) connecting a strip chart recorder(s) to the amplified output of the analog photomultiplier detector(s) and continuously recording the analog signal versus time; and (3) acquiring and storing $S_{em}$ at regular intervals with a computer. Subsequent data processing is easiest using the latter method. An example of computer-acquired kinetic data is shown in Fig. 10. These data were recorded during competitive hybridizations between fixed amounts of DNA probes (F, Fluorescein) and various amounts of target DNA (*Escherichia coli* plasmid). At the low probe concentrations used in these assays, the background fluorescence was large and the sample-to-sample variation in the background fluorescence was significant. To correct for this background variation, each data set was offset by a different intensity value to equalize the initial $S_{em}$ of each data set. Before equalization of the initial $S_{em}$ values, the final values of the fluorescein emission intensities could not be correlated with the target DNA concentrations.

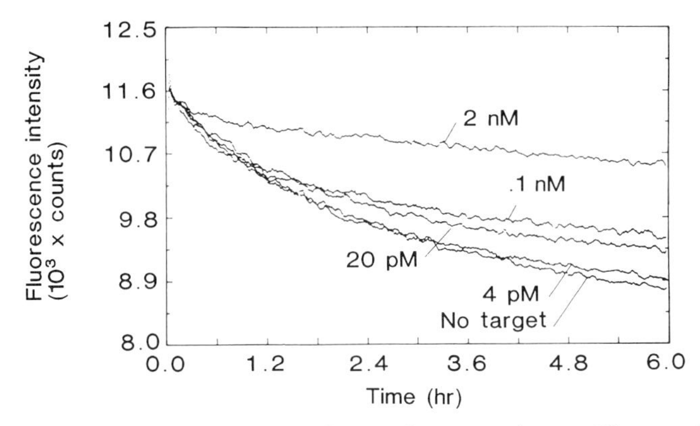


**Fig. 10** Competitive hybridization assay for *E. coli* enterotoxin gene. Five complementary pairs of labeled probes were used simultaneously to increase sensitivity at a concentration of 20 pM per pair. Each strand was labeled with both a 5′-fluorescein and a 3′-pyrenebutyrate. Samples consisted of the *E. coli* enterotoxin gene fragment and 10 $\mu$g $\lambda$ DNA/ml. Samples were heated in a boiling water bath for 10 min, after which time the probe pairs were added and heating continued for another 2 min. The samples were then diluted with buffer to a final concentration of 1 *M* NaCl, 10 m*M* EDTA, and 10 m*M* *tris*(hydroxymethyl)aminomethane, pH 8.0. Fluorescence was monitored using photon counting over 2-sec intervals separated by 5 sec for 6 hr following dilution of the denatured probe and target solution. The fluorometer sample compartment was maintained at 42°C throughout the experiment. The data were smoothed by averaging each time point with the five time points on either side (11 points total, approximately 1 min time resolution). From Morrison *et al.* (1989) (Fig. 6, p. 240).

Following the offset procedure, the final fluorescence values decreased with decreasing target concentration, as expected in a competitive hybridization assay, and a sample containing 4 p*M* target DNA was distinguishable from the sample containing no target DNA.

## 3. Chemiluminescence Measurements

Chemiluminescent reactions are initiated by the addition of a chemical coreagent or catalyst. I shall describe here only the luminol reaction since it has proved useful in energy-transfer assays. Other chemiluminescent compounds can be used with the limitation that they be attached to the DNA probe by a linkage that remains intact at the time of light emission. Likewise, the chemiluminescent or bioluminescent catalyst can be conjugated to the DNA only if the chemiluminescent compound remains associated with the catalyst at the time of light emission.

Under conditions amenable to DNA hybridization, the luminol reaction requires an oxidant and a catalyst for appreciable rates of light emission. A number of oxidant/catalyst combinations have been examined, and the most sensitive include hydrogen peroxide as the oxidant and microperoxidase as the catalyst (Schroeder and Yeager, 1978). Solution pH values near 7 are acceptable; however, higher rates of light emission are obtained at higher pH values, such as pH 8, where double-stranded DNA is still stable.

1. If the luminometer accepts emission filters, select the filter from the information provided in Table III.
2. Prepare a 10 $\mu$g/ml solution of microperoxidase (MP-11, Sigma) in the luminescence reaction buffer (e.g., 25 m*M* sodium phosphate, 100 m*M* NaCl, pH 7.2).
3. Prepare the sample in the luminescence reaction buffer and add 10 $\mu$l of the 10 $\mu$g/ml microperoxidase solution per ml of the reaction buffer.
4. Place the sample in a cuvette or glass tube appropriate for the luminometer and allow the sample to equilibrate to the temperature of the thermostated sample holder. Room temperature is acceptable if it does not interfere with hybridization.
5. Prepare a 400-fold dilution of 30% $H_2O_2$ using demineralized water.
6. Inject 100 $\mu$l of the diluted $H_2O_2$ solution per 1 ml of sample to initiate chemiluminescence.
7. Measure the signal(s) produced by the emission detector(s) at regular intervals as the chemiluminescence increases and then begins to decrease.
8. Repeat steps 2 through 6 for a *background* sample in which the luminol-labeled probe is omitted, measuring chemiluminescence over the same period as for the other sample.
9. Determine the value of the maximum signal produced by the first sample and the time at which the maximum occurred. Subtract the signal recorded on the background sample at the same reaction time from the first sample to obtain the net signal, $S_{em,\ net}$. The percentage quenching and enhancement are calculated using Eq. (8) where $S^0_{em,\ net}$ and $S^{\infty}_{em,\ net}$ are obtained using the same sample compositions described in the section on static fluorescence measurements.

## B. Reagent Characterization

Some degree of labeled probe characterization should be performed before using the newly labeled probes in a hybridization assay. The minimum characterization includes a determination of the probe concentration and the labeling percentage, both of which can be determined from an absorbance spectrum. When probes of a new length or containing a new label are first prepared, further characterization may be necessary to determine the temperature range in which the probes hybridize best and to determine whether the new label's emission properties are adequate.

### 1. Composition and Concentration of Labeled Probes

The concentration of probe is determined from the probe absorbance at 260 nm, $A_{260}$, after subtracting the absorbance contribution of the label. This is done by recording the $A_{260}$ and the absorbance at the long wavelength maximum of the label, $A_{\lambda max}$, and solving the following two simultaneous equations:

$$A_{260} = \varepsilon_{260,\ label}\,[label] + \varepsilon_{260,\ DNA}\,[probe] \qquad (9)$$

and

$$A_{\lambda max} = \varepsilon_{\lambda max,\ label}\,[label] + \varepsilon_{\lambda max,\ DNA}\,[probe] \qquad (10)$$

where $\varepsilon_{260,\ label}$ is the absorbance extinction coefficient ($M^{-1}cm^{-1}$) of the label at 260 nm, $\varepsilon_{260,\ DNA}$ is the absorbance extinction coefficient of the DNA strand at 260 nm, $\varepsilon_{\lambda max,\ label}$ is the absorbance extinction coefficient of the label at the long wavelength maximum of the label, $\varepsilon_{\lambda max,\ DNA}$ is the absorbance extinction coefficient of the DNA strand at the long wavelength maximum of the label, [label] is the concentration (*M*) of the label attached to the DNA strand, and [probe] is the concentration of the DNA strand (*M*) composing the DNA probe. For labels absorbing in the visible portion of the spectrum, $\varepsilon_{\lambda max,\ DNA}$ is negligible and [probe] can be calculated according to Eq. 11:

$$[probe] = \{A_{260} - (\varepsilon_{260,\ label}/\varepsilon_{\lambda max,\ label})A_{\lambda max}\}/\varepsilon_{260,\ DNA} \qquad (11)$$

$\varepsilon_{260,DNA}$ can be approximated by a value of 10,000 n$M^{-1}$ $cm^{-1}$, where n is the number of nucleotides in each probe strand. Extinction coefficients of oligomers can be calculated within 10% certainty using tables of values for nearest neighbor interactions (Borer, 1976). Some experimentally determined values of $\varepsilon_{260,\ label}$ and $\varepsilon_{\lambda max,\ label}$ are provided in Table III for several common labels.

The concentration of conjugated label is likewise calculated from the simultaneous solution of Eqs. 9 and 10. For labels absorbing in the visible portion of the spectrum, the calculation reduces to

$$[label] = A_{\lambda max}/\varepsilon_{\lambda max,\ label} \qquad (12)$$

The probe composition, determined as the percentage of probe labeled, is then calculated as 100[label]/[probe]. The Q-labeled probe should be labeled as close to 100% as possible to obtain maximal quenching or energy transfer. The F-labeled probe should be labeled as close to 100% as possible to obtain the maximal fluorescence intensity.

## 2. Titration of Labeled Probes

Probe concentrations determined by the above procedure may be in error as much as 12%, 16%, and 17% for 20-, 12-, and 10-nucleotide long oligomers, respectively (Morrison *et al.*, 1989), leading to even greater uncertainty in the relative concentrations of two probes used together in an assay. The proportions in which labeled probes should be combined in a hybridization assay can be determined better by titrating one probe with the other. For titrating complementary probes, use the following procedure.

1. Prepare solutions containing the probe labeled with F at a fixed concentration and the probe labeled with Q at concentrations ranging between 0 and 5 times the fixed probe concentration.
2. Allow the probes sufficient time to hybridize.
3. Measure the amount of light emitted by one or both labels.
4. Plot the amount of emitted light versus the ratio of the two probes [(Q-labeled probe)/(F-labeled probe)].
5. Fit the data points recorded on samples with excess F-labeled probe with one straight line and fit the data points recorded on samples with excess Q-labeled probe with a second straight line.
6. Determine the point of equivalence as the intersection of the two straight lines.
7. Recalculate the concentration of the Q-labeled probe by dividing the previously estimated Q-labeled probe concentration by the ratio of the two probes at the equivalence point. If it is believed that the initial estimate of the Q-labeled probe concentration is the more reliable value, then recalculate the concentration of the F-labeled probe as the product of the previously estimated F-labeled probe concentration and the ratio of the two probes at the equivalence point.

An example of the equivalence point determination is shown in Fig. 11. Below the equivalence point, the light emission should decrease linearly with increasing probe ratio, and above the equivalence point, the light emission should be invariant. For probes used in the noncompetitive assay

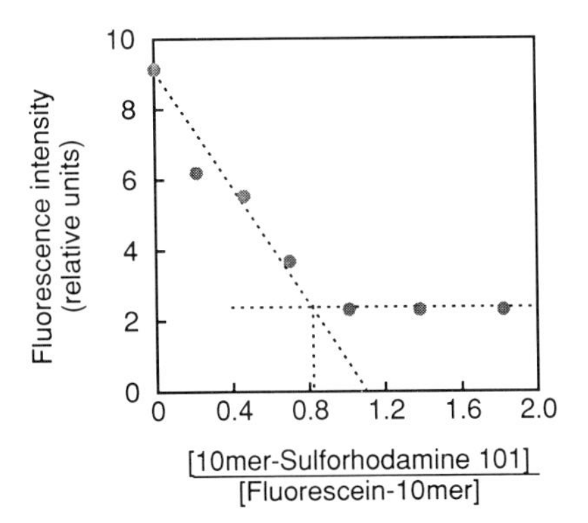


**Fig. 11** Titration of 5′-fluorescein-TTGGTGATCC with 3′-sulforhodamine 101 (Texas Red)-labeled complementary oligomer to determine the equivalence point. A fixed concentration of the fluorescein-labeled oligomer was mixed with increasing amounts of the sulforhodamine 101-labeled complement and the relative fluorescein emission intensity plotted versus the ratio of the two labeled oligomers. The linear decrease in fluorescein emission with increasing concentration of the sulforhodamine-labeled complement below the equivalence point is fit with a straight line, as is the constant fluorescence measured above the equivalence point. The intersection provides the optimal concentration ratio for use in competitive hybridization assays. The equivalence point of 0.82 indicated in this plot is within the predicted uncertainty in the ratio of the two oligomer concentrations of the expected value of 1.0. The buffer contained 1 *M* NaCl and 10 m*M* Tris at pH 8.0.

format, the above procedure can be used, except that several titrations of the F-labeled probe with the Q-labeled noncomplementary probe are performed, each titration using a different target DNA concentration that brackets the estimated F-probe concentration. Several data sets similar to those in Fig. 11 are generated with the one displaying the lowest light intensity in the horizontal region of the data (high probe ratio) indicating equivalence between the target and the F-labeled probe concentration. The equivalence point in this data set is used to recalculate the probe concentrations relative to one another.

## 3. Determination of Hybridization Conditions

Hybridizations should be performed at a temperature low enough to afford nearly complete association between the probe and target DNA strands, but high enough to provide rapid hybridization. Likewise, the length of the hybridization period should be long enough to ensure nearly complete hybridization. This requires on the order of about $10 \cdot t_{1/2}$. The optimal hybridization temperatures and times are strongly dependent on the composition and concentration of the DNA probe as well as the solution ionic strength. Several relationships have been derived for predicting the $T_m$ values (the temperature at which 50% of the possible base pairs are formed) and the hybridization rates of polynucleotides. These relationships and several rules of thumb for hybridization conditions have been reviewed (Wahl *et al.*, 1987). For short oligomers a temperature about 5°C below the $T_m$ is optimal for hybridization, whereas a temperature between 18 and 32°C below the $T_m$ is optimal for high-molecular-weight DNA.

Labeling of the DNA probes can alter their hybridization characteristics, and a more accurate determination of the $T_m$ and hybridization rate may be desired. The $T_m$ can be measured by recording the fluorescence of a solution containing the probes and buffer salts at the concentrations intended for the hybridization assay. The target is omitted for complementary probes (competitive assay format) and included at a concentration equal to the F-labeled probe for noncomplementary probes (noncompetitive assay format). The temperature of the solution is increased slowly to allow equilibration of the DNA strands and the fluorescence recorded at fixed temperature increments (Morrison *et al.*, 1989). Plotting the fluorescence of F and/or Q, where applicable, provides the melting curve with midpoint equal to $T_m$. Hybridization rates can be measured using the same sample compositions as used for the $T_m$ measurements. Hybridization is initiated by either combining the complementary probes or rapidly cooling the complete hybridization mixture from a temperature well above the $T_m$ to the hybridization temperature. Some $T_m$ and $t_{1/2}$ values measured in my laboratory are listed in Table II and may be used as a first approximation for the $T_m$ and $t_{1/2}$ values of oligomers of similar length in 1 *M* NaCl.

## C. Nonamplified DNA Hybridization Assays

Hybridization assays can be performed once the probe stock concentrations have been determined and the $T_m$ and $t_{1/2}$ values have been measured or estimated for the probes under the conditions selected for the assay. The level of fluorescent impurities should be carefully controlled by pretreatment of the samples (e.g., *spun column* purification or ethanol precipitation) and by using high-purity buffer components. Particularly important is the use of high-purity NaCl, such as J. T. Baker (Phillipsburg, New Jersey) Ultrex grade or Johnson Matthey Puratonic grade distributed by Alfa Products (Danvers, Massachusetts).

### 1. Competitive Hybridization Assay

1. Prepare a stock solution of 1:1 complementary labeled probes in water at a concentration 12 times greater than the final concentration desired in the assay.
2. Combine 50 $\mu$l of the sample with 10 $\mu$l of the 1:1 probe solution and 60 $\mu$l buffer containing 2 *M* NaCl, 1 m*M* EDTA, 4–20 $\mu$g/ml carrier DNA (e.g., $\lambda$-DNA), and 20 m*M* $NaH_2PO_4$ at pH 8.0 in a 500-$\mu$l conical plastic tube.
3. Place the tube in a boiling water bath for 2 min.
4. Immediately place the tube in a water bath or thermostated sample holder at the hybridization temperature and begin timing.
5. Measure the amount of emission from either or both labels periodically over the incubation period (kinetic measurement) or when the incubation period is complete (static measurement).
6. Calculate the $\%\Delta f_{max}$ for each sample using Eq. 8. Calculate the target DNA concentration according to Eq. 13:

$$[T_1 \cdot T_2]_0 = (100 - \%\Delta f_{max})[P_1 \cdot P_2]_0 / \%\Delta f_{max} \quad (13)$$

Note that the two labeled probe reagents may be added separately to the sample and the denaturation step (step 3) omitted if the target polynucleotide is already in single-stranded form, as is the case when the target polynucleotide is RNA.

### 2. Noncompetitive Hybridization Assay

The noncompetitive hybridization is performed by exactly the same procedure as the competitive assay, except that the labeled complementary probes are replaced by the two labeled probes that hybridize to contiguous sequences on the target polynucleotide.

## D. Amplified DNA Hybridization Assays

The amplification example provided below uses PCR to amplify double-stranded DNA targets. A number of experimental parameters may be varied in the amplification steps, and other sources should be consulted in determining primer oligonucleotide lengths, temperatures, the number of amplification cycles, etc. (Mullis and Faloona, 1987). Primers that exactly flank the sequence recognized by the labeled probes provide the shortest amplified DNA possible, thereby minimizing the consumption of PCR reagents and reducing possible problems associated with strand displacement. The length of the amplified DNA is the sum of the two primer lengths and the probe length. The following PCR protocol is adapted from the Perkin-Elmer Cetus GeneAMP Amplification Reagent Kit (Norwalk, Connecticut) and is combined with the competitive fluorescence-quenching assay (Morrison *et al.*, 1989).

1. Mix the following together in a 500-$\mu$l conical plastic tube: 10 $\mu$l 10X PCR reaction buffer (500 m$M$ KCl, 15 m$M$ $MgCl_2$, 0.1% (w/v) gelatin, and 100 m$M$ *tris*(hydroxymethyl)aminomethane at pH 8.3), 16 $\mu$l dNTP solution (1.25 m$M$ each of dATP, dTTP, dCTP, and dGTP neutralized to pH 7 with NaOH), 5 $\mu$l primer solution (20 $\mu M$ of each of the two primers in demineralized water), 68 $\mu$l or less sample DNA, demineralized water to a final volume of 100 $\mu$l, and 0.5 $\mu$l Taq Polymerase (5 units/$\mu$l, Perkin-Elmer Cetus).
2. Incubate the sample at the following temperatures and times using a programmable temperature block (Perkin-Elmer Cetus DNA Thermal Cycler) or three water baths: 1 min at 94°C, 2 min at 37°C, and 3 min at 72°C.
3. Repeat step 2 for the desired number of amplification cycles (e.g., 25 cycles).
4. Complete the amplification with a 7-min incubation at 72°C.
5. Proceed with the nonamplified detection procedure using 50 $\mu$l of the PCR reaction solution as the sample.

An example of the data obtained from amplified hybridization assays detected by fluorescence quenching is shown in Fig. 12. Application of Eq. 13 to the PCR-amplified data determines the amount of amplified DNA present in the sample. The amount of DNA present in the original sample before amplification can only be estimated since the amount of amplification will vary from sample to sample and is dependent on a number of factors, including the original amount of target DNA present in the sample and the number of PCR cycles performed. Amplification can be estimated by repeating the assay using known amounts of the target DNA. The data plotted in Fig. 12 were obtained using known quantities of DNA target at the start of the PCR reactions; an amplification factor of 750,000 was obtained after 25 cycles of PCR (Morrison *et al.*, 1989).

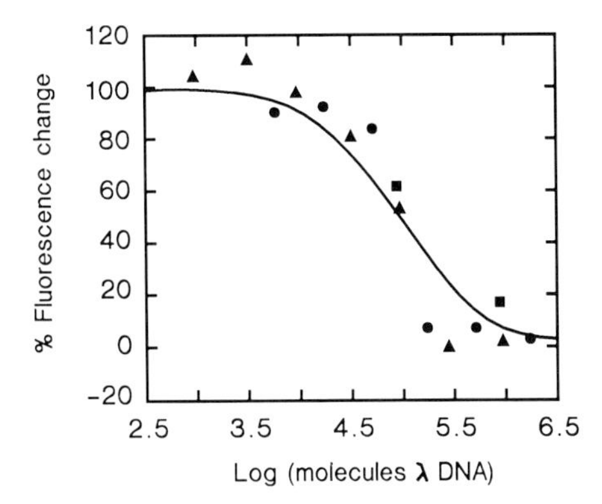


**Fig. 12** Detection of PCR-amplified DNA by competitive hybridization with labeled probes. Samples containing various amounts of λ DNA were carried through 25 cycles of PCR, and half of each sample (50 μl) was denatured with 1 n*M* each of fluorescein-(ethylenediamine)GATCCCACCTCATTTTCATG and CATGAAAATGAGGTGGGATC(aminohexylaminoadenosine)-pyrenebutyrate in a final volume of 120 μl. Fluorescence was recorded after 15 min of reassociation at 40°C. Fluorescence is plotted as $\%\Delta f_{max}$ =% fluorescence change, and the solid line was calculated using Eq. (13) with $[T_1 \cdot T_2]_0$ =751,000[λ DNA]. The abscissa displays the number of λ DNA molecules present in each 50 μl of sample before PCR amplification. Different plotting symbols represent data from experiments performed on different days. From Morrison *et al.* (1989) (Fig. 7, p. 241).

# REFERENCES

Borer, P. N. (1976). Measured and calculated extinction coefficients at 260 nm, 25°C and neutral pH for single-strand RNA and DNA. *In* "Handbook of Biochemistry and Molecular Biology" G. D. Fasman, ed.), 3rd Ed., Vol. I, p. 589. CRC Press Cleveland, Ohio.

Cardullo, R. A., Agrawal, S., Flores C., Zamecnik, C., and Wolf, D. E. (1988). Detection of nucleic acid hybridization by nonradiative fluorescence resonance energy transfer. *Proc. Natl. Acad. Sci. U.S.A.* **85,** 8790–8794.

Chu, B. C. F., Wahl, G. M., and Orgel, L. E. (1983). Derivatization of unprotected polynucleotides. *Nucleic Acids Res.* **11,** 6513–6529.

Conrad, R. H., and Brand, L. (1968). Intramolecular transfer of excitation from tryptophan to 1-dimethylaminonaphthalene-5-sulfonamide in a series of model compounds. *Biochemistry* **7,** 777–787.

Eftink, M. R., and Camillo, G. A. (1981). Fluorescence quenching studies with proteins. *Anal. Biochem.* **114,** 199–227.

Förster, T. (1959). Transfer mechanisms of electronic excitation. *Disc. Faraday Soc.* **27,** 7–17.

Heller, M. J., and Jablonski, E. J. (1987). Fluorescent Strokes shift probes for polynucleotide hybridization assays. *European Patent Application* 229 943.

Heller, M. J., and Morrison, L. E. (1985). Chemiluminescent and fluorescent probes for DNA hybridization systems. *In* "Rapid Detection and Identification of Infectious Agents" (D. T. Kingsbury and S. Falkow, eds.), pp. 245–256. Academic Press, Orlando, Florida.

Heller, M. J., Morrison, L. E., Prevatt, W. D., and Akin, Cavit (1983). Light-emitting polynucleotide hybridization diagnostic method. *European Patent Application* 070 685.

Lakowicz, J. R. (1983). "Principles of Fluorescence Spectroscopy." Plenum Press, New York.

Morrison, L. E. (1987). Competitive homogeneous assay. *European Patent Application* 232 967.

Morrison, L. E. (1988). Time-resolved detection of energy transfer: Theory and application to immunoassays. *Anal. Biochem.* **174,** 101–120.

Morrison, L. E., and Stols, L. M. (1989). The application of fluorophore-labeled DNA to the study of hybridization kinetics and thermodynamics. *Biophys. J.* **55,** 419a.

Morrison, L. E., Halder, T. C., and Stols, L. M. (1989). Solution-phase detection of polynucleotides using interacting fluorescent labels and competitive hybridization. *Anal. Biochem.* **183,** 231–244.

Mullis, K. B., and Faloona, F. A. (1987). Specific synthesis of DNA *in vitro* via a polymerase-catalyzed chain reaction. *Methods Enzymol.* **155,** 335–350.

Patel, A., Davies, C. J., Campbell, A. K., and McCapra, F. (1983). Chemiluminescence energy transfer: A new technique applicable to the study of ligand–ligand interactions in living systems. *Anal. Biochem.* **129,** 162–169.

Schroeder, H. R., and Yeager, F. M. (1978). Chemiluminescence yields and detection limits of some isoluminol derivatives in various oxidation systems. *Anal. Chem.* **50,** 1114–1120.

Telser, J., Cruickshank, K. A., Morrison, L. E., and Netzel, T. L. (1989a). Synthesis and characterization of DNA oligomers and duplexes containing covalently attached molecular labels: Comparison of biotin, fluorescein, and pyrene labels by thermodynamic and optical spectroscopic measurements. *J. Am. Chem. Soc.* **111,** 6966–6976.

Telser, J., Cruickshank, K. A., Morrison, L. E., Netzel, T. L., and Chan, C. (1989b). DNA duplexes covalently labeled at two sites: Synthesis and characterization by steady-state and time-resolved optical spectroscopies. *J. Am. Chem. Soc.* **111,** 7226–7232.

Ullman, E. F., Schwarzberg, M., and Rubenstein, K. E. (1976). Fluorescent excitation transfer immunoassay. *J. Biol. Chem.* **251,** 4172–4178.

Wahl, G. M., Berger, S. L., and Kimmel, A. R. (1987). Molecular hybridization of immobilized nucleic acids: Theoretical concepts and practical considerations. *Methods Enzymol.* **152,** 399–407.

# INDEX

## E



## F



## G



## H



## I